Biology and Politics

Publications of the International Social Science Council

19

Mouton · The Hague · Paris

ALBERT SOMIT

Biology and Politics
Recent Explorations

Mouton · The Hague · Paris

Published with the support of the Maison des Sciences de l'Homme, Paris

ISBN 2-7193-0439-5 (Mouton)

Printed in France

Contents

Preface

I WANT TO EXPRESS my appreciation to the Trustees of the Harry Frank Guggenheim Foundation for their generous grant in support of the Paris Conference on Biology and Politics. Without that sagacious philanthropy, neither the conference nor this book would have been possible. Thanks are due, as well, to the Executive Committee of the International Political Science Association whose cosponsorship was an essential aspect of both undertakings.

Anyone who organizes a conference of this sort incurs a number of personal obligations. I am indebted to all those who participated in the Paris meeting for their unstinting cooperation and ready assistance. Special recognition is due Jean Laponce for help, in all stages of the conference, far and above the normal call of duty.

I would be remiss if I did not identify the three gracious ladies whose services were literally indispensable in organizing the conference and completing this manuscript—Marty Fiorella, Rose Levin and Charlotte Poole. Steven Peterson, as on several previous occasions, gave generously of his time and talents in a variety of capacities. And, finally, William Richardson, my graduate assistant, has efficiently accomplished that myriad of tasks without which no manuscript can be satisfactorily completed.

Several internationally known social scientists—John Crook, David Easton, William Mackenzie, Lionel Tiger and John Wahlke—served as discussants and their perceptive comments and criticisms are an invaluable feature of this volume. Albert Somit has provided an overview essay and has reprinted a recent review of the literature. Finally, Steven Peterson has contributed a most useful annotated bibliography for those who wish further to explore this engrossing area.

ALBERT SOMIT

ALBERT SOMIT

Introduction

Although it may not be readily apparent from the literature, political scientists pursue a relatively small number of objectives. They seek: *a)* to explain past and present political phenomena; *b)* to predict, using the term rather loosely, future political developments; *c)* to find solutions for existing or anticipated political problems; and *d)* to propound, and perhaps achieve acceptance of, some vision of the "good" political society. To be sure, not all of them would agree upon the legitimacy of all the foregoing objectives. But, taking the profession collectively, these can reasonably be offered as embodying a disciplinary consensus.

In the pursuit of these goals, political scientists have relied both on their own ingenuity and on the creativity of scholars in other fields. Many of the theories, models and methods employed have been developed by political scientists themselves; others have been borrowed, with striking open-mindedness,[1] from such diverse sources as philosophy, physics, jurisprudence, social psychology, psychology, economics, sociology, anthropology, geography, and even mathematics.

I. Previous Efforts at a Biologically-Oriented Political Science

This volume, and the conference whose proceedings it reports, deals with the latest such effort at scientific cross-fertilization—an attempt to draw upon the intellectual and methodological armory of modern biology. Admittedly, this enterprise is not altogether novel, for the notion that there are meaningful similarities between the body politic and a biological organism, and the sibling conviction that biology can cast light on political behavior, has a long if not always distinguished history in political science.[2]

The best known and probably most unfortunate previous attempt to effect a marriage betwen biology and politics took place, of course, during the nineteenth century. As Darwin achieved scientific acceptance, the concepts of evolution, natural selection, and survival of the fittest were promptly pressed into service both by those who resisted and those who sought the extension of governmental authority into economic and social spheres. The fact that advocates of diametrically opposed political philosophies could find persuasive support in evolutionary doctrine eventually tended to discredit what we now call Social Darwinism. Regrettably, it also tended to discredit the parallel premise that biological concepts could be fruitfully employed in the study of social and political behavior. Worse, it cast into temporary disrepute the assumption underlying this premise—i.e., that human behavior was meaningfully influenced by man's biological makeup.

Looking back, the earlier quest seems almost foredoomed to failure. The biology of the era was still unable adequately to explain biological, let alone social, phenomena. Second, many of those who rushed to apply evolutionary doctrine to social issues were not aware of the limitations and weaknesses of the concepts they were so enthusiastically borrowing; frequently, they did not even grasp the ideas involved. Third, as so often happens when those in one discipline seize upon developments in another, Darwinian theory was put to purposes for which it was never intended and forced to bear a burden of argument for which it was manifestly inadequate. The coup de grâce was the embrace of "biology" by the proponents of racism, a group whose scientific ignorance was equalled only by the certainty with which its members proclaimed their views.

For these and other reasons, the inevitable reaction occurred. By the 1920's and '30's, most social scientists drew a sharp line between man and all other forms of life. The behavior of other species was understood entirely or largely in terms of biological inheritance (instinct); that of Homo sapiens was viewed as almost totally learned and having little, if any, relationship to our genetic legacy. Political man was treated as if he had no behavioral predispositions whatsoever (back to *tabula rasa*) and political and social behavior were explained as the product of learning and social conditioning, leavened by "intelligence" and spiced with a dash of "personality." So held the conventional wisdom.

II. The Revival of Interest in Biology

By the 1950's and 1960's this view came under increasingly severe attack. The way for the change was prepared by Freudian psychology. Although many of his disciples were slow to admit it, Freud's emphasis on the inherently irrational (or non-rational) nature of human behavior demanded, in the final analysis, a direct link between biology and human action, since any innate behavioral tendency has necessarily to be rooted in man's genetic makeup.

More important, though, were major advances in biology itself. Since these have been dealt with at considerable length elsewhere,[3] they need only be noted here. Research in neurology, psychoneurology, and psychopharmacology demonstrated beyond a doubt that human behavior—intellectual and emotional, as well as physical—can be influenced and even controlled by altering the physiological functioning of the body.[4] Developmental biology forged a conceptual apparatus increasingly capable of explaining animal evolution, natural selection, and speciation. Genetics was enriched by discoveries which simultaneously clarified the process whereby genetic instructions are transmitted from generation to generation and opened the way for human intervention in, and modification of, that process.

Perhaps the most decisive factor was the transformation of ethology from a relatively obscure to an almost indecently glamorous specialization. Two ethologists, Konrad Lorenz and Nikolaas Tinbergen, became Nobel Laureates; several ethologically-based books made the best-seller lists. Without exception, these books stressed the manner in which an animal's behavior is influenced by its genetic legacy. Although the ethological evidence was based almost entirely on studies of sub-human species (ranging from earthworms to chimpanzees), the moral, not always merely implied, was clear: human behavior, too, has its roots in our biological makeup.

These developments, reinforced by an upsweep of concern over man's relationship to his environment, rekindled among many social scientists a desire to explore the possibility of utilizing biological concepts and techniques.[5] Within political science, this experiment has been underway for a decade or so. Similar ventures have been launched, over much the same time span, in sociology, social psychology, geography, anthropology, and economics. In short, the pendulum seems to be well along its return journey. One can only hope that the lessons of the earlier debacle will not be entirely forgotten.

III. Major Avenues of Exploration

The resurgence of interest in biology among political scientists has already generated a sizable literature and achieved sufficient importance to earn "panel" recognition at International Political Science Association Congresses and at many national and regional professional conventions. In 1974, it received the ultimate *imprimatur*—a foundation grant to hold a special conference. Before turning to the papers produced for and at that conference, it might be useful briefly to describe the general configuration of this emerging field.[6] I say "briefly" for two reasons: first, I undertook this task in a recent review article, reprinted herein,[7] and there seems little point in restating or paraphrasing views which have not significantly altered in the interim; second, some of this ground is covered, albeit in sketchy fashion, in the conference papers.

As is often the case with a newly developing area, interests range widely and irregularly; practitioners and literature alike resist precise classification; any organizing schema must necessarily have a certain arbitrary quality. The requisite caveat delivered, I will employ here the four categories utilized in the above mentioned article. They are probably as valid today, to put it somewhat ambiguously, as they were then.

1. *The Case for a Biologically-Oriented Political Science*

Quite naturally, scholars essaying a novel approach to their subject are eager to persuade others of the benefits, present or prospective, promised by that approach. While all of the papers prepared for the Paris conference have this as one of their objectives, I think it fair to say that none springs primarily from a proselytizing desire.

2. *Ethology and Political Behavior*

What is the nature of political man? Is man inherently good or inherently selfish? Peaceful or agressive? To what degree is human nature fixed? To what degree is it, or can it be, molded by society? Whether stated in naive or sophisticated fashion, these have been central issues in political philosophy from Greek speculation to the present.

Contemporary ethology, we noted above, seeks to account for animal behavior in terms of the interplay among *a)* genetically transmitted behavioral patterns (acquired over literally millions of years); *b)* learned behavior; and *c)* the environment in which the organism functions. Though still troubled by methodological and theoretical disagreements, ethology has made swift strides forward in the past few decades. For reasons which are almost self-evident, political scientists have seized upon ethological concepts and, to a lesser degree, upon ethological techniques. The lure of ethology has not been limited to political philosophers but has also been felt, again for obvious reasons, by those working in politics, organizational behavior, international relations, and public policy. The multifaceted aspects of this attraction are reflected in the papers by Corning, Schubert, Masters, and Schwartz. Each explores the manner in which ethological concepts can be employed by political scientists and/or seeks to apply those concepts to specific political problems.

3. *Physiological Aspects of Political Behavior*

Our genetic heritage, to be sure, profoundly influences our physiological functioning. The difference between these two categories is, therefore, a matter of emphasis. Political scientists interested in ethology are primarily concerned with the manner in which genetic programming might manifest itself in political and social behavior; those who are more "physiologically" oriented look at the way in which that behavior can be altered by modifying the biological functioning of the human body.

Among the physiological/biological variables which might, or seem to, have some bearing on political behavior are nutrition, health, disease, fatigue, exercise, neurological excitation (or deprivation), age, sex, psychopharmaceuticals, pain, age at puberty—even, in one imaginative essay, hair. Three of the papers presented at the Paris conference explore this type of variable: David Schwartz suggests that "health" is a significant element in political participation; Jean Laponce examines the relationship between left-handedness and political behavior; and James Davies argues that revolution, or other forms of rapid political change, might be explained, in significant measure, by neurological over-excitation.

Physiological reactions can also be treated as dependent variables and utilized, directly or indirectly, to measure attitudes and/or behavioral potential. Among the "bio-polimetrics," to adopt Easton's term, so employed have been galvanic skin response, heart beat rate, blood pressure, facial expression, body posture, energy level, hand squeeze, muscular tension—i.e., much of what we call "non-verbal" behavior. As can be seen, the physiologically-oriented approach may be conceptual; it may be methodological; it may incorporate elements of both. The paper by Davies exemplifies the first, that by Lodge/Tursky/Cross the second, and Schwartz's the third.

4. *Public Policy Applications*

For whatever reasons, and these may include the modest esteem in which applied research has been held by many political scientists, this area has attracted comparatively few investigators. Among this handful, two major foci have emerged: *a)* the moral, ethical, and political issues posed by our expanding ability, via "biological engineering," to influence human behavior; and *b)* biologically-related social issues—environmental pollution, food supply, radiation levels, population control, and the regulation of drug and alcohol usage. Of the papers presented at Paris, Peter Corning's is the best example of this genre, though the problems explored in several of the others also have significant, if less direct, implications for public policy.

IV. The Paris Conference on Biology and Politics

The first panel specifically devoted to "biology and politics" was held, I believe, at a meeting of the Southern Political Science Association in 1967. Thereafter, similar panels were frequently organized at international, national, and regional professional meetings.

However welcome these sessions, they had serious limitations. The time allotted was too brief to permit adequate consideration of the larger problems encountered in an admittedly exotic cross-disciplinary venture; the panel format did not readily lend itself to the task of discussing, let alone planning, broad research strategy; and, despite determined efforts to achieve a "mix" of participants—that is, of skeptics as well as true believers—the panelists were usually persons sympathetic, if not already committed, to the notion of a biologically-oriented political science.

The idea for a conference devoted entirely to biology and politics grew out of these several shortcomings. Initially suggested (as I recall) by Jean Laponce, proposed by me, sponsored by the International Political Science Association, and supported by a grant from the Harry Frank Guggenheim Foundation, the conference was held in Paris, January 6-8, 1975. The "regular" conferees[8] were augmented, from time to time, by members of the IPSA Executive Committee attending an overlapping IPSA function. Among those who participated in this latter fashion were Klaus von Beyme, Karl Deutsch, Serge Hurtig, John Meisel, Stein Rokkan, and John Trent.

The Paris meeting sought some half-dozen objectives: *a)* to assess what had been accomplished in biopolitics generally and in the major sub-areas thereof; *b)* to subject these assessments to the scrutiny and criticism of "non-committed" political scientists and "outside" specialists; *c)* to identify key theoretical and methodological problems and to suggest how these might best be attacked; *d)* to consider, in more systematic fashion than had so far been possible, the actual or potential benefits to political science from utilizing biological concepts and techniques; *e)* to develop a strategy for acquainting political scientists with what had been accomplished in biopolitics to date; and *f)* on the basis of the foregoing, to plan the "biology and politics" panels scheduled for the 1976 IPSA (Edinburgh) Congress.

In keeping with these goals, participants were recruited from three groups: political scientists working in biopolitics; "non-committed" political scientists chosen for their knowledge of the discipline, critical acuity, and receptivity to new ideas; and "outsiders" with an appropriate biological expertise.

Seven papers, collectively crosscutting the major sub-areas of biopolitics, were prepared and presented by "committed" political scientists. After each paper, there was a fairly extended comment by a discussant who was either a "non-committed" political sientist or an "outside" expert and, then, a general discussion of issues which had been raised. Of the five conference sessions, four were devoted to a presentation and analysis of formal papers. The final meeting was reserved for what the program, with dubious Latin scholarship, termed a "*quo vadis*" session. This began with "overview" statements by the three non-committed political scientists (Messrs. Easton, Mackenzie and Wahlke) and was followed by a far-ranging and freewheeling interchange among all the participants.

All seven papers and the overviews delivered at the closing session are contained in the following pages. Regrettably, limitations of space made it impossible to include—with a single exception—the discussants' comments. The papers printed here differ slightly from those given at the conference in that they have been polished for publication and, in some instances, revised to take cognizance of criticisms and suggestions expressed during the conference. These revisions in no way significantly alter their original substantive thrust.

To what extent did the conference achieve its intended objectives? So far as the conferees were concerned, the assessments were uniformly positive, ranging from a Delphic "far more successful than I had dared hope" to "tremendous." Part of the answer can be found in the papers contained in the following pages; part, I trust, will emerge in the course of the 1976 IPSA "biology and politics" sessions. My own sense—and I am not an altogether neutral observer—was that, by the time that the Paris sessions ended, the major issue was not *whether* a more biologically-oriented approach could enrich political science but rather *how* that could best be accomplished.

Altough each author has provided an abstract of his paper, I trust that these will serve only as a point of departure. Most of the papers are rather lengthy; in most, the argument is complex rather than simple; in most, the data are suggestive, rather than conclusive. I speak for all the contributors to this volume in stating that we want the reader to sit in personal critical judgment on what has been written and said. As almost every scholar knows, there is a feel and a texture to an argument which can only be felt first-hand. The abstract is but the skeleton; the paper itself provides the intellectual flesh and muscle.

V. Concluding Remarks

In this final section, I would like to make some personal observations and, in a couple of instances, to comment on points made in the conference papers.

1. Biology—more precisely, a number of biologists—has developed a general systems theory which, some have urged, should be applied to the study of social and political behavior. At first hearing, it seems like an attractive idea, especially in light of the many apparent similarities between biological systems theory and the systems approach currently favored by many political scientists.

Still, the proposal is not altogether persuasive. There have been very few successful attempts to transplant conceptual structures from one discipline to another. Moreover, *a priori* considerations aside, man is profoundly different from other species, including his closest primate relatives. A theoretical framework appropriate to the study of other organisms is not necessarily well suited to the study of Homo sapiens. My own sense is that we are likely to make better progress by borrowing selectively rather than in a wholesale fashion.

2. The proposition that our behavior is influenced, to greater or lesser degree, by our genetic endowment is central to most of the essays in this book. But that endowment was "selected for" under conditions profoundly different from those in which modern man currently exists, a difference inexorably widened by our unique capacity to make our surroundings increasingly unsuitable for human existence. We may now have reached a point where we are no longer equipped, as species, successfully to survive in the environment we have created for ourselves.[9]

The standard response, of course, is that the same intelligence which got us into our present predicament will somehow get us out. We all hope so. But the ethological viewpoint is not completely reassuring. "Rationality" is seen as a limited rather than an absolute characteristic of the species. As Konrad Lorenz lately observed, ". . . man, even while endeavoring to plan intelligently and flattering himself that he is a rational being, is still subject to unreflected motivations which force him to commit acts which are obsolete even at the time they are committed."[10]

Further, there may be inherent structural limits to human intelligence, quite apart from the non-rational impulses to which Lorenz refers. The same process of evolution and natural selection which produced the human brain, Lionel Tiger warns, could have built into that brain "design defects" which literally make it incapable of grasping—or effectively coping with—certain critical aspects of its environmental reality. The history of the past three or four millenia is, from one perspective, the record of repeated failures of this sort. Whether they are in any way related to genetically fixed limits on our intellective capacities we do not know. It is a possibility, though, which we should not blithely brush aside.

3. As the conference papers testify, several of the participants are not altogether pleased with the current state of political science.[11] This dissatisfaction was perhaps most vigorously voiced by Messrs. Wahlke and Schubert and was specifically addressed by Wahlke in response to his own question "What political science questions should we bring biology to?"[12]

What is it that political scientists should be studying? how can they best go about that inquiry? Here, the history of ethology may be instructive. The field of animal behavior made relatively slow progress so long as ethologists limited their observations primarily to animals in zoos, experimental laboratories, and other man-made settings. A near quantum jump in knowledge was achieved when the realization dawned that, adequately to understand the behavior of an organism, it must be studied in the environment for which it has been designed by evolution—its natural habitat. This forced some drastic changes in the life style of ethologists, but the results were most impressive.

Much of what we deal with in political science is, at best, only indirectly related to political behavior; much of our inquiry takes place in what is, at best, a simulated political environment. To date, the greatest impact of ethology on political science has been in terms of concepts and explanatory theory. We might profit most, at least in the short run, if the ethologists' example persuades us to look at actual political behavior in an actual political arena.[13]

4. The contents of this book will, I trust, convince some readers that the subject merits further exploration. With this in mind, we have provided what is probably a surfeit of bibliographic assistance.

For those interested in the topics covered by the conference papers, the authors have provided extensive footnotes. For those who want a general overview of biopolitics, I have reprinted the best (and only) review article. It covers the literature, item by item, to September, 1971. Finally, Steven Peterson has put together an excellent critical bibliography for those who seek a broader perspective. What Peterson has done, in effect, is to provide both a "do it yourself" guide and a reading list for a basic course in biology and politics.

I hope that the reader will take advantage of these materials. For those with little background in biology, I would urge that reading be complemented by discussions with knowledgeable biologists. I know of no better way for a layman to achieve an understanding of the issues on which the biologists themselves are in disagreement. It also does wonders for one's morale to be reminded that our colleagues in the "harder" disciplines are riven by the same type of disputes which we too often regard as characteristic only of the social sciences.

5. Professor Mackenzie cogently argues, *inter alia,* that those interested in the relationship between biology and politics should not seek to establish a separate *field* within political science and that we should find some descriptive term other than "biopolitics."[14]

There was near unanimity among the Paris conferees that a separatist movement in biopolitics [*sic*] would be counter productive. We want to convince other political scientists that biology can contribute to many, if not all, of the fields into which the discipline is presently organized. That objective would in large part be defeated if biopolitics were to be established simply as another "field." The strategy followed must win acceptance for biopolitics without setting up shop, literally and figuratively, in competition with already established sub-specializations. The diversity of research interests represented at the Paris Conference, and at earlier meetings as well, gives reason to believe that this can successfully be accomplished.

I agree with Professor Mackenzie that the term "biopolitics" leaves much to be desired. But, as the foregoing pages testify, I have not always been able to heed his admonition. True, "biology and politics" is more accurate; regrettably, it is far more cumbersome. The merits of the longer term notwithstanding, I fear that even those who have not already drifted into slothful habits will occasionally find the shorthand phrase too convenient to resist. Professor Mackenzie has, nonetheless, succeeded in making me feel guilty every time I err in this respect.

Notes

1. Cf. *infra* David Easton's comments on this point, pp. 237-247.
2. S. A. Peterson, "Biological Basis of Student Unrest," State University of New York at Buffalo, 1974 (unpublished Ph. D. dissertation), chap. 2: "Biopolitics: Origin and Development," pp. 21-50.
3. See below Peterson's critical bibliography, Appendix, pp. 279-291.
4. These findings cast a good deal of light, *inter alia,* upon another phenomenon which had engrossed and baffled social scientists—brain-washing.
5. This coincided, Steven Peterson reminds me, with the popularity of a futurist literature which, in both fictional and non-fictional form, described the deliberate control of human behavior by biological and chemical means.
6. As noted below (p. 10-11) the term field is both undesirable and difficult to avoid.
7. See Appendix, pp. 277-298.
8. Peter C. Corning, John Crook, James C. Davies, David Easton, Jean A. Laponce, Milton Lodge, William J. M. Mackenzie, Roger Masters, Steven A. Peterson, Glendon Schubert, David C. Schwartz, Albert Somit, Lionel Tiger, and John Wahlke.
9. The term "environment" can be used, in this context, to include the political, economic, and social institutions, through which we seek to solve our problems.
10. *Science* 187, March 14, 1975, p. 907.
11. While most of the criticism focused on the discipline's conceptual shortcomings, the Lodge/Tursky/Cross paper, interestingly enough, expresses grave reservations about research techniques currently being utilized by many in the profession.
12. See below, pp. 253-259.
13. A leading psychologist, Dr. Urie Bronfenbrenner, has recently criticized research in child psychology because it focuses on ". . . the strange behavior of children in strange situations with strange adults." Instead, Dr. Bronfenbrenner argues, investigators should attend to the interaction between "the growing organism and the enduring environment or context in which it lives out its life," *Science and Government Report* 5 (9), May 1, 1975, p. 4.
14. See below, pp. 249-251.

Conference Papers

DAVID C. SCHWARTZ

*Somatic States and Political Behavior: An Interpretation and Empirical Extension of Biopolitics**

THIS CHAPTER IS AN ATTEMPT to interpret, evaluate and improve the study of biopolitics by addressing several unmet needs which threaten to inhibit research progress. These are: a) *the need to make explicit the basic assumptions which are, or ought to be, common to biopolitical studies—in order to guide new research cumulatively and to facilitate general evaluation of the biopolitics field;* b) *the need to state precise theoretical linkages between somatic variables and the major empirical paradigms in Political Science—in order to encourage the regular inclusion of somatic variables in most studies of political behavior; and* c) *the need to document the important influence of somatic states on political behavior in studies on representative populations.*

Three basic assumptions are identified as appropriately involved in all biopolitics research. Taken together, these assumptions hold that those somatic characteristics which influence the individual's capacities and/or predispositions to respond to the sociopolitical environment will tend, in interaction with specified environmental conditions, to have significant causal influence on political behavior; these causal associations will tend to be significantly, but not totally, mediated by processes already studied in political science (socialization, personality, the arousal of politically predisposed individuals, etc.)—such that an important portion of the somatic influence on political behavior is exerted in a manner which requires the inclusion of somatic variables in most political science studies of political behavior.

* The national survey reported in this chapter was supported by grants from the H. F. Guggenheim Foundation, Rutgers Research Council, Rutgers Center for Computer and Information Services and the N.I.H.-Rutgers BioMedical Faculty Support Grant Program. The regional study was conducted at the University City Science Center of Philadelphia with the cooperation of the Jefferson Medical University, under the general direction of L. Peterson, M.D. and the day-to-day supervision of Nicholas Zill. I am grateful to these institutions and colleagues for their generous support. The following persons deserve acknowledgement, and much more, for the encouragement, advice and assistance they have given to my research: Lionel Tiger, Robin Fox, James C. Davies, Albert Somit, Michael D'Amato, Julian Wolpert, Peter Corning, Mary Etazady, Dan Landis and, not least, Sandra Kenyon Schwartz. Special acknowledgement is due to my research associate, Margaret Gaboury, for her outstanding management of our data analytic efforts. These individuals and institutions are, of course, in no way responsible for any errors in this paper.

A. Somit, Biology and Politics, pp. 15-44.

The importance and validity of these assumptions are illustrated in a detailed analysis of the relationships between somatic variables and the major political science explanations of political behavior—an analysis which draws heavily upon previous literature. The assumptions are then tested, rather directly, in two new studies concerning the influence of health on political behavior. This chapter contains the first published reports on these large-scale studies.

The chapter concludes with an optimistic prediction about the future of biopolitical research.

Introduction

In the past seven years, more than fifty articles, reviews and scholarly papers on biopolitics have appeared in the political science literature (see, for example, Schubert, 1974). Taken together, these writings suggest that a very wide range of biopsychological variables may be causally related to a number of important, diverse political phenomena. The list of biopsychological variables which have been putatively linked to political behaviors includes health status, arousal level, physical development patterns, nutritional factors, hormonal variation, drug-medicine use, central nervous system functioning, genetic differences and a variety of body images which have been shown to be associated with specific somatic states.* Among the political behaviors for which biopsychological correlates have been sought are: citizen participation in electoral politics (Dearden, 1973; Jaros, 1972; Schwartz, 1975; Schwartz / Zill, 1971); participation in protests and demonstrations (D. C. Schwartz, 1970; Peterson, 1973), participation in violent political events (Hummel, 1970; Davies, 1963; Corning, 1971), identification with charismatic leaders (Hummel, 1970), and the making of perceptual discriminations between political options (Jaros, 1972). Basic political attitudes—such as political interest, efficacy, alienation, reformism and cosmopolitanism—have also been related to somatic states in biopolitical theorizing and research (all of the pieces cited in note below, touch upon one or another aspect of the linkage between somatic states and political behavior).

The preceeding paragraph indicates, I think, that those of us interested in biopolitics have been rather busy and reasonably productive of late. Clearly, we have not shied away from the central theorical concerns in contemporary political science (e.g., the causes of political violence, the

* Health and politics is the subject of D. C. Schwartz' work (1975, in press). The relationship between health status and politics is the topic of a large scale research program presently being conducted by the author; some of the data generated in this program is reported in a later section of this paper. Arousal levels and their political consequences are explored in: D. C. Schwartz / N. Zill (1971); S. A. Peterson (1973). Physical development as a factor in political socialization is discussed in: Leroy Ferguson / *et al.* (1970). On the relevance of nutrition, see: J. C. Davies (1963), R. Stauffer (1969). Hormonal variation is considered in the following biopolitical studies: R. H. Hummel (1970), J. Dearden (1973). Drug-use studies include: R. B. Stauffer (1971); D. Jaros (1972). J. C. Davies has charted some of the ways in which the central nervous system's functioning influences political behavior. See Davies (1970); P. A. Corning (1971). Additional bibliographical citations to Corning's work on this topic are available in: Schubert (1974). The relevance of body images for political study is shown in: Schwartz (1975) and in: P. Shubs (1973).

wellsprings of citizen participation in political events and movements). Rather, we have articulated a number of plausible theories linking important political behaviors to antecedent somatic states, in interaction with specified environmental conditions, and have conducted a few small scale pilot studies testing selected aspects of those theories.

But if there has been some modest progress achieved in our recent efforts to relate somatic phenomena to political behavior, we must also acknowledge—and seek to rectify—some important present shortcomings in biopolitics research. Stated briefly, I believe the following problems, shortcomings or next steps must be addressed.

Initially, the fundamental assumptions which are, or ought to be, common in biopolitics research must be made more explicit—so that we, our colleagues and our students can examine their general validity and utility. Such explication of our assumptions can also be expected to guide new research in a useful, integrative, cumulative fashion. If, on the other hand, the common assumptions underlying biopolitics research are not rendered clearer and more persuasive, I fear that our findings will remain too largely unintegrated and therefore the discipline as a whole will not be encouraged to confront man's somatic nature as it bears upon the study of human political behavior.

Secondly, we must state, with precision, how biopsychological and/or biopolitical variables are related to the variables and processes in the major empirical paradigms in political science. I believe we can show that the effective employment of each of these paradigms either requires or is significantly aided by biopsychological inquiry. When that argument is more persuasively made, when our variables are clearly related to more familiar political science paradigms, we may hope that substantial numbers of political scientists will begin regularly to include biopsychological variables in their studies. *To be maximally useful, biopsychological inquiry must either change the prevailing paradigms in political science or be incorporated into those paradigms; it cannot be unrelated to those paradigms.*

Finally, and perhaps most importantly, it seems necessary to advance beyond the articulation of plausible theories and the administration of pilot studies on small unrepresentative populations to the empirical documentation of consistent, significant relationship between biopsychological and political variables *in politically relevant populations.* We must forcefully demonstrate, not merely argue or suggest, that biopsychological variables do add significantly to our ability to explain political phenomena. Basic theoretical work and focused small scale research must, of course, continue (and continue to be highly valued) but at least on those biopolitical topics where the groundwork has been carefully done, large scale interdisciplinary studies are now manifestly required.

This chapter represents my effort to address these fundamental problems in biopolitics. In the sections which follow, I try to make explicit some assumptions which are, or ought to be, common in biopolitics research, to state some of the ways in which biopolitical variables relate to the major empirical paradigms in political science, to demonstrate that one set of somatic variables, the individual's physical health status, does have significant and arguably causal impact on political behavior.

Three Assumptions Which Are, or Ought to Be, Common in Biopolitics

Studies in biopolitics have taken their independent and intervening biopsychological variables from a broad array of disciplines in the life sciences, including neurology, endocrinology, ethology, behavioral genetics, psychophysiology, nutritional science and clinical medicine. These disciplines of course differ in the levels of analysis to which they typically relate (e.g., species, populations, individuals), in the degree to which human variability on the factors of interest may be observed (e.g., from strong individual differences to remarkable species-wide similarities) and in the degree to which the somatic variables are influenced, and their relationship to sociopolitical behavior are mediated, by environmental (especially cultural) conditions (i.e, while some mediation may almost always be assumed, marked differences in the degree of mediation seem clear). As indicated above, the life sciences which have been drawn upon suggest a wide variety of biopsychological variables which may have causal association with human sociopolitical behavior. Accordingly, previous reviews have correctly emphasized the rich, variegated and unintegrated character of the biopolitics literature (see Introductory Chapter, this volume, and Bibliography).

I believe, however, that three, linked, general assumptions are or ought to be common to the whole biopolitics enterprise, assumptions which are linked not by the disciplines or theories from which they derive but the kinds of relationships they posit to exist between somatic states and political behavior. These assumptions, I should stress, are not the untested axioms of biopolitical inquiry but rather are generalized propositions, selected specific aspects of which have been confirmed.

> *Assumption 1*
> Those of the individual's non-transient somatic characteristics and recurrent somatic states which influence his capacity and/or predispositions to respond to the sociopolitical environment, will tend, in interaction with specifiable environmental conditions, to have causal association with the degree and/or type of his political activities. *Ceteris paribus,* such associations should be equally observable across individuals, within-individuals-across-time, and/or across populations (where the somatic characteristic is based upon some averaging of individuals within the populations). *Corollary: Ceteris paribus,* the greater the somatic variation across units, or within-units-across-time, the greater or more likely is the sociopolitical influence of the somatic variables.

This formulation is clearly the most basic assumption in biopolitical research and on its validity hinges the viability of the biopolitics efforts as a whole (though not the validity of individual studies). The validity of this assumption, then, and improving our ability to assess its validity, are the larger, ultimate subjects of this chapter (i.e., my purpose is to encourage greater precision in the expression and testing of biopolitical propositions and broader political science employment of biopolitical variables in order to better assess the general validity of this assumption). The chapter as a whole, then, bears on the validity of this assumption and I will concentrate here on explicating the assumption and showing its centrality to the major strands of biopolitical inquiry.

Perhaps it would be well to begin explicating the assumption by making clear what it does *not* mean (since errors of incautious overgeneralization of propositions are not uncommon in our discipline and may be more within an author's control than are errors of underutilization). First, the assumption does not hold that all observed variation in political behavior is attributable, even in part, to variation in somatic characteristics. We recognize that political behavior need not covary with somatic states and that, even in instances of systematic covariation, causality is not always appropriately attributable to the somatic states. The limited assumption we make is that there are instances in which important political behaviors are explicable by somatic antecedents in interaction with specified environmental conditions.

Second, the assumption does not mean that two individuals or populations that differ on a basic somatic dimension will therefore differ in all political behaviors, even holding environmental conditions constant. Third, the assumption does not deny that some environmental conditions may be "controlling" with reference to a given political behavior. Just as we assume that great somatic differences have stronger and/or more likely political influence than smaller differences, we recognize that the same relationship probably obtains for environmental conditions. Fourth, the assumption does not deny the important sociopolitical influence of short-term somatic states (e.g., sudden increases in secretion of a given hormone) where such states are recurrent or typical under specified conditions (as in stress). It is true that the assumption, as stated, does not cover short-term, non-recurrent somatic states. Much needed future research on the consequences of short-term health traumas, therefore, may force a broadening of the assumption.

Probably few if any students of biopolitics will disagree that the assumption stated above is consistent with, and derivable from, the major strands of biopolitical inquiry. Nonetheless it may be worthwhile here to briefly demonstrate that this is the case in order to document the centrality of this assumption to biopolitical research and to illustrate the many forms in which the assumption has been cast. To do this, I will transpose into the terms of the assumption, the major arguments and/or findings from representative articles in five biopolitical areas: the political influence of hormonal variation; ethological research on politics; drug use and politics; health and politics; the political implications of genetic processes.

Perhaps the most general case for the relevance of hormonal variation to political behavior is made by John Dearden (1973) who begins by citing evidence that variation in a non-transient somatic characteristic (male-female differences in such hormone configurations as androgen levels) is causally related to variation in aggressive predispositions (including aggressivity as a general personality trait). His basic theory is that the confluence of a somatic state (high androgen level), a resulting psychological condition (aggressivity) and an (internalized) environmental condition (differential sex-role socialization) accounts for the greater observed male activism and aggression in politics. Clearly, this theory is entirely compatible with the assumption being explicated here. So, too, is the reasoning of Ralph Hummel (1970) who has a more situationally-specific notion of the relevance of hormones to political behavior. Hummel contends that individuals most characterized by a recurrent somatic state (increased adrenaline secretion in conditions of high social stress) and most influenced by a specific environmental condition (the presence of one or more aggression-justifying, charisma-claiming leaders) will tend to be most politically aggressive and most attracted to charismatic leaders.

The assumption under reviews seems to be equally central in ethological research on politics. The works on crowding and political aggression, by Singer / Luterbracher (1970) and by Welch / Booth (1973) for example, assert that conditions of crowding initiate similar biochemical processes in humans and in nonhuman primates. These biochemical processes are seen to have behavioral effects which include increased irritability, an activation of territorial defense needs and overt violence. It should be noted that in these papers, as in Hummel's, the biopsychological state is treated primarily as an intervening variable: an environmental condition (crowding, stress) predicts to a situationally-specific but recurrent somatic state (biochemical processes including specific hormone secretions) that predicts to aggressive predispositions which, when activated by environmental stimuli, yield a politically violent response. Dearden's theory, *per contra,* treats the biopsychological variable of interest (hormone levels) primarily as an independent variable. Despite this difference in the form of the explanatory proposition, in the type of hypothesized link between somatic and political variables, it can be seen that the assumption being explicated here operates in all these studies.

Studies on drug-alcohol use and political behavior show the same general pattern. Jaros (1972) reports experimental evidence that the ingestion of sizable but safe quantities of depressants had the short term effects of: *a)* reducing the individual's ability to make perceptual discriminations among political options in a standardized stimulus presentation; *b)* interfering with previously held participatory orientations. Stauffer's (1971) extensive review of the drug use literature concludes that long-term, heavy and/or repeated use of depressants, amphetamines and hallucinogens tends to be associated with passive personality configurations and with social withdrawal, both of which appear to be inconsistent with the behavioral requirements of sustained political activity. Stauffer concludes that the psychophysiological effects of such drugs, in confluence with the sociocultural consequences of prolonged or heavy drug use, tend to reduce probable political involvements. In the terms of our general assumption, Jaros and Stauffer assert that a drug-related somatic state is causally associated with a diminished capacity and/or readiness to respond to sociopolitical stimuli and hence with diminished political activity.

My own research (see above, p. 16, note*) concerning the influence of people's health status on their political attitudes and behaviors also illustrates the basic biopolitical assumption. I find that: *a)* the individual's health status, a non-transient somatic characteristic, is significantly and consistently related to his political attitudes and behaviors; and *b)* some of these relationships are significantly mediated by environmental variables (e.g., by demographic variables). These findings are discussed at length in a later section of this paper.

The assumption that antecedent somatic states, in interaction with environmental conditions, have causal association with political behavior derives most clearly from that area of biopolitics informed by behavioral genetics. Indeed one may note the similarity between our basic assumption and this behavioral genetics conclusion recently quoted by Corning (1971, pp. 323-324; bracketed phrase added): "Neither the genetic parameter nor the environmental parameter alone can account for more than a portion of behavioral variability. With the development of [. . . a thorough going] interactional approach, a revitalized interest in the genetic basis of social behavior has been witnessed." More specifically, students of biopolitics have noted the evidence that such influences on behavior as intelligence, emotionality and aggressivity have considerable heritability. Applying our basic assumption, we recognize that these traits, and the genetic processes involved in their transmission, can have profound influence on people's capacity and predisposition to respond to the sociopolitical environment and hence on human political behavior.

Assumption 2

Many of the individual's non-transient somatic characteristics and recurrent somatic states have causal influence not only on political behavior but on the basic psychosocial processes which political scientists have most emphasized as yielding explanations of political behavior. Among the basic psychosocial processes which can be shown to be influenced by biopolitical variables are: political socialization; the linkage between personality and political attitudes; the elicitation of political behavior from politically predisposed individuals (via stimuli which 'arouse' or 'activate' the individual); and the frustration-aggression-displacement process underlying much political violence. It is also recognized that psychosocial processes can significantly mediate the relationships between biopsychological variables and political behavior.

The validity, as well as the centrality to biopolitics, of this set of assumptions deserves examination at length because, if valid, these assumptions suggest considerable change in the contemporary theory and practice of political science. If we can show that the predominant individual-level paradigms in political science either require or are significantly aided by biopsychological inquiry, then the case will have been made for a substantial broadening of basic theory in our discipline. Moreover, *if we can show that somatic variables significantly influence not only the dependent political behavior variables typically studied in political science but also the psychosocial processes ordinarily treated as the independent variables which underly and explain political behavior, then, I think, the case will have been made for the regular incorporation of somatic variables in much, if not most, political science research.*

To illustrate that major individual-level models in political science can be usefully informed by somatic variables, I will review in depth one such model, the predisposition-activation model which underlies almost all of the research literature on political participation. My review explicates the model as applied in political science, and summarizes the evidence from biopolitical theory and research which suggest that somatic factors influence the variables and subprocesses in this model.

Somatic Variables and the Predisposition-Activation Model

The vast preponderance of political science research on political participation (voting, election campaign involvement, etc.) proceeds from one or another variant of predisposition-activation theory in psychology (Milbrath, 1965; Verba / Nie, 1972). The central core of this theory, as applied in political science, states that a given behavior is to be explained as a combined function of the strength of the individual's predispostion to engage in the behavior and the strength of the stimuli which arouse the individual to act—presumably to act in accordance with his predispositions (for a fuller consideration of predisposition activation notions, see Schwartz / Zill, 1971). Using voting behavior as an example, we typically measure the strength of the respondent's sense of civic duty and previous turnout behavior as indications of his predisposition to vote; we measure the strength of his party identification, candidate preference and perceived issue agreement with the preferred candidate as indicators of this predispositions toward voting for a specific party or candidate; and we measure the strength (frequency) of the respondent's exposure to political stimuli (as in newspapers, television and the like) as a means of assessing stimuli-strength (Campbell / *et al.*, 1964). Our measures of the strength of these predispositions and our indirect measure of stimuli strength are then combined to yield a prediction concerning the probability and direction of the respondent's vote.

The political science application of predisposition-activation theory, which I will criticize below as too narrow, nonetheless has been sufficiently successful or popular to give rise to the questions: *a)* where do these participatory predispositions originate ?; and *b)* what processes account for the stability and/or change of these predispositions ? The response to these questions has been quite varied but, in the main, political scientists have come increasingly to see the origins of basic political predispositions in the socialization process (Easton / Dennis, 1969; Greenstein, 1965; Hess / Torney, 1967) and to explain the stability of such predispositions in terms of the predispositions' centrality, reinforcement, mutual constraint and personality relevance. In other words, we believe that people learn, evolve and change their political predispositions largely as a result of family, school, peer, media, community and workplace contacts; that these predispositions are further reinforced by the relative homogeneity or redundancy of such contacts; and that such predispositions will tend to remain stable to the degree that they are early learned, frequently reinforced, important to the individual, central to a mutually coherent or constrained attitude set and linked to the individual's basic self image or identity (Rohter, 1975).

Theory and research in biopolitics suggest that the individual's somatic characteristics influence each and every process involved in the predisposition-activation model. More specifically, the influence of somatic states on each of the following variables and processes has been indicated in biopolitical studies: socialization processes and outcomes; the formation and functioning of personality; the content of political attitudes or predispositions; the degree to which the individual exposes himself to political stimuli; the degree to which the individual is able or likely to respond to such stimuli as he does perceive; the probability that the predisposed individual will participate in political activities.

In briefly reviewing this complex literature, I will concentrate on the political influence of five somatic phenomena: hormonal variation, nutrition, drug-alcohol use, health status, and genetic processes.

Socialization and Somatic States

Socialization processes and outcomes have been found to be significantly related to each of the five somatic phenomena mentioned above, albeit to markedly different degrees and with profound differences in the extent to which explicitly political learning has been investigated. On hormonal variation, Dearden (1973) cites a number of studies relating male-female differences in hormone configurations to childhood learning. After summarizing research which demonstrates that androgen levels are strongly correlated with aggressivity in young children, Dearden describes an important study comparing male-female differences in aggressivity over a range of preadolescent years. This study found that the greatest differences in aggressivity showed up in the youngest age groups, before sex role socialization had been strongly reinforced in school situations, although substantial gender-related differences in aggressivity remained apparent in the older children. While not ignoring the influence of sex role socialization, Dearden interprets this finding as highlighting the important influence of hormone configuration on early learned response tendencies in children. Given our emphasis, in political socialization, on the importance of early learning in structuring later learning, this kind of finding should be at least thought provoking to political scientists.

Nutritional factors have been far more extensively related to socialization processes and outcomes. Birch and Gussow (1970), for example, pulled together a vast array of evidence on the psychosocial influence of nutritional deficiences and attendant health difficulties, evidence taken from national and cross-cultural studies as well as investigations on regional and smaller samples. Birch and Gussow document the existence of a coherent syndrome of nutritional, somatic, personality and attitude elements. I have elsewhere characterized this syndrome as involving malnutrition, gross physical underdevelopment, severe learning problems, social withdrawal, depression, high degrees of anxiety, a low sense of self-esteem and low or diminished capability to handle stress. Several of these elements—withdrawal and low self-esteem—have already been shown to have important political consequences.

The one somatic variable which has been linked both to socialization processes and to socialization outcomes, to general socialization and to specifically political socialization, is health status. On socialization processes, there is a wealth of data from large-scale studies in several countries showing unambiguously that health status significantly affects the relationship between socialization agencies and socializee (Pless / Douglas, undated; Pless / Roghmann, 1970). Less healthy children have been found to have consistently more limited and less satisfactory relationships to school, family and peer groups than have their healthier counterparts. Not surprisingly, then, health status also influences the outcomes of general socialization processes: sickly children—who comprise 5% to 20% of the children in developed counties and probably a higher proportion in less developed nations *(ibid.)*—tend to adopt appreciably more passive and negative attitudes toward both self and society than do healthier youngsters. The process and content of political socialization has also been shown to vary with the individual's health status: in a study of 2,100 American high school students (Schwartz, 1975), I found that less healthy students were significantly less involved in school activities, less knowledgeable about politics, less vicariously participant in politics and less likely to participate in sociopolitical activities *even if attitudinally predisposed to do so*. Health status was shown to have important influence on aspects of the individual's personality and these personality elements, in turn, were shown to have significant influence on political attitudes and behaviors.

Drug and alcohol use, and genetic processes, have also been linked to socialization phenomena. Briefly, Stauffer (1971) reviewed a sizeable body of literature showing that prolonged or heavy use of depressants, amphetamines and hallucinogens tends to depress the influence of family and workplace contacts on the individual; Jaros (1972) showed that one of the psychophysiological effects of depressants is to "desocialize" the individual, at least temporarily. Genetic processes have been empirically related to socialization only inferentially—as in a number of studies showing (White, 1972) that intelligence, a seemingly heritable trait, does have significant impact on political attitudes. As indicated above, the inherited psychological characteristics of emotionality and aggressiveness (summarized in Dearden, 1973) should also have major socialization effects such that, again, genetic processes may ultimately be linked directly to socialization. Thus far, however, the well validated twin methodology has not been employed in reported studies on political attitudes and the consistently low correlations between parents' and children's political attitudes (Jennings / Niemi, 1968) suggest caution in the interpretation of the inferential materials presently available.

Personality and Somatic Phenomena

It is, of course, quite arbitrary to separate personality from socialization because socialization processes obviously influence key personality elements and such elements, at least when they are well established, influence subsequent socialization processes. In the predisposition-activation model, personality is expected to influence the content of the individual's predispositions (people are presumed to adopt and maintain political attitudes which are consistent with their personalities: see Lane, 1972) and the stability of political predispositions (those attitudes which are most strongly linked to the individual's self image or identity are, *ex hypothese,* most likely to be stable over time, *ibid.*). Both of these personality functions are clearly related to socialization processes, as they both involve the learning, reinforcement and extinction of attitudes. We treat the somatic influences on personality as distinct from the somatic influences on socialization, here, solely for reasons of presentational clarity.

Again, each of the five somatic phenomena we are considering appear to have interesting associations with personality; again, the strength and scope of somatic influences varies widely over the five different somatic phenomena. Of these five, three are primarily related to the formation or content of personality; only nutrition and health status seem clearly to influence the political functioning of personality elements.

Hormonal variation has been linked to the content of personality structures in two ways: *a)* as we have seen, androgen levels—within and between the sexes—has been repeatedly shown to be associated with the personality trait of aggressiveness (Dearden, 1973); *b)* hormonal variation within-individuals-across-time has been shown to be associated with the content and oscillation of moods (Wessman / Ricks, 1966). We have also seen above that extensive drug-alcohol use is regularly associated with the personality trait of passivity (Stauffer, 1971). Several personality traits have been found to be at least partly transmitted genetically. In a study cited by Corning (1971, especially p. 350), 76% of the inter-individual variance in aggressiveness measured as a personality trait was attributed to genetic inheritance; similarly, Eysenck (1967) has found a tendency for introversion-extroversion to be an inherited personality characteristic.

Nutritional factors and health status influence both the content of personality structures and the ways in which personality operates in politics. Concerning the influence of nutrition on the content of personality structures, I have already described Birch and Gussow's findings on the personality syndrome typically associated with nutritional deficiences. Concerning nutritions role in the influence of personality on politics, I need only refer the reader to James C. Davies' pathbreaking work (1963) relating Maslow's hierarchy of personality needs to political behavior. Davies convincingly demonstrates that people whose energies must be primarily devoted to the satisfaction of basic physical needs, including nutritional needs, simply cannot engage in sustained political behavior. While not as focused on personality as is Davies' work, Stauffer (1969) documents an empirical relationship between nutritional sufficiency and political phenomena across a range of nation-states.

Investigations of the influence of health status on personality yields similar results. A national survey based study on American adults shows that health status is significantly associated with the individual's self images or identity and that health status also significantly mediates the relationships between self images and political behavior (data available on request).

Specific Political Predispositions and Somatic Phenomena

As somatic phenomena influence the socialization and personality processes which underly the formation and stability of political predispositions, they must indirectly influence the specific content of the individual's political predispositions. To my knowledge, however, only the health status variable has been directly linked to specific political attitudes. The national study mentioned above was designed, in part, to test the hypothesis that poor health would be significantly associated with passive and negative attitudes toward the polity. Different health status measures were found to have just such associations: various indicators of poor health were linked to political alienation, powerlessness, reformism, low deference to political authority, low civic duty, low belief in the effectiveness of elections as a political linkage mechanism, and high expectations political violence (see below).

Exposure to Political Stimuli and Somatic Phenomena

In political science applications of the predisposition-activation model, the frequency with which the individual exposes himself to political stimuli is a crucial variable in that it is a surrogate measure for the strength of political stimuli. (It will be recalled that the greater the strength of the stimuli, the more likely is the predisposed individual to be roused or activated to political activity.) Three of the somatic phenomena we are considering here have been linked in theory the to individual's predisposition to expose himself to, or be aware of, political stimuli: nutrition, drug use and health status. Davies (1963) has shown how preoccupation with nutritional needs can so strongly influence the individual's perception of the world as to make him block out stimuli not immediately related to these needs (or to substantially distort such stimuli). Stauffer (1971) has shown that extensive drug-alcohol use leads to a generalized withdrawal from the normal social contacts which might be expected to transmit and interpret political stimuli; my studies on health provide some support for the hypothesis that low health status tends to diminish attention to, and/or the impact of, political stimuli (some of these data are presented below).

*The Response to Political Stimuli and Somatic Phenomena**

Political science studies employing the predisposition-activation theory have virtually ignored a crucial intervening variable in that theory, the individual's arousal level or readiness to respond to political stimuli. Such studies have tended to assume, tacitly, that people whose survey responses yield identical predisposition scores (say on strength of party I.D., candidate preference, intention to vote, etc.) are in fact identical in their readiness to respond to political stimuli. This assumption seems quite dubious not only in light of the relative discriminatory weakness of survey techniques (Wahlke / Lodge, 1970) but because a great variety of somatic, psychological and social factors have been shown to influence both the individual's general or characteristic arousal level and his situationally-specific readiness to respond to external stimuli. Here, of course, we are interested in the somatic influence on such readiness or arousal.

Hormonal factors seem to have very clear, strong influence on the individual's readiness to respond to external stimuli. Indeed, increasing the individual's readiness to respond seems to be a major function served by adrenaline secretion. Davies (1970) has reasoned that the stable adrenaline

* See Schwartz / Zill, 1971; Schwartz, 1974, especially pp. 121-123.

balance (i.e., adrenaline/nonadrenaline ratio) in different species may well convary with the tendency to aggression among these species. Hummel (1970) has theorized that individual differences in adrenaline secretions under conditions of high stress help to account for individual differences in the degree of emotionality and aggressiveness of response to such stress. Empirical work relating either general or situational differences in adrenaline secretion to differences in human political behavior has yet to be accomplished but, in light of the promising empirical work in other disciplines and on other animals, such research clearly seems called for.

Nutritional deficiency and the extensive use of depressants and hallucinogens also seem to influence the individual's readiness to respond to external stimuli. Chronic malnutrition is associated with low energy, a characteristic inconsistent with high reactivity to stimuli (Birch / Gussow, 1970); most depressants and hallucinogens tend to induce passivity in the face of otherwise activating stimuli (Stauffer, 1971).

Finally, health status also impacts upon the individual's readiness to respond to external stimuli. In our national study on health and politics, we found that health status consistently mediated the relationships between exposure to political stimuli and political participation. People who were exposed to a great deal of political stimuli, and who were highly predisposed to engage in political activities, nonetheless tended to be quite nonparticipant if they experienced a variety of even relatively mild health difficulties (some of these data are presented below).

The reasoning and evidence adduced above is, I think sufficient to illustrate that many of the individual's basic somatic characteristics have influence, and arguably causal influence, on several of the psychosocial processes which political scientists have most emphasized as explanations of political behavior. These somatic states, then, may be said to have at least indirect influence on political behavior. Our data on health status and political behavior, presented in a later section of this chapter, tends to suggest some rather direct influences of somatic states on political behavior.

If space permitted, it could be illustrated that somatic variables have similar influence on the processes underlying political violence and elite decision-making. But I think the point has been made strongly enough: biopsychological variables appear to merit a far more complete incorporation into political science theories and research than they have yet received.

Assumption 3

Part of the influence of somatic characteristics on political behavior is exerted directly, or at least through mechanisms not presently studied in political science. Alternatively stated, not all of the somatic influence on political behavior is encapsulated or mediated by the underlying psychosocial processes—like socialization, personality functioning or the activation of political predispositions—now included in political science research.

This assumption may already be tacitly accepted by some researchers but it has not been commonly expressed in biopolitics research. Nevertheless, it is crucial that this assumption be made explicit, examined carefully and, if valid, insisted upon—if biopsychological variables are to be regularly incorporated into political science research. Why is this so?

Very simply, *the consequences of a successful subdiscipline of biopolitics will be to require the introduction of large numbers of new variables and new methods into political science and to force a dramatic broadening, if not fundamental restructuring, of political science theory.* The dictates of parsimony, and the appropriate skepticism of most political scientists, oppose such changes unless and until the necessity of change is persuasively documented by plausible new theory confirmed in new, strong, relevant data. If, therefore, all of the influence of somatic states on political behavior can be assumed to be indirect, can be assumed to be wholly encapsulated or mediated by variables already studied in political since, then the biopsychological theories, variables and methods can be ignored or minimized—at least as regards studies of political behavior *per se.* (Presumably, the evidence offered above on belahf of assumption 2 will require students of political socialization, personality and politics, etc., to include some somatic variables in their studies.)

The assumption stated here, of course, holds that biopsychological variables cannot properly be ignored or minimized because at least some of their influence on political behavior is direct, is not entirely encapsulated or mediated by processes already studied in political science.

Unfortunately, the general validity of this assumption cannot now be unambiguously assessed because many of the most promising somatic variables have not been operationally measured in studies which include political behavior measures as dependent variables and measures on socialization or personality or other putatively intervening variables. All too often, students of biopolitics have used the following fragmentary or inferential strategy of research:

a) Cite literature from one or more of the life sciences to suggest a linkage between somatic factors and some psychological or environmental variable.

b) Test an hypothesized relationship between political behavior and that psychological or environmental variable.

c) From the observed significant association between the political variable and the psychological or environmental variable, infer an association between the unobserved somatic variable and the political variable.

To be sure, almost all of the researchers who have pursued this strategy have entered all the appropriate *caveats,* have fully understood that they were developing plausible theories not documenting the relationships posited in their theories. *I am not faulting these scholars in any way.*

But the present state of biopolitics, and of political science, will suffer grievously if we do not move beyond this state of the science! And we *can* quickly move beyond this state because all of the somatic variables which have been discussed in this chapter, all of the somatic factors whose political impact has been explored only inferentially, can be measured rather directly by well validated techniques. What is required is the specification of appropriate interdisciplinary research designs and the effort/resources to to execute these designs.

The next section of this chapter is devoted to a research report on two interdisciplinary studies exploring the influence of the individual's physical health status on his/her political attitudes and behaviors. These studies were designed to elaborate and test the assumptions identified above and therefore, in addition to such substantive interest as they may have, the studies may be helpful to other biopolitical investigators in developing appropriate research designs.

The Influence of Health on Political Behavior: Two Corroborative Studies

In recent years, I have undertaken a series of five large-scale, survey-based efforts to examine the influence of the individual's physical health status on his political attitudes and behaviors. The first two of these efforts are reported here. The first study was conducted in 1970 on a population of 198 American adults participating in a regional health maintenance program in the urban Northeast. This program provided the interdisciplinary team of investigators with a large body of health data based upon repeated clinical examination of each participant by staff physicians as well as a standardized health history and a record of each participant's use of drugs and medicines. These physician-based health assessments, in various analytic combinations described below, were then related to political attitude and behavior measures obtained in a mailed survey (after telephone contact with the participants). The second study was conducted by the author in 1973 on a national area-probability sample of 827 American adults.* This study used an extensive battery of self-report items, adopted from health assessment inventories used by the United States Public Health

* The study was conducted by leaving a self-administered questionnaire with a national probability sample of 3,019 American adults. This sample was constructed by Audits and Surveys, Incorporated, by: 1) combining the 3103 counties and county equivalents into 1900 primary sampling units; 2) stratifying these primary sampling units by region, metropolitan-nonmetropolitan, size and other standard demographic criteria; 3) selecting blocks and final area segments at random from the primary sampling units; 5) selecting households by including every "n[th]" unit from a random starting point. The respondent to be surveyed, in each household, was selected at random by the interviewer using guidelines as to randomization by age and sex established by Audits and Surveys.
Because of the highly personal character of many of the survey items, and to maximize cost effectiveness, we chose a self-administered, mailed-back design for the completion of the survey instrument. Recognizing that such a procedure would inevitably lead to a lower and less certainly representative response rate than a personal interview strategy, we also conducted two studies designed to validate our procedures as to representativeness. First, a systematic random sample of 78 non-respondents were interviewed by telephone; secondly, a national random sample of 113 adults were interviewed in person. No statistically significant differences were observed between the responses of the 827 persons returning the questionnaire and either of the other samples. These findings tend to lend support to the general validity of our data by indicating an absence of systematic bias in our procedures. In order to project the sample results to the population it purports to represent, weights were applied in the form of ratio estimates to ensure that certain basic statistics in the sampling operation coincided with known population characteristics. Ratio estimates were applied for 9 separate categories including religion, sex, status and locale-type. A separate validation effort to test the validity of our weighting procedures revealed no significant differences between the weighted population distributions by age and sex in our study and those in the Federal Census.

Service and various regional public health surveys, to assess the respondent's present and previous health statues.

Measures of the respondent's political attitudes and behaviors—and of several demographic personality and life stress variables—were included in the survey instrument, the instrument having been distributed in person by an interviewer in the respondent's home and mailed back after self-administration.

It should be stressed that all of our respondents were healthy enough to be functioning in jobs, schools or as homemakers, none of them were confined to their homes or to health care institutions. We are studying, here, the influence of variations in health status on political participation among persons physically capable of such participation.

The specific measures and findings in each study are described separately below. To avoid repetition, however, the three basic hypotheses which were common to these studies are stated below:

a) The individual's present health status and/or average adult status will be significantly related to the strength of his political attitudes and the level and type of his political participation.

b) The relationships between health status and political behavior will be significantly, but not totally, mediated by the strength of political attitudes and by environmental variables (e.g., demographic variables locating the individuals on several socio-environmental dimensions).

c) The relationships between the strength of the individual's political attitudes and his level of political participation will be significantly mediated by his health status. Alternatively stated, health status will significantly influence the probability that politically predisposed individuals will act on their predispositions.

A Regional Study on Health and Politics: Measures and Data

Seven analytically distinct measures of the individual's health status were created from the health records described above. These were:

a) Life Health Status: an index composed of three equally weighted components—the individual's total number of different clinical diagnoses, the ratio of organ systems in which an illness was discovered to the total number of organ systems examined and the total number of self-reported health difficulties.

b) Age-Corrected Life Health Status: the above reported index standardized by age.

c) Perceived Present Health: a single-item indicator used primarily to construct indices relating clinical and perceptual assessments of health.

d) Age Corrected Perceived Present Health Status: #3 standardized for age.

e) Health Perturbation: the relationship between perceived childhood health status and perceived present health status (ranging from poor childhood and present health through several intermediate or mixed categories to good childhood and adult health).

f) Perceived Health Development: an index composed of two equally weighted variables—health perturbation and a variable expressing the relationship between perceived present health and several clinical assessments of health.

g) Illness Level: a simple count of all diagnosed illness and abnormalities (using physician-based data only, excluding all self-reports not corroborated by physicians).

Our political participation measures yielded the following indexes:

a) General Political Participation: the number of different types of political participation in which a respondent reported himself ever to have engaged (from a list of six types of activities including 'work for political candidates,' 'to go political meetings,' 'sign petitions,' 'write Congressmen,' 'go to neighborhood meetings,' etc.)

b) Recent Political Participation: the total number of such activities in which the respondent engaged during the year prior to our study.

c) Recent Voting Participation:* the frequency with which respondents reported themselves to have voted in general and/or primary elections during the year prior to our study.

d) Voting 1964-1968: the frequency with which respondents reported themselves to have voted in general and/or primary elections in the Presidential election years of 1964 and 1968.

Our measures of the strength of political attitudes were multi-item indices constructed by aggregating the responses to closed-ended, "agree-disagree" type items after assuring ourselves of the appropriateness of such aggregation by close inspection of the inter-item correlations among the items in each desired index. We constructed indexes on a wide range of attitudes, including political interest, alienation, inefficacy, reformism and withdrawal. The items used to construct these indexes are of the form familiar to most political scientists, many of them were adapted from previous studies such as those of the Michigan Survey Research Center; a complete listing of the items in each index, and their inter-item correlations, are available on request.

Findings

The data resulting from the procedures described above were analysed twice: once permitting the health indices to vary across their natural spectrum (by dividing the indices, at natural "breakpoints," generally approximating quartiles) and a second time by dividing each index so as to contrast the relatively less health subpopulation (averaging 22% on our seven indices) with the relatively healthier subpopulation (avering 78%). These analyses are presented separately here.

* Several of our studies have revealed the utility of separating voting-type participation out from other forms/types of political participation because very different causal patterns between the forms of participation often exist which are obscured by combining them. Reader's interested in learning about the relationship which emerge in our data when voting-type participation is analytically combined with the other forms are invited to request this information from the author.

The Spectrum of Health Statuses and Political Behavior

The data arrayed in Table 1 provide at least modest support for the hypothesis linking health status to level of political participation. Almost all of our health status measures have two or more significant, consistent associations with the political participation measures; there are twenty such associations, in the .20 to .41 range, out of thirty five possible cases. Good health then, does tend to be associated with political participation: poor health does appear to be a barrier to political involvement.

The health status measures in this study tend to bear significant and consistent association with few political attitudes, however. While poor health does seem to be linked to political inefficacy and to diminished political interest, it is regular associated with no other political attitude.

Table 1 should also cue us to the potential importance of the individual's self-perceptions of health (and perhaps of self-images generally) because: *a)* the one health measure which entirely excludes the respondent's self report on health ("illness level") has the fewest significant associations with the political variables; and *b)* the general absence of significant associations between health status and political attitudes should suggest that the processes by which health influences political behavior are not primarily through political attitudes.

Our second hypothesis, that political attitudes and demographic variables would significantly but not totally mediate between health and political behavior, was tested by introducing each putatively mediating factor as a control on the bivariate relationship between each pair of relevant health and political behavior variables. A significant mediation was defined, as is typical, as one in which the control variable results in a difference equal to or greater than .10 in the level of association between the initial variables (or in "reducing" the relationship to statistical insignificance). Tables 2 and 3 summarize the results of these analyses, using two attitudinal mediators (political interest and efficacy) and two demographic mediators (age and sex).

Inspection of Table 2 reveals that the political attitudes did significantly mediate the relationships between health and political participation in a majority of cases and that the link between health and political behavior seems to be strongest among persons whose political interest and sense of efficacy are of moderate strength. This later finding seems to indicate that poor health tends to be an insufficient barrier to participation among highly interested, efficacious respondents and that good health, alone, will not lead to political involvement among persons who are very disinterested or feel powerless, in politics—a finding which intuitively appears reasonable. More detailed analysis of the tabular data which underly the summary statistics presented in Table 2 is clearly called for to aid in the interpretation of this finding but, frankly, limitations of space prohibit further discussion of this matter here. It should be noted, however, that about half of the significant mediations in Table 2 are associations in the "nonmoderate" cells suggesting, as hypothesized, that attitudes do not totally intermediate the health-to-politics linkage.

The data in Table 3 suggests that both age and sex tend to influence the relationship between health and political participation. As might be expected, health difficulties appear to have more pronounced impact in

TABLE 1. *The relationship between health status and political variables in a regional study of health and politics (associations expressed as gamma coefficients)*

	Life health status	*Age corrected life health status*	*Perceived present health*	*Age corrected perceived health*	*Health perturbation*	*Health development score*	*Illness level*
General political participation	.40 (.02)	.37 (.006)	.37 (.02)	.41 (.008)	.41 (.04)	*	.31 (.04)
Recent political participation	.37 (.001)	.25 (.08)	.34 (.07)	.34 (.03)	*	*	*
Recent voting participation	*	*	.28 (.02)	.31 (.002)	.33 (.02)	.29 (.001)	*
1968 voting participation	*	*	.25 (.04)	.38 (.001)	.36 (.004)	.33 (.009)	*
1964 voting participation	*	*	*	.27 (.01)	.21 (.04)	*	*
Political interest	*	*	*	.29 (.003)	*	.21 (.09)	*
Political efficacy	.26 (.07)	*	.29 (.02)	*	*	.29 (.08)	*
Political alienation	*	*	*	*	*	*	*
Political withdrawal	*	*	*	*	*	.24 (.06)	*
Political reformism	*	*	*	*	*	*	*

* Not significant, otherwise significance levels given in parentheses.

N.B. All health indices range from poor to good health; all political variables range from high to low.

TABLE 2. *The effect of political attitudes on the relationships between health status and political participation (associations expressed as gamma coefficients)*

	Control variable	High values control variable	Moderate values	Low values
Life health/general participation	Political interest	.27 (ns)	.52 (.03)	.44 (.06)
Life health/recent participation	Political interest	.22 (ns)	.53 (.08)	.40 (.06)
Age linked life health/general participation	Political interest	.16 (ns)	.54 (.002)	.64 (ns)
Age linked life health/recent participation	Political interest	.05 (ns)	.59 (.002)	.45 (.07)
Perceived health/general participation	Political interest	.39 (ns)	.37 (ns)	.38 (ns)
Perceived health/recent participation	Political interest	.50 (ns)	.30 (ns)	.21 (ns)
Perceived health/voting	Political interest	.12 (ns)	.42 (.04)	.12 (ns)
Perceived health/voting 1968	Political interest	.31 (ns)	.20 (.08)	.33 (ns)
Age linked perceived health/general participation	Political interest	.29 (ns)	.37 (ns)	.34 (ns)
Age linked perceived health/recent participation	Political interest	.31 (ns)	.29 (ns)	.16 (ns)
Age linked perceived health/voting	Political interest	.24 (ns)	.39 (ns)	.22 (ns)
Age linked perceived health/ voting 1968	Political interest	0.6 (ns)	.14 (ns)	.46 (.05)
Age linked perceived health/ voting 1964	Political interest	.05 (ns)	.19 (ns)	.07 (ns)
Health perturbation/general participation	Political interest	.24 (ns)	.28 (ns)	.54 (ns)
Health perturbation/voting	Political interest	.37 (ns)	.40 (.03)	.05 (ns)
Health perturbation/voting 1968	Political interest	1.0 (ns)	.18 (.03)	.56 (.08)
Health perturbation/voting 1964	Political interest	1.0 (ns)	.24 (.08)	.20 (ns)
Health development/voting	Political interest	.30 (ns)	.20 (.06)	.06 (ns)
Health development/voting 1968	Political interest	.07 (ns)	.21 (ns)	.39 (ns)
Illness level/general participation	Political interest	.07 (ns)	.40 (.002)	.55 (ns)
Life health/general participation	Efficacy	.33 (ns)	.54 (04)	.16 (ns)
Life health/recent participation	Efficacy	.53 (.001)	.44 (06)	.17 (ns)
Age linked life health/general participation	Efficacy	.36 (ns)	.50 (.002)	.11 (ns)
Age linked life health/recent participation	Efficacy	.38 (ns)	.38 (ns)	.30 (ns)
Perceived health/general participation	Efficacy	.30 (.03)	.56 (.02)	.12 (ns)
Perceived health/recent participation	Efficacy	.18 (ns)	.39 (.07)	.13 (ns)
Perceived health/voting	Efficacy	.48 (ns)	.20 (ns)	.10 (ns)
Perceived health/voting 1968	Efficacy	.67 (ns)	.20 (ns)	.09 (ns)
Age linked perceived health/ general participation	Efficacy	.43 (.06)	.54 (.01)	.01 (ns)
Age linked perceived health/recent participation	Efficacy	.16 (ns)	.43 (.03)	.18 (ns)
Age linked perceived health/voting	Efficacy	.49 (ns)	.35 (.02)	.09 (ns)
Age linked perceived health/ voting 1968	Efficacy	.80 (.04)	.39 (.03)	.09 (ns)
Age linked perceived health / voting 1964	Efficacy	.48 (ns)	.27 (.06)	.07 (ns)
Health perturbation/general participation	Efficacy	.12 (.03)	.72 (.02)	.74 (ns)
Health perturbation/voting	Efficacy	.12 (ns)	.33 (ns)	.43 (ns)

TABLE 2. *(Continued)*

	Control variable	*High values control variable*	*Moderate values*	*Low values*
Health perturbation/voting 1968	Efficacy	.45 (ns)	.29 (ns)	.17 (ns)
Health perturbation/voting 1964	Efficacy	.17 (ns)	.43 (ns)	.28 (ns)
Health development/voting	Efficacy	.13 (ns)	.29 (.001)	.38 (ns)
Health development/voting 1968	Efficacy	.32 (ns)	.29 (ns)	.22 (ns)
Illness level/general participation	Efficacy	.24 (ns)	.45 (ns)	.15 (ns)

TABLE 3. *The effects of age and sex on the relationships between health status and political participation (associations expressed as gamma coefficients)*

Relationships	*Male*	*Female*	*Age 19-44*	*Ages 45 +*
Life health/general participation	.39 (ns)	.38 (ns)	.37 (ns)	.36 (ns)
Life health/recent participation	.19 (ns)	.43 (.02)	.40 (.03)	.26 (ns)
Life health/voting	.04 (ns)	.47 (.03)	.44 (.09)	.07 (ns)
Life health/voting 1968	.12 (ns)	.36 (ns)	.27 (ns)	.02 (ns)
Life health/voting 1964	.09 (ns)	.05 (ns)	.08 (ns)	.16 (ns)
Age linked life health/general participation	.38 (ns)	.32 (.07)	.41 (.009)	.33 (ns)
Age linked life health/recent participation	.10 (ns)	.26 (ns)	.33 (.04)	.13 (ns)
Age linked life health/voting	.05 (ns)	.21 (ns)	.23 (ns)	.03 (ns)
Age linked life health/voting 1968	.27 (ns)	.25 (ns)	.08 (ns)	.09 (ns)
Age linked life health/voting 1964	.39 (.05)	.21 (ns)	.11 (ns)	.25 (ns)
Perceived health/general participation	.45 (.001)	.33 (ns)	.24 (ns)	.64 (.09)
Perceived health/recent participation	.51 (ns)	.30 (.07)	.20 (ns)	.49 (ns)
Perceived health/voting	.50 (ns)	.53 (.001)	.20 (.06)	.39 (ns)
Perceived health/voting 1968	.40 (ns)	.46 (.001)	.02 (ns)	.37 (.03)
Perceived health/voting 1964	.05 (ns)	.22 (.05)	.11 (ns)	.45 (.07)
Age linked perceived health/general participation	.40 (.002)	.38 (ns)	.26 (ns)	.59 (.001)
Age linked perceived health/recent participation	.57 (.07)	.26 (.0)	.23 (ns)	.46 (.09)
Age linked perceived health/voting	.02 (ns)	.46 (.002)	.30 (.009)	.31 (ns)
Age linked perceived health/voting 1968	.12 (ns)	.50 (.001)	.15 (ns)	.57 (.002)
Age linked perceived health/voting 1964	.36 (ns) o	.25 (.05)	.01 (.05)	.53 (.03)
Health perturbation/general participation	.68 (.001)	.24 (ns)	.18 (ns)	.63 (.09)
Health perturbation/recent participation	.09 (ns)	.42 (.08)	.08 (ns)	.54 (.07)
Health perturbation/voting	1.0 (ns)	.54 (.001)	.43 (.08)	.54 (.07)
Health perturbation/voting 1968	.43 (ns)	.53 (.001)	.20 (ns)	.41 (.04)
Health perturbation/voting 1964	.28 (ns)	.21 (.02)	.04 (ns)	.29 (.02)
Health development/general participation	.34 (.001)	.06 (.02)	.08 (ns)	.39 (.01)
Health development/recent participation	.03 (ns)	.32 (ns)	.06 (ns)	.38 (.07)
Health development/voting	.06 (ns)	.41 (.001)	.30 (.01)	.29 (.03)
Health development/voting 1968	.22 (ns)	.39 (.004)	.37 (ns)	.30 (.05)

TABLE 3. *(Continued)*

Relationships	*Male*	*Female*	*Age 19-44*	*Ages 45 +*
Health development/voting 1964	.08 (ns)	.15 (ns)	.12 (ns)	.13 (.01)
Illness level/general participation	.43 (ns)	.19 (.08)	.25 (ns)	.42 (ns)
Illness level/recent participation	.11 (ns)	.23 (.06)	.28 (ns)	.05 (ns)
Illness level/voting	.05 (ns)	.09 (ns)	.1 (ns)	.08 (ns)
Illness level/voting 1968	.06 (.09)	.15 (ns)	.02 (ns)	.04 (ns)
Illness level/voting 1964	.05 (ns)	.11 (ns)	.14 (ns)	.26 (ns)

limiting the political participation of older respondents. In this data, the relationships between health and political behavior tended to be stronger among women than among men *but this finding was not corroborated in our national data (reported below).*

Our third hypothesis, suggesting that health factors mediate between political attitudes and political participation, was tested in much the same fashion as was described immediately above (i.e., by introducing health status measures as controls on the relationships between political attitudes and political participation, with the same standards applied as the definition of a significant mediation).

The data arrayed in Table 4 clearly demonstrate that health status can and does function as an intervening variable between political attitudes and behaviors. The individual's health does influence the degree to which he acts on one important variable, political interest. In Table 4, health status mediates the relationships between political attitudes and behaviors in a majority of cases; in most of these, the link between attitudes and behavior is strongest among persons in relatively poor health.

Several complementary interpretations of these data can be offered. First, it may well be that persons in relatively poor health must husband their energies to a relatively greater degree than do their healthier counterparts, such that political interest is a better predictor of political participation among these people than among healthier persons whose excess energies can be expended even on activities in which they are not strongly interested. Second, it could be argued that poor health should reinforce the nonparticipatory orientations of persons who are not very interested in politics. Finally, it might be contended that poor health should interfere with even the strong predispositions to participate of those who are keenly interested in political affairs.

Inspection of the tabular data underlying the summary statistics presented here gives the clear impression that all three of these processes are operative among our respondents. Among respondents enjoying ebullient good health, almost everyone participates—whether or not strongly interested in politics—hence the relatively low attitude-to-behavior associations in this subpopulation. In contrast, moderate health difficulties suffice to drive many less interested citizens to non-participation but does not substantially depress the participation of those keenly interested in public affairs. In other words, respondents in the middle ranges of physical health status tend

TABLE 4. *The effects of health status on the relationships between political interest and political participation (associations expressed as gamma coefficients)*

Participation variable	*Health variable*	*Poor health*	*Moderate health*	*Good health*
General participation	Life health	.39 (ns)	.39 (ns)	.20 (.06)
Recent participation	Life health	.23 (.09)	.37 (.03)	.11 (ns)
General participation	Age linked life health	.64 (ns)	.50 (.06)	.24 (ns)
Recent participation	Age linked life health	.56 (.03)	.42 (.05)	.38 (ns)
General participation	Health development	.39 (.09)	.52 (.04)	.39 (ns)
Recent participation	Health development	.59 (ns)	.56 (ns)	.39 (ns)
Voting	Health development	.17 (ns)	.41 (ns)	.41 (ns)
Voting 1968	Health development	.86 (.001)	.56 (.04)	.63 (.05)
General participation	Perceived health	.29 (ns)	.33 (ns)	.46 (ns)
Recent participation	Perceived health	.19 (ns)	.12 (ns)	.48 (ns)
Voting	Perceived health	.18 (ns)	.40 (.10)	.29 (ns)
Voting 1968	Perceived health	.77 (.005)	.58 (.007)	.61 (ns)
Voting 1964	Perceived health	.58 (.03)	.42 (.07)	.66 (.05)
General participation	Age linked perceived	.24 (ns)	.36 (ns)	.38 (ns)
Recent participation	Age linked perceived	.20 (ns)	.15 (ns)	.34 (ns)
Voting	Age linked perceived	.43 (ns)	.35 (ns)	.12 (ns)
Voting 1968	Age linked perceived	.88 (.02)	.52 (.02)	.70 (.03)
Voting 1964	Age linked perceived	.63 (.03)	.41 (.07	.71 (.02)
Recent participation	Health perturbation	.29 (ns)	.66 (ns)	.30 (ns)
Voting	Health perturbation	.18	.44 (ns)	.42 (.03)
Voting 1968	Health perturbation	.77 (ns)	1.0 (.001)	.53 (.008)
Voting 1964	Health perturbation	.59 (ns)	.70 (ns)	.47 (.01)
General participation	Illness	.56 (.06)	.24 (ns)	.28 (.08)

to participate less frequently than do healthier respondents, but the lower levels of participation in moderate health ranges are attributable largely to the politically disinterested citizens. Finally, high levels of health difficulties (i.e., poor health status) begin to depress the participation of even strongly interested citizens. Poor health status, however, typically drives almost all of the less interested citizens out of participatory activities (while only beginning to diminish the activity levels of more interested persons), hence the higher gammas between political interest and participation in the poor health subpopulation.

Contrasting the Relatively Less Healthy Subpopulation with the Healthier Population: A Brief Discussion

As might be expected from both substantive and statistical reasoning, dichotomizing our health indexes to contrast the least healthy subpopulation (22%) with the rest of the population, reveals a stronger pattern of associations between health status and political participation (here the range of bivariate associations tends to be between .30 and .60 rather than the previous range of .20 to .40). Some differences in the nature of the pattern are also revealed. Contrasting Table 5, below, with Table 1, for example we find that indexes including physician-based health assessments become better predictors of behavior and some of the more perceptual and more

Table 5. *The relationships between dichotomized health status indexes and measures of political participation and political attitudes (associations expressed as gamma coefficients)*

Political variables	*Life health*	*Age linked life health*	*Perceived health*	*Age linked perceived health*	*Health perturbation*	*Health development*	*Illness level*
General participation	.41 (.06)	.50 (.01)	.59 (.008)	*	*	*	.35 (.03)
Recent participation	.55 (.01)	*	.52 (.02)	*	.41 (.04)	*	.33 (0.8)
Voting	*	*	.52 (.01)	*	.33 (.02)	*	*
Voting 1968	*	*	.55 (.003)	*	.36 (.004)	*	*
Voting 1964	*	*	.42 (.04)	*	.21 (.04)	*	*
Political interest	*	*	.35 (.01)	*	.22 (.05)	*	*
Efficacy	*	*	*	*	*	*	*
Alienation	*	*	*	*	*	*	*
Withdrawal	*	*	*	*	*	*	*
Reformism	*	*	*	*	*	*	*

N.B. Starred entries indicate associations which are below .20 gamma and/or statistically insignificant.
Otherwise significance levels are given in parentheses.

age-linked health indexes become worthless as predictors. Note, too, that virtually none of the associations between health indexes and attitudes which obtained in Table 1 are observable in Table 5, a finding which strongly suggests that in this population, poor health depresses political participation in ways which are essentially independent of the political attitudes held by the respondents.

I believe that the findings of our regional study are, on the whole, clear and consistent: health status tends to have a significant, general and arguably causal influence on the level of an individual's political participation. This influence, moreover, is reasonably direct (e.g., is only partly mediated by political attitudes) and, operates, in part, by mediating the degree to which the individual acts upon his political predispositions.

A National Study on Health and Politics: Measures and Data

Of the more than thirty indexes of health status which were generated from the forty seven separate health items on the national survey instrument, we have selected five indexes to report here. These have been selected as representing the broad diversity of alternative conceptualizations of health status in the rich literature of the public health field (for an excellent review of the rich and varied literature on health status assessment, see Patrick / *et al.,* 1972). These variables, discussed below, were constructed by setting the magnitude of present clinical health indicators equal to that magnitude which the respondents identified as characteristic of an average adult year. More specifically: *a)* a difference score for each respondent was computed between the individual's perceived present health status and average adult health status as represented on a self-anchoring scale; *b)* this difference score was then employed to adjust each clinical variable from its initial level (representing present health) to new, analytically-defined "average" or "characteristic" levels (a procedure justified, *inter alia,* by the very high association between perceived and clinical health (Gamma = .89). The specific components of each index so constructed are listed below.

a) Characteristic chronic illness experience: a summative index of fourteen medical conditions which are chronic in nature.

b) Characteristic symptom experiences: derived from a checklist of eleven common symptoms.

c) Characteristic comprehensive health status: based on disability in social functioning, because of health conditions, impaired hearing or vision, chronic medical conditions, the symptom checklist, and energy level items.

d) Characteristic health restrictions: an index based on the number of days on which the respondent reported himself to have limited normal activities (e.g., working) due to health difficulties.

e) Scope of health restrictions: a measure of limited social functioning derived not from the days of restricted activity but from the behaviors eschewed for reasons of health.

Our political participation measures derive from a thirty three item battery in which the respondent was asked to indicate whether or not he had engaged in each of eleven types of activities for three different time periods. Using factor analyses and inter item correlation matrices as guides in constructing indexes matched to our theoretical purposes, we constructed the following nine political participation indexes.

a) General voting participation: frequency of voting in elections and primaries, all time periods.
b) Recent voting participation: #a only for 1972.
c) Presidential year voting: #a for 1968 and 1972.
d) General impersonal participation: frequency of petitioning and writing representatives, all time periods.
e) Recent impersonal participation: #d only for 1972.
f) General interpersonal participation: frequency of campaign activities for all time periods.
g) Recent interpersonal participation: #f only for 1972.
h) General protest participation: frequency of protest activity self-reported for 1972 + "usually."*
i) Recent protest participation: #h only for 1972.

Our political attitude indexes, constructed as were those in the regional study, included measures on political interest, alienation, civic duty, reformism, expectations of future political violence in America, efficacy, and withdrawal. Again, scholars interested in the specific items in each index and/or the inter-item correlations are invited to request these materials.

Findings

The summary statistics in Tables 6 and 7 are the results of analyzing the national data after dichotomizing the health indexes to contrast the least healthy subpopulation (averaging 15% of the sample on our five health dimensions) with the healthier population (the remaining 85%). We choose to present this subset of our findings because those analyses in which we permitted health status to vary more or less continually over its full spectrum reveal a similar pattern of findings but one which is much weaker in the strengths of association (hence less interesting and less useful for more refined analysis).

Inspection of Table 6 indicates, again, that health status does tend to influence an individual's degree of political participation; there are some twenty significant, if generally quite modest, associations between health and participation in that table. Table 7 clearly indicates, once again, that characteristic health status tends to have very limited, if any, general influence on political attitudes. (Present health status, health status assessed at the same time as were the attitudes, has somewhat greater influence.) The bivariate national data, then, seems broadly corroborative of our regional findings.

As in the regional study, we examined the degree to which the associations between health and political behavior were mediated by demographic variables and by political attitudes. Eschewing tabular presentations of these analyses for reasons of space limitation, we found: *a)* social class tends

* We exclude 1968 protest activity because preliminary analyses reveal dramatically different causal patterns for protest in that year, patterns requiring explication at greater length then can be made here.

TABLE 6. *The relationship between dichotomized characteristic health status variables and political behavior in a na. survey (associations expressed as gamma coefficients)*

General voting populations	*Characteristic comprehensive health status*	*Characteristic chronic illness experiences*	*Characteristic symptom experiences*	*Characteristic health restrictions*	*Scope o health restrictio*
General voting participation	*	*	.20 (.004)	*	*
Recent voting participation	.23 (.001)	*	.25 (.04)	*	*
Presidential voting participation	.20 (.001)	*	.24 (.001)	.21 (.004)	*
General impersonal participation	.21 (.002)	.28 (.04)	*	*	*
Recent impersonal participation	.22 (.001)	*	*	*	.22 (.00
General interpersonal participation	.24 (.001)	*	*	.27 (.001)	.27 (.00
Recent interpersonal participation	.27 (.001)	*	*	.20 (.02)	.22 (.00
General protest participation	.33 (.09)	*	*	*	
Recent protest participation	.73 (.004)	*	*	.36 (.09)	.47 (.04

* Not significant gammas .20 and/or not statistically significant; otherwise significance level is given in pa theses.

TABLE 7. *The relationship between dichotomized characteristic health status measures and political attitudes in a national survey (associations expressed as gamma coefficients)*

	Characteristic comprehensive health status	*Characteristic chronic illness experiences*	*Characteristic symptom experiences*	*Characteristic health restrictions*	*Scope of health restrictions*
Political interest	*	*	*	*	*
Civic duty	.25 (.01)	*	*	.22 (.03)	*
Alienation	*	*	*	*	*
Efficacy	*	*	*	*	*
Withdrawal	*	*	*	*	*
Expectation of political violence	.25 (.006)	*	*	.28 (.001)	
Reformism	*	*	.23 (.003)	*	*

* Not significant; otherwise significance level is given in parentheses.

not to mediate the linkages between health and political behavior; *b)* there is no evidence to suggest that health is a stronger influence on political participation among women than among men (if anything, something of a countertrend to that regional finding may exist in our national data); *c)* there may be a tendency for the health-to-politics linkage to be stronger among persons thirty five and older than it is among younger adults; *d)* political attitudes do mediate the relationships between health status and political participation in a majority of cases but do so in an inconsistent, rather patternless fashion. One exception to the patternless nature of this mediation is the link between health status and voting where the linkages tend to be strongest among persons whose attitudes least predispose them toward participation. Further exploration of this finding seems warranted and is planned.

We examined, too, the degree to which health mediates the relationships between seven basic political attitudes and political participation. As in the regional study, health factors, more often than not, significantly intervene between attitudes and behavior. The influence of health on the linkages between attitudes and behavior is most consistent, most patterned, with reference to the more active, energy-requiring forms of participation—protest and interpersonal activity. In these domains, poor health very consistently depresses participation—even among respondents who are strongly predisposed toward participation. To parallel our analysis of the regional data, we intensively examined the influence of health status on the relationships between political interest and participation. We found that poor health functions to depress voting primarily among less politically interested respondents, that health difficulties depress protest activity even among the most keenly interested respondents. Poor health has less powerful impact on the linkages between attitudes and interpersonal and/or impersonal political participation.

I believe that these findings of our national study, like those in the regional research, strongly suggest the utility of including health indicators in most studies of political behavior. First, health factors do bear rather consistent and significant associations with political participation. Second, the attitudinal and demographic variables so often included in political research do not entirely encapsulate the influence of health on political participation, do not so mediate the link between health and politics as to eliminate the influence and/or importance of health factors. Finally, health factors do consistently mediate the attitude-to-behavior link, do influence the degree to which politically predisposed individuals act upon their attitudes. Failing to include health factors in our studies, therefore, must deleteriously, and needlessly, weaken our understanding of human political behavior.

A Concluding Note

I mean to end this chapter not with a bang and not with a whimper and certainly not with a summary of this already overlong piece. Perhaps, however, my indulgent reader will permit me one paragraph of prediction.

The assumptions of biopolitical research will prove to be valid and useful. The relationships between somatic variables and political attitudes and behavior will prove to be intimate, important and very complex. The confrontation between the paradigms now predominant in political study and those in the life sciences will prove to be exciting and worthwhile, improving the theory and methods of the former and extending the range and relevance of the latter. The interpretation, evaluation and extension of biopolitics offered here—though hopefully useful now—will soon be superseded.

Bibliography

Birch, H. / Gussow, J. D.
1970 *Disadvantaged Children,* New York, Harcourt, Brace and World.

Campbell, A. / *et al.*
1964 *The American Voter,* New York, Wiley.

Corning, P. A.
1971 "The Biological Bases of Behavior and Some Implications for Political Science," *World Politics* 23, pp. 321-370.

Davies, J. C.
1963 *Human Nature in Politics,* New York, Wiley.

Dearden, J.
1973 "Sex Linked Differences of Political Behavior," Paper read to the 9th World Congress of the International Political Science Association, Montreal.
1970 "Violence and Aggression: Innate or Not," *Western Political Quarterly* 23, pp. 611-623.

Easton, D. / Dennis, J.
1969 *Children in the Political System,* New York, McGraw-Hill.

Eysenck, H. J.
1967 *The Biological Basis of Personality,* London, Thomas.

Ferguson, Leroy / *et al.*
1970 "An Attempt to Correlate Rate of Physical Maturation with Attitudes Toward Politics," Paper read to the 8th World Congress of the International Political Science Association, Munich.
1965 *Children and Politics,* New Haven, Conn., Yale University Press.

Harburg, E. / *et al.*
1970 "A Family Set Method for Estimating Heredity and Stress — I," *Journal of Chronic Diseases* 23, pp. 69-81.

Hess, R. D. / Torney, J. V.
1967 *The Development of Political Attitudes in Children,* Chicago, Ill., Aldine.

Hummel, R. H.
1970 "Charisma: the Biological Dimension," Paper read to the 8th World Congress of the International Political Science Association, Munich.

Jaros, D.
1970 "Biochemical Desocialization," Paper read to the Annual Meetings of the American Political Science Association, Los Angeles.
1972 "Biochemical Desocialization," *Midwest Journal of Political Science* 16, pp. 1-28.

Jennings, M. K. / Niemi, R. G.
1968 "The Transmission of Political Values from Parent to Child," *American Political Science Review* 62, pp. 169-184.

Lane, R. E.
1972 *Political Man,* New York, The Free Press.

Milbrath, L. W.
1965 *Political Participation,* Chicago, Ill., Rand McNally.

Patrick, D. L. / *et al.*
1972 "Toward An Operational Definition of Health," San Diego, Calif., University of California, Department of Community Medicine, unpublished manuscript.

Peterson, S. A.
1973 "The Effect of Physiological Variables upon Student Protest Behavior," Paper read to the 9th World Congress of the International Political Science Association, Montreal.

Pless, I. B. / Douglas, J. W. B.
n. d. "Chronic Illness in Childhood," Rochester, NY, University of Rochester School of Medicine and Dentistry, unpublished manuscript.

Pless, I. B. / Roghmann, K. J.
1970 "Chronic Illness and Its Consequences," Paper read to the Meeting of the American Public Health Association.

Rohter, I.
1975 in press "A social Learning Approach to Political Socialization," in: Schwartz / Schwartz, *op. cit.*

Schubert, G.
1974 "Bibliography... in Biopolitics," Honolulu, Hawaii, University of Hawaii, mimeo.

Schull, W. J. / *et al.*
1970 "A Family Set Method for Estimating Heredity and Stress — II," *Journal of Chronic Diseases* 23, pp. 83-92.

Schwartz, D. C.
1970 "Perceptions of Personal Energy and the Adoption of Basic Behavioral Orientations to Politics," Paper read to the 8th World Congress of the International Political Science Association, Munich.
1974 "Toward a More Relevant and Rigorous Political Science," *Journal of Politics* 36, pp. 103-137.
1975 in press "Health, Body Images and Political Socialization," in: Schwartz / Schwartz, *op. cit.*

Schwartz, D. C. / Schwartz, S. K.
1975 in press *New Directions in Political Socialization,* New York, The Free Press.

Schwartz, D. C. / Zill, N.
1971 "Psychophysiological Arousal as a Predictor of Political Participation," Paper read to the annual meetings of the American Political Science Association, Chicago.

Singer, J. D.
1970 "Crowding and Combat in Animal and Human Societies," Paper read to the 8th World Congress of the International Political Science Association, Munich.

Shubs, P.
1973 "Political Correlates of Body Self Image," Paper read to the 9th World Congress of the International Political Science Association, Montreal.

Stauffer, R. B.
1969 "The Biopolitics of Underdevelopment," *Comparative Political Studies* 2, pp. 361-387.
1971 *The Role of Drugs in Political Change,* Morristown, NJ, General Learning Press.

Verba, S. / Nie, N. H.
1972 *Participation in America,* New York, Harper and Row.

Wahlke, J. C. / Lodge, M.
1970 "Psychophysiological Measures of Political Attitudes and Behavior," Paper read to the 8th World Congress of the International Political Science Association, Munich.

Welch, S. / Booth, A.
1973 "Crowding as a Factor in Political Aggression," Paper read to the 9th World Congress of the International Political Science Association, Montreal.

Wessman, A. E. / Ricks, D. F.
1966 *Mood and Personality,* New York, Holt, Rinehard and Winston.

White, E.
1972 "Genetic Diversity and Political Life," *Journal of Politics* 34, pp. 1203-1242.

J. A. LAPONCE

The Left-Hander and Politics

THE FIRST PART OF THIS ARTICLE shows that the number of left-handers (defined by the hand used in writing) can be used as a social indicator measuring the level of permissiveness, tolerance and/or liberalism in a society. The data used in the study comes mostly from the United States, Great Britain and Canada and spans the last seventy years.*

The second part of this article tests the hypothesis that left-handers are likely to have the characteristics of minority groups and that in the Canadian context, where the hypothesis was tested, they are more likely to be oriented politically to the left; the findings are inconclusive.

Neither the pleas for the emancipation of the left hand, by Benjamin Franklin for example, nor the creation of associations of left-handed individuals, such as that founded by Hakozaki—effective as they may be in changing social attitudes toward the "wrong side"—will make the world ambidextrous.[1] The data accumulated in the past two centuries is beginning to show conclusively that biology as well as culture must be used to explain the universal dominance of the right hand. Barring unforeseen genetic and social controls, the left hander is bound to remain in the minority in all communities larger than the family and the small village. Are there political consequences? Can the imbalance between the two hands be used for political analysis?

The Dominance of the Right Hand: A Survey of Explanations and Findings

To the dominance of the right hand Aristotle gave cultural as well as physiological explanations: the organs of the right side are both better trained and by nature more powerful than those of the left. Plato prefered

* This paper has been written while the author was on a research leave supported by the Harry Guggenheim Foundation, whose help he gratefully acknowledges.

a simple cultural cause: nurses play with and look after babies in such a way that the baby's right hand is freer to develop than his left.[2] Subsequent theories offered many variations on the nature-nurture theme. It has been theorized that when men stood out of the monkey, it was not only, as Giraudoux supposed, to pin more easily and display medals on the left side of his chest, but, more immediately, in order to take a shield in one hand and a weapon in the other, an action which, according to the theory, gave a greater chance of survival to the right handers who, naturally, would protect their heart and aggress more effectively than their left-handed opponents. An assumed better blood irrigation of the left side of the brain, the rotation of the earth, the position of the fœtus in the uterus, the social taboos associating the notion of left with the notion of wrong, the location of speech functions predominantly on one side of the brain, the assumed existence of one or more Mendelian recessive genes have also been proposed, either singly or in combination, to explain that, if some men prefer to use their left hand, man is right-handed.[3] However, the gathering of systematic statistical evidence began only in the late nineteenth century; and that evidence—though still unconclusive—favors Aristotle rather than Plato: neither exclusively biological nor exclusively cultural explanation can suffice to explain the unbalance between the two hands, and culture appears to be a reinforcing rather than a primary cause.

The data gathered thus far show that the incidence of left-handedness varies markedly *a)* across cultures and *b)* according to the test of handedness used to determine either preference, strength or "adroitness." The Wyle compilation of twenty-five different findings obtained before the mid-1930s shows variations in left-handedness ranging from 1% to about 30%, the lowest percentage describing the dominant "tool" hand of military recruits in the Pre-World War I German army, the highest describing the stronger hand on the dynamometer test among groups of children and adults surveyed by Galton in the 1880s and by Parson in the 1920s.[4] More recent studies have produced similar variations, from very low percentages such as the 0.5 reported among Katangan school children in 1959-1960 on a test of writing, cutting a circle with scissors and screwing a cap bottle, to high percentages such as the 25% reported by Bryngelson.[5]

To make sense of the many statistics available, let us first of all consider separately the data pertaining to infants and that pertaining to children and adults.

Infants and Handedness

The observations made in Gesell's clinic indicate frequent changes in hand preference during the first years of life; clear dominance does not appear until the age of two and does not begin to stabilize until the age of four.[6] The fact that the baby appears ambidextrous has been used to support the theory that handedness is an acquired characteristic, that the infant is made to grow into a right-handed child in order to meet the demands of parents, playmates, teachers and "right handed" objects. Contrariwise this evolution from ambidexterity to right-handedness is also used to support the theory that hand unbalance is linked to brain unbalance and that right hand dominance is linked to the development of speech, a faculty primarily located in the left hemisphere of the brain, the hemisphere that controls the movements of the right hand. The latter interpretation, that which fits best the data available, distinguishes man from other animals by both speech and right-handedness, two characteristics linked to brain hemispheric specialization.[7]

Children and Adults

The perception by the child of the conceptual difference between left and right comes relatively late, much later than his recognition of up and down or front and behind. The explanation for the lateness of this conceptual mastery is that, in man as in nature, the symmetry between left and right is much greater than the symmetry between either front and behind or up and down. It is not until he is about 10 to 12 years old that the child has mastered conceptually the distinction between left and right and is able to distinguish verbally each of the two sides in himself as well as in others.[8] But long before he can conceptualize the differences between his two hands the child specializes their functions, a specialization which is part of a more general division of labor between sides of the body. The right side of the brain, the right eye, the right hand and the right foot become better at doing certain things, while the left is better at doing other things. This division of tasks leads to the formation of a hierarchy between a dominant and a helping side. For example, the left hemisphere of the brain is normally better at speech than the right side, which is better at music and at spatial perceptions; the right hand is better at precise manipulation while the left hand, by contrast, is better at holding and at pinpointing within a context.

Here again, it could be and it has indeed been argued that the functions creates the organ, that a culturally-acquired specialization leads to physiological imbalance. If one is taught to wink frequently and always with the same eye, one will develop a wrinkle on that side. If one is taught, like some Ibo women, not to use the left hand in such frequent activities as cooking and eating, one will develop a weak and idiotic collection of fingers on the passive side. True. But then how can we explain that no society for which data is sufficiently abundant and reliable has ever produced a majority of left-handed individuals? Because, it will be argued, the data comes from contemporary societies which are all under the influence of a few cultural centers that happen to be right-handed. As if the European and the Chinese scissors had made the world right-handed. That is difficult to argue. Such argument would not explain that nearly all primitive societies have taboos affecting the left side, taboos written in myths and rituals much older than the Indo-European influence. One can of course point to a few photographs by Ratzel showing Australian aborigines holding their spears in the left hand or to some prehistoric cave drawings showing as high a frequency of left hand as of right hand grips.[9] But overwhelmingly, the data on primitive societies indicates right hand dominance.

Left-Handedness as a Social Indicator

The wide variations reported earlier in the incidence of left-handedness, variations ranging from 1 to over 30% in cross-sections of a normal population[10] might suggest that, unlike suicide or birthrate, level of left-handedness is an unstable characteristic difficult to use as a measure of social change. Such an impression would be entirely mistaken. Once we control for the test by which handedness is measured and control also for the culture of the subjects, we obtain remarkably stable statistics. To illustrate such stability let us take "the hand used in writing" as our criteria of left-handedness. Consider the following measures obtained in the last twenty years in Britain, Canada and the United States. Marian Annett reports 11.63% males (N=1928) and 10.62% females (N=920) writing with their left hand among undergraduates at the University of Hull in the 1960s, and 9.37% (N=630) among military recruits.[11] A few years earlier, in the mid-50s, Enstrom recorded in Eastern United States primary schools, levels of left handwriting of 12.5% among boys (N=48009) and 9.7% among girls (N=44647).[12] In 1971 a survey made by Laponce among Canadian school children in the Vancouver area found an incidence of left-handedness of 12.1% among boys (N=21429) and 9.75% among girls (N=20233).[13] Thus in three countries which are roughly similar in their level of industrial development, in their political system, in their religious traditions and in their policy of school permissiveness, the levels of left-hand writing are remarkably close to one another. By contrast, a study made in a different culture, that of Lebanon, found in the mid-50s, 5% left-handed boys (N=1430) and 4.9% left-handed girls (N=226).[14] To these contrasted statistics, compare now a measure obtained in Israel among 14-18 years-olds: for native Israelis the level of left-handedness is 12.7% while for the foreign-born it is only 2.2%.[15] Clearly these percentages indicate variations "in space" that make sense culturally. The incidence of left-handedness appears to be the result of a cultural pressure of varying intensity on what looks like a constant biological counterpressure. It is tempting to line up percentages of left-handedness along the Y axis of a Cartesian space and to search for the variable which would, when put on the X axis, produce a good and meaningful regression line. Degree of religiosity? Degree of anomia? Degree of permissiveness? Tolerance of deviance? Democracy? Equality? Freedom? Importance of the use of teamwork with one-sided tools? A world map of left-handedness might give us the answer.

An answer can also be sought in the history of a single culture. Take the United States for example: in the 1910s Laura Smith found 5.5% left-handed boys and 4.5% left-handed girls among school children of the Eastern United States public schools.[16] In the mid-twenties Wilson and Dolan found 4.6% left-handed boys (N=1147) and 2.6% left-handed girls (N=1181) among junior high school pupils in Oklahoma City while Chamberlain found 4.5% left-handers in a sample of male Ohio state undergraduate (N=2177) and 4.1% among the fathers of these undergraduates.[17] In the following decade, in the 1930s, an overall 7% was reported in Kansas schools (N=622) and 6.4% in Detroit.[18] About ten years later, in the 1940s, a Michigan school study recorded 8.2% and yet another ten years later, the

Eastern United States study already mentioned, obtained an overall level of 11.6% (N=13.438).[19] In England the evolution has been identical (see Table 1): in 1913 Burt reported 3.8% for school boys and 2.1% for school girls; in 1923 he reported 4.9% for boys and 2.7% for girls.[20] By the mid-sixties the overall percentage had risen to slightly over 10%.[21] By locating in time and in space the cultures surveyed for left-handedness we should have, eventually, data as rich as that which was used by Durkheim when he characterized societies by their rate of suicide. I have not tried to do such systematic locating for lack of enough valid observations, but, fortunately, statistics on left-handedness are now collected and reported with a care lacking in many of the older studies; in a few years time, we might have a reliable world map of left-handedness.

TABLE 1. *Percentage of male students, in schools or universities, who write with the left hand**

	1900 to WW I	*1920s*	*1930s*	*Post 1950*
United States	4.1 (Chamberlain)	4.5 (Chamberlain)	7.8 (Pyle)	12.5 (Enstrom)
	5.5 (Smith)	4.6 (Wilson)		
Great Britain	3.8 (Burt)	4.9 (Burt)		11.6 (Annett)
Canada				12.1 (Laponce)

* The average age of the students varies from study to study and is thus a slightly distorting factor to be kept in mind when comparing two nearby decades. I estimate, on the basis of inadequate information, that in the above studies the average age was very roughly: Chamberlain (about 19), Smith (10 to 12), Pyle (about 9), Wilson (about 13), Burt (8 to 12), Enstrom (about 8), Annett (about 19), Laponce (about 12).

In the meantime, I have tried to verify, in a single locality, the hypothesis implied by my tentative naming of the hypothetical X axis of the hypothetical regression life; I sought to verify that left-handedness is related to sex, social class, and religiosity. On the basis of a questionnaire filled out by teachers in the Greater Vancouver area, I obtained confirmation that the incidence of left-handedness is higher among areas of the city characterized by high socio-economic standing; higher in lay than in religious schools, and, as expected, higher among boys than girls (see Figure 1). Deviant behavior is a characteristic of males rather than females[22] and the tolerance of such deviant behavior appears to increase with income, education and secularism. This finding matches those obtained in an entirely different context by Stouffer in his study of the tolerance of communism and communists in the United States;[23] it fits also the data on primitive cultures recorded by Hertz who hypothesized that man, having used his body to explain the cosmos, had assigned to the right side the function of symbolizing the pure and the religious, and to the left the function of symbolizing the dangerous and the secular, with the result—we can infer—that the higher the degree

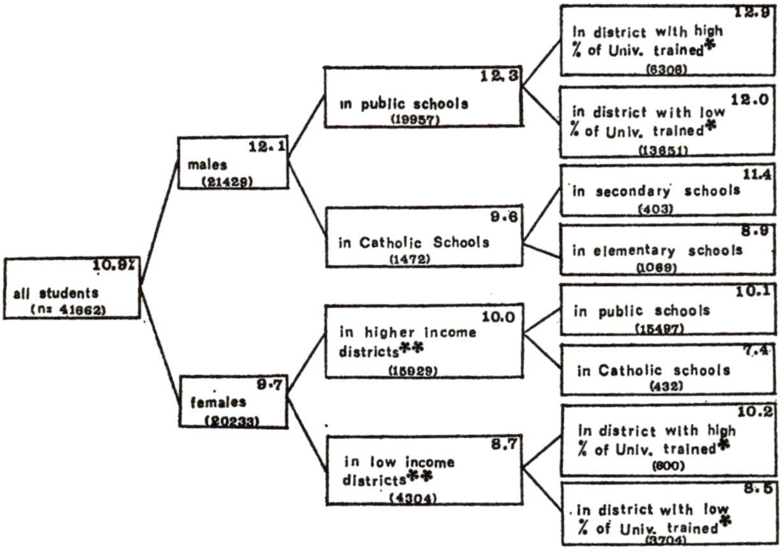

N.B. In all splits reported here the difference between means of the two subgroups is significant at .05 on the t test. The automatic interaction detector (AID) was used to build the tree.

* Higher education is defined as a characteristic of the census districts where the school was located and where the percentage of individuals having attended university among the people no longer in school was higher than 15%. Low education was defined as the characteristic of the district where the percentage was lower than 15%. The census used is that of 1961.

** High income districts were defined as those where the percentage of males in the labor force with an income of $ 6 000 or more was over 20%; low income districts are those where the percentage was under 10%.

FIG. 1. *Tree analysis of the incidence of left-handedness as reported by the teachers of selected schools of the Greater Vancouver area in 1971 (grade one to grade twelve)*

of religiosity and social control the more likely the left to be subjected to taboos since its opposite, the right, is the side of God and of social order.[24]

POLITICAL BEHAVIOR AND THE LEFT HAND

The left-handed individual, often studied by neurologists and child psychologists has been ignored by political scientists, understandably so since we have rarely gone beyond sex and age in the relating of biological factors to politics and since there are, I admit, more obvious and more important bio-political correlations to be sought.[25]

In seeking a relationship between biological and political leftism, as I propose to do, I may well be tricked into entering the realm of the absurd by what may be no more than a fortuitous semantic similarity. The following pages would then be the illustration of a mischievous hypothesis born exclusively from the right side of the brain, that of fantasy, rather than from its more scientific counterpart. I like to think however that there were "reasons" for risking this possibly absurd enterprise; the following reasons: even if left-handedness had purely cultural causes, it would still remain that the left-handed individual—an individual who cannot escape living in a right-handed world—could be expected to develop mental attitudes and to be led to kinds of behaviors different from those of people who, being right-handed, move their hands and their minds in harmony with the society at large. Left-handed people live in a more or less hostile or at least in a more or less unfriendly surrounding. Of course many will overcome the difficulty and the awkwardness resulting from their being wrongly "bent," and this very overcoming may well strengthen their ego, may even give them a certain sense of superiority, but we should expect nevertheless that, on the average, the left-hander would be at a constant disadvantage, a disadvantage leading him to be relatively dissatisfied with the present and hopeful of a more equalitarian future.

Furthermore, if in addition to having cultural roots, left-handedness has also biological determinants, then the likelihood of there being specific political correlates to left-handedness is even greater if, as appears to be the case, the left-hander is not simply a mirror reversal of a right-handed person but is a different individual with different reactions to self, others and environment.[26]

In short, it seems reasonable to expect that either culture by itself or more likely culture and biology reinforcing each other create between left and right-handers a tension of the kind observable between dominant groups and minorities. Whether the minority rejects a dominant group that could and wants to absorb it[27] or whether, on the contrary, the minority is rejected by the dominant culture we have a type of situation which, in the Canadian context at least, justifies hypothesizing that the left-hander will be inclined to the left politically.

I am aware that these hypotheses and assumptions are a little like bridges over obstacles that are not there. Whatever their worth I sought to verify them on a sample of 686 Canadian students.

Is the Left-Hander Politically to the Left?

The subjects on whom the hypothesis is tested came from the Universities of British Columbia, Waterloo and Queen's. Each respondent was asked to fill out a questionnaire that took about ten minutes to complete. In addition to questions on political preferences, the questionnaire included a question asking the subject to indicate his writing hand.[28] The proportion of left-handed individuals in the population studied was 12.8%; consequently we have only 88 left-handers, a number which is uncomfortably small but can serve for what is not more than a preliminary and very tentative report inviting verification on larger populations.

Findings—The Left-Handers: A Socio-Political Profile

Social Characteristics

We found the expected sexual and social class differences in the incidence of left-handedness; but in most other respects our respondents were a disappointment to the hypothesis.

Compared to 11.25% of women (N=240), 13.48% (N=423) of men were left-handed. The difference is not statistically significant at .05 on the chi square test, but it is in the direction and of the magnitude the literature had led us to predict.

Among the respondents who described the milieu in which they were brought up as being either "very well off" or "well off" 14% (N=237) were left-handers while those who described their family milieu as ranging from "above average" to "poor" had only 12.2% left-handers (N=437). This slight difference is again not statistically significant at the .05 level on the chi square test but, as in the case of sex, it is in the expected direction. Using three instead of two classifications of SES suggests that the percentage of left-handers is higher in the upper and in the lower categories of the social hierarchy: 14.3% (very well + well off: N=24) at the top and 20% at the bottom (below average + poor + very poor: N=44) but only 11.2% in the middle group (above average + average: N=378). The differences are not statistically significant and the number of cases very small in some of the categories; I report them however as an invitation to further testing since it makes sense that the middle-middle class would enforce social norms of "proper" behavior to a greater extent than either the lower or the upper social strata.

A question intended to measure religiosity did not produce the differences expected and neither did a question recording the educational backgrounds of the parents of the respondent. Among male respondents who said that they had attended religious services no more than twice during the previous twelve months, the percentage of left-handers was 13% (N=338) while among those saying that they had attended such services at least once a month, the percentage was 15.4% (N=84). This small difference is not in the expected direction. Equally disappointing was the lack of correlation between a respondent's handedness and the educational level of his parents. Among the subjects whose two parents had attended university, the percentage of left-handers was 13.2% (N=128) while among those whose two parents had not attended university it was 13.1% (N=373).

Psychological Characteristics

Our questionnaire included a series of questions asking the subject to indicate whether he was more oriented toward his physical than toward his social environment, whether he was introverted or extroverted and whether he was optimistic or pessimistic. The answers to these questions produced very small and statistically non-significant differences which did not form a consistent pattern: 24.5% of left-handers (N=86) said that they were more oriented to their physical than to their social environment while 27% of right-handers gave a similar answer (N=587). More left-handers than right-handers described themselves as extroverted; 55% (N=80) compared to 48% (N=560). The question asking the subject to indicate whether, on the whole, he was optimistic or pessimistic showed 72% of left-handers to classify themselves as optimistic (N=86) compared to 77% of right-handers (N=584). These differences are not statistically significant and much to small to warrant our giving much attention to them in the absence of comparable findings in the literature.

Political Characteristics

The hypothesis that left-handers are more likely to be to the Left politically is not verified when we define a left preference as a preference for the Socialist Party (NDP). Only 25% of left-handers (N=85) indicated a preference for that party compared to 30% of right-handers (N=581). But a marked difference appears in the support given by right- and left-handers to the Liberal Party; 63% of left-handers supported that party while only 54% of right-handers did so.[29] One could possibly rescue our hypothesis (the hypothesis that left-handedness and political leftism are positively correlated) by claiming that there are two kinds of Left in the Canadian political system: one a socio-economic Left represented by the New Democratic Party (a party which distinguishes itself from its opponents by the support it receives from trade unionists); the other, a cultural Left represented by the Liberals, the party of ethnic and religious minorities such as Catholics, Jews, and French Canadians.[30] Maybe the left-handers have the feeling of being deviants in relation to the dominant group in a way similar to that of Catholics or Jews. However, on such a small sample, such an explanation must be treated as no more than pointing to an unlikely possibility.

Slightly more rewarding was a question asking the respondent to locate himself, his father, his mother and selected other concepts between "extreme left" and "extreme right" taken in a political sense. 69% (N=84) of left-handers put themselves on the left but only 60% of right-handers (N=567) did so.[31] This difference cannot be explained by the technique used for locating concepts in the Left-Right continuum (a technique that required drawing an arrow as long as one wanted it to be, an arrow going from the middle of the page, where the words to be classified had been listed in one column, toward either the left or the right margin) since on other concepts than "self" left-handers and right-handers used similar classifications. They did not differ for example in the locating of either their fathers or mothers. 25% of left-handers (N=86) put their fathers on the left and 28% of right-handers (N=659) did so; 28% of left-handers put their mothers on the left and 31% of right-handers did so. Even better than "father" or "mother" at showing that there is no systematic tendency of left-handers to give a "left" identification is the classification given for the words "banker" and "worker;" 74% of left-handers put "worker" on the left; 75% of right-handers did so; 2% of left-handers put "banker" on the left and 4% of right-handers did so. The tendency of left-handers to "see" themselves on the left does not appear to be accidental. A possible explanation of that tendency is that there is an automatic reaction leading the left-handed respondent to see himself on the left irrespective of his party preference. The semantic similarity between "left" in its physical sense and left in its political sense might trigger such reaction. Many people, children and adolescents especially, are very much influenced by signs; if these signs are found in one's own body, all the more reason for they being powerful. A name, an odd characteristic can be taken as evidence of a force pushing the self to incarnate itself in preordained and fitting behavior. The Liberal Party would then have this great political advantage of being left at the level of perception without being markedly such at the level of behavior and policy. A tenuous explanation, I admit.

Conclusion

Our findings are both intriguing and disappointing. Intriguing is the consistency in the differences and variations in the incidence of left-handedness when we compare large normal populations on such criteria as sex or social permissiveness; disappointing because our attempt at finding attitudinal and behavioral correlates of left-handedness that would be politically significant was most inconclusive. I am not however quite prepared yet to say that the search for such correlates is bound to lead to spurious or at best trivial associations, and that for three reasons.

1. Whenever we obtained what appeared to be a stable correlation between left-handedness and social or physical characteristics, such as sex or state of economic, political, and religious evolution, we found the correlation to be very small. If we had, in this study, reported for the first time the relationship between sex and hand used in writing, we might have dismissed, indeed we would have most surely dismissed the correlation as uninteresting because of its lack of statistical significance and because of the smallness of the gap in the percentage of left-handers when males are compared to females. To establish the importance of such a small correlation, large samples and many tests on many different populations were needed. This argues for not dismissing too quickly what appeared as disappointingly small differences.

2. The links between left-handedness, brainsidedness, and personality are not yet sufficiently well established in the biological literature for us to have been able to theorize properly. I may have used wrong tools and hypotheses, especially when it came to linking handedness to brainsidedness.

3. The criteria of handedness that I used—the hand normally used in writing—provides an easy measure but is far from perfect. A more complex operationalization of handedness and sidedness providing measures that would take into account a variety of manual tasks other than handwriting and which would have linked hand to eye and foot sidedness might have been more useful; they would have enabled us to line up our subjects on a continuum rather than force us to put them in dichotomized categories. The work of Annett and Burt,[34] among others, have well established that

one is not simply left-handed or right-handed, but that one is more or less a mixture of the two. Furthermore, the measure I used did not enable me to identify the interesting category of subjects who would naturally have been left-handed but who had been forced to convert to right-handedness for writing. Populations much larger than those I studied would have been required by these more refined classifications.

However, even if more complex studies of the relation between biological handedness, cultural handedness, political attitudes and political behavior were to prove disappointing, it would still remain that the global incidence of left-handedness is an interesting, and very likely, an important measure of the state of a social system, a measure providing a balance point between biology and culture; a measure akin, in that respect, to such useful indicators as suicide or birth rates.

Left-handedness may well appear unrewarding as an independent but very useful as a dependent variable.

Notes

1. Benjamin Franklin, "A petition to those who have the Superintendency of Education." The text is given in full in I. S. Wile, *Handedness : Right and Left,* Boston, Mass., Lothrop, Lee and Shepard, 1934, pp. 356-357. A note on the Japanese Association of left-handers is in *Time* magazine of January 7, 1974. Soichi Hakozaki, a psychiatrist, has published a work widely read in Japan called *Warnings Against Rightist Culture.*
2. Plato, *The Laws,* Book 7; for Aristotle the left-right division is characteristic of the animate world, the inanimate having only an up-down contrast. For him the heavenly bodies have a left and a right because they are animate, and they move to the right which is the better side; thus man is in harmony with the universe, his right hand being the stronger. Aristotle concedes however that one can train the left hand to become as strong as the right. See his *Ethics,* v. 10. A discussion of these two theories is in V. Fritsch, *Links und Rechts in Wissenschaft und Leben,* Stuttgart, Kohlhammer, 1964, translated as *Left and Right in Science and Life,* London, Barrie and Rockliff.
3. A review of the various theories used to explain right hand dominance is in A. Blau, *The Master Hand: A Study of the Origins and Meaning of Right and Left Sidedness and its relation to Personality and Language,* New York, American Orthopsychiatric Association, 1946; see also G. Hildreth, "The Development and Training of Hand Dominance: Origins of Handedness and Lateral Dominance," *The Journal of Genetic Psychology* 75, 1949, pp. 255-275. Recent findings, contradicting Plato's suppositions, show that infants cry less if they are carried on the left side of the holder's body and that 75% of mothers given their baby to hold in their arms within twenty-four hours of delivery tend indeed to hold them on the left (after twenty-four hours they show no side preference). See Salk, "The Role of the Heartbeat in the Relations between Mother and Infant," *Scientific American,* May 1973.
4. See Wile, *op. cit.,* p. 68.
5. P. Verhaegen / A. Ntumba, "Note on the Frequency of Left-Handedness in African Children," *Journal of Educational Physiology* 55 (2), 1964, pp. 89-90. B. Bryngelson, "Stuttering and Personality Development," *Nervous Child* 2, 1943, pp. 162-171; and B. Bryngelson / T. B. Clark, "Left-Handedness and Stuttering," *Journal of Heredity* 34, 1939, pp. 387-390. More recent estimates of left-handedness in the absence of cultural controls given by Bryngelson put the percentage as high as 34; see T. B. Clark, *Left-Handedness,* Edinburgh, 1966.
6. See A. Gesell, *The First Five Years of Life: A Guide to the Study of the Pre-School Child,* New York, Harper, 1940.
7. A survey of the findings on brain asymmetry and laterality is in Maya Pines, *The Brain Changers: Scientists and the New Mind Control,* New York, Harcourt, Brace, Jovanovich, 1974.
8. See J. Piaget / B. Inhelder, *La représentation de l'espace chez l'enfant,* Paris, Presses Universitaires de France, 1948.
9. See Wile, *op. cit.,* p. 69; Paul Sarasin reports that the Stone age implements found around

Moustier were intended almost equally for both hands: 135 for the left, 146 for the right, and 145 for either. In the Bronze age the dominance of right-handed implements become overwhelming. See P. Sarasin, *Über Rechts- und Linkshändigkeit in der Prehistorie, und die Rechtshändigkeit in der historischen Zeit,* Naturforschende Gesellschaft in Basel, Verhandlungen 29, pp. 122-196. See also D. G. Brinton "Left-Handedness in North American Aboriginal Art," *American Anthropologist* 9, May, 1896, p. 175. It could be that early man fought and worked with both hands and that it took him time to discover the paradox that unlike animals he could fight and work better by giving a leading role to only one hand.

10. In sub-groups such as the offspring of "homozygote" left handers studied by Chamberlain one obtains even higher percentages (50% in the Chamberlain study). See H. D. Chamberlain, "The Inheritance of Left-Handedness," *The Journal of Heredity,* 1928, pp. 557-559.
11. The N and % are in a letter to the author. The analysis of the data is in M. Annett, "A Classification of Hand Preference by Association Analysis," *British Journal of Psychology,* 1970, pp. 313-321, and "The Distribution of Manual Asymmetry," *British Journal of Psychology,* 1972, pp. 343-358.
12. E. A. Enstrom, "The Extent of the Use of the Left Hand in Handwriting," *Journal of Educational Research,* February, 1962, pp. 234-235.
13. The data is deposited in machine readable form with the Data Bank of the University of British Columbia.
14. W. Dennis, "A Note on Sex Equality in the Incidence of Left-Handedness", *Journal of Educational Psychology,* 1958, pp. 209-210. Unfortunately, the study does not control for religion.
15. Joseph Weiser, "Handedness—Leggedness—Eyedness," *Journal of Experimental Education,* Summer 1965, p. 347 ff.
16. L. G. Smith, "A Brief Summary of Left-Handedness," *Pedagogic Seminary,* 1917, pp. 19-35. Everything in the article points to the author having used the 'writing hand' as the test of handedness but this is not stated unambiguously. The children studied are described as public school pupils. The ages are not recorded.
17. See M. O. Wilson and L. B. Dolan, "Handedness and Ability," *American Journal of Psychology* 43, 1931, pp. 261-268; and Chamberlain, *op. cit.*
18. For Kansas see Wyle, *op. cit.*, p. 66. For Detroit see W. H. Pyle / A. Drouin, "Left-Handedness: Experimental and Statistical Study," *School and Society* 36 (121), August 20, 1932, pp. 253-256. The authors obtained data on 13,438 children. They do not give the breakdown by sex. I have assumed an equal number of boys and girls.
19. The Michigan study, done by Palmer, is quoted by Enstrom. I have not been able to verify the source. For the Pennsylvania study see Enstrom, *op. cit.*
20. C. Burt, *The Backward Child,* London, Lowe and Brydone, 1937. The author does not report the N.
21. See Annett, *op. cit.*
22. These findings contradict those reported by Cesare Lombroso at the end of the nineteenth century. Among 1,029 operatives and soldiers he found 4% left-handers among men and from 5 to 8% among women. Among lunatics the proportion was not much different; but among criminals, he found thirteen percent left-handed men and twenty-two percent left-handed women. Lombroso does not indicate his operational definition of handedness, nor does he give the number of cases by sex. He attributes handedness to lack of proper lateralization in the brain, a characteristic, as he sees it, of lower forms of life. C. Lombroso, "Left-Handedness and Left-Sidedness", *The North American Review,* 1903, pp. 440-444.
23. In his summary, Stouffer finds the less educated to be less tolerant of communists and communism. See S. A. Stouffer, *Communism, Conformity and Civil Liberties,* New York, Doubleday, 1955.
24. R. Hertz "La prééminence de la main droite: étude philosophique," 1909, 553-580. This splendid article has been retranslated and republished in Rodney Needham (ed.), *Right and Left: Essays on Dual Symbolic Classification,* Chicago, Ill., Chicago University Press, 1973.
25. On the lack of attention given by political science to biological factors see J. A. Laponce, "Of Gods, Devils, Monsters and One-eyed Variables," *The Canadian Journal of Political Science,* June 1974; an excellent survey of the literature on bio-politics is in A. Somit, "Bio-politics," *British Journal of Political Science,* April 1972, pp. 209-238.
26. According to Jerry Levy and her colleagues, there are two kinds of left-handers: less than half of them have the speech center in the right hemisphere of the brain, the remainder, like right-handers, have it on the left. A brief summary of these and other related findings is in Pines, *op. cit.*, and in M. Pines, "We Are Left-Brained or Right-Brained," *The New York Time Magazine,* September 9, 1973.

27. The hypothesis that the left hander is essentially a negative person who expresses his negativeness through his handedness is in Blau, *op. cit.* Interestingly this theory has been adopted by Benjamin Spock in his *Baby and Child Care,* 3rd ed., New York, Simon and Shuster, 1968, p. 233.
28. The questionnaires were administered between January and March of 1974 during regular class hours. The data is deposited with the UBC data bank.
29. The chi square test comparing Liberals on the one hand to Conservatives and New Democrats on the other is not significant at .05.
30. For the characteristics of the Canadian party electorates see in particular J. Meisel, *Working Papers on Canadian Politics,* Montreal, McGill Queen's Press, 1973; and J. A. Laponce, "Post-dicting Electoral Cleavages in Canadian Federal Elections, 1919-68," *Canadian Journal of Political Science,* June 1972, pp. 270-286.
31. The chi square is barely significant at .05 ($\chi^2 = 3.84$, df $= 1$).
32. See Annett, *op. cit.,* and Burt, *op. cit.*

BERNARD TURSKY / MILTON LODGE / DAVID CROSS

*A Bio-Behavioral Framework for the Analysis of Political Behavior**

IT HAS NOT BEEN CLEARLY DEMONSTRATED whether the poor correlations found between political attitudes and behavior are due to poor theory, weak methodology, or a combination of both factors. This paper is directed at presenting a cross-modal, bio-behavioral approach to the evaluation of the cognitive and evaluative components of political attitudes. Psychophysical and psychophysiological approaches and methodology are discussed, and examples of ongoing research are described. Evidence is presented from this research that cross-modally valid, bias-free psychophysical rating scales can be developed to improve the evaluative measure of attitude and that these techniques combined with appropriate psychophysiological measures can be employed to accurately assess the cognitive component of attitudes. It is hoped that this improvement in measurement will be helpful in answering the questions of the strength of the relationship between attitude and behavior.

From the perspective of the biological and behavioral sciences, behavior is defined as any measurable and observable movement of an organism, including both internal and external movements and their effects, as well as glandular secretions and their effects. Under proper conditions, an eyeblink, fidgeting, change in brain wave activity, rise in blood pressure, nod of the head, the checking of a form, squeeze of a dynamometer, an expletive or monologue and so on across a range of overt and covert, voluntary and involuntary human responses are measureable behaviors. Although physiological, physical, and verbal responses are all sensitive to environmental conditions, and commonly take place concurrently, the relationships among them are neither fully integrated nor completely understood (Shapiro / Crider, 1969).

Unless and until one can establish lawful functional relationships among behavioral responses, the most seemly posture for the behavioral scientist is what we shall call a "cross-modal, multiple indicator" approach. What this involves is measuring behavior, whenever feasible, in each response

* Research supported by National Science Foundation Grant #S041125.

mode (verbal, physical, and physiological) by means of more than one instrument, scale, or indicator. The power of this cross-modal approach goes well beyond the simple correlation of multiple measurement within and across independent modes of response. Rather, the cross-modal approach suggested in this paper relies on theory-derived measurements of response with known, predictable interrelationships. By way of illustration, Lacey (1959, 1967) has demonstrated in his work on *physiological* response specificity that the autonomic nervous system is not a simple activational system but is patterned according to the type and intensity of coping behavior adopted by the individual to specific environmental demands. In the area of *psychophysical* measurement, Stevens' (1975) procedures for validating the relationship between the intensity of a stimulus and its perceived magnitude require the matching of two or more physical responses with known power functions against social stimuli. By using selected indicators from more than one response mode, we bid fair to improve the measurement of attitude in a fashion that will enhance our ability to test the predictive and explanatory capabilities of attitudinal theories.

For the political scientist whose research paradigms rely heavily on attitudes, the utility of the cross-modal, multiple indicator approach is especially commanding. Although variously defined, an attitude generally designates a relatively stable and well-organized set of beliefs and feelings that predisposes response to external stimuli. Attitude, however, like intelligence, motivation, and personality, is a construct which scholars cannot observe or measure directly. Rather they must infer it from some behavior, most commonly verbal response, that can be directly observed and measured. The usual paradigm for attitudinal research undertaken at a single point in time is illustrated in Figure 1.

As is well recognized, the validity of a construct such as attitude is formidably difficult to demonstrate. Since an attitude has no physical referent with a conventional metric, e.g., weight in pounds, length in meters, or sound in decibels, there is no direct way of determining whether the indicator used to measure an attitude is really measuring what it is supposed to be measuring. Hence, we tend to fall back on what is called construct validity: the empirical testing of theory-derived predictions about relationships between attitudes and observable behaviors. Unfortunately, predictions from measurements of attitude to action are seldom impressive, despite rigorous controls over situational variables (DeFleur / Westie, 1958) and increasingly sophisticated methodological advances (Warner / DeFleur, 1969; Wincher, 1969; Schwartz / Tessler, 1972). After a careful review of the literature in social psychology, Tittle and Hill (1967) concluded that "only a moderate degree of correspondence between measured attitude and other behavior can be observed," and that Campbell's earlier conclusion (Campbell, 1963) is inescapable: "The degree of correspondence is, for the most part, yet to be discovered." Weak theory may, of course, be the root cause of our relative lack of success in predicting behavior from attitude. But poor prediction may also result from the failure of any single verbal indicator to incorporate fully and precisely the major cognitive, evaluative, and conative dimensions of the attitude.

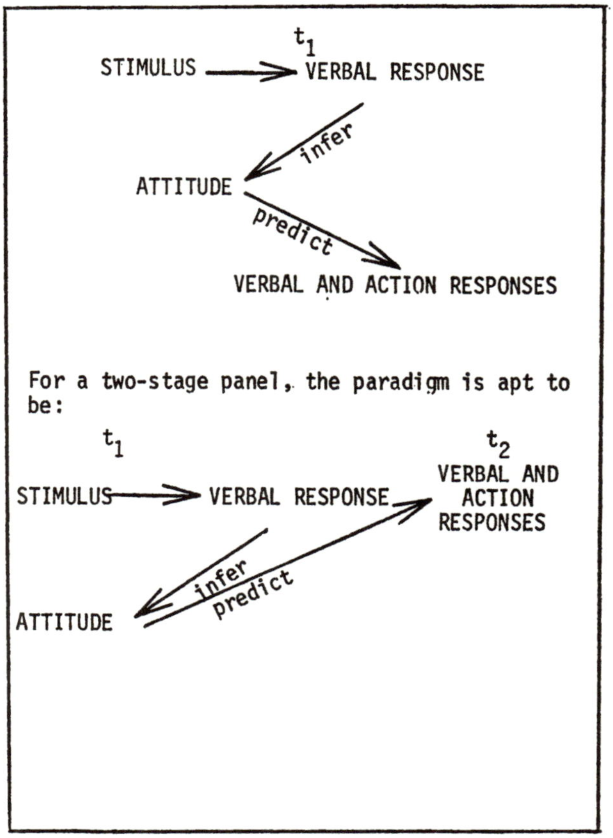

FIGURE 1. *Conventional paradigm in attitudinal research*

Multiple verbal indicators can reduce the hazards of testing predictions derived from attitudinal theory with incomplete and imprecise measures of attitudinal variables. However, even multiple verbal indicators of an attitude are hardly ideal, especially when dealing with political attitudes inferred from survey data. The question-answer format of the typical survey instrument probably promotes an intellective response bias (Lazarus, 1962, 1965) and, in all likelihood, generates in the respondent ratiocinations about the political world (Crespi, 1971). Thus, relying on verbal indicators tends to elicit evidence of attitude structures that may well be inappropriately rationalistic and cannot, with the partial exception of routinized behaviors such as voting, reliably predict many varieties of non-verbal behavior (Wahlke / Lodge, 1972).

What is more, relying solely on verbal indicators assumes that human beings can reliably discern, assess, and verbalize the cognitive as well as the evaluative components of their perceptions. That these assumptions do not always hold is strikingly demonstrated by Schachter (1971) who administered sympathomimetic drugs (adrenaline and ephedrine) to subjects to induce a pattern of physiological (autonomic nervous system) reactivity similar to, if not identical with, that of strong emotional states. One group of subjects was not informed concerning the reaction they would experience, another group was accurately informed, and a third group was misinformed. The uninformed subjects, having no appropriate explanation for their physiological arousal, were readily manipulated by the experimenters who cued one subset to feel angry, another amused, the third euphoric. Those subjects with an explanation for their arousal were unaffected by the experimental manipulations. Summarizing the relationship between subjective emotional experience and physiological state, Schachter (1964) advanced three propositions:

a) Given a state of physiological arousal for which an individual has no immediate explanation, he will "label" this state and describe his feelings in terms of the cognitions available to him. To the extent that cognitive factors are potent determiners of emotional states, one might anticipate that precisely the same physiological arousal could be labeled "joy" or "fury" or any of a great number of emotional labels, depending on the cognitive aspects of the situation.

b) Given a state of physiological arousal for which an individual has a completely appropriate explanation, . . . no evaluative needs will arise, and the individual is unlikely to label his feelings in terms of the alternative cognitions available.

c) Given the same cognitive circumstances, the individual will react emotionally or describe his feelings as emotions only to the extent that he experiences a state of physiological arousal.

Synoptically, then, "emotion is composed of some combination of physiological activation and cognition. In order to measure emotion, it thus becomes necessary to measure both of these factors" ***with independent measurements from more than one response mode.*** Although all types, degrees, and intensities of emotion may have differential physiological patterns, these are not reliably discernible by the individual. Such feelings as anger, fear, mirth, hate, rage, trust, love, or hunger are readily manipulated in experimental settings devoid of cues for the labeling of the emotion.

The underlying rationale, in fine, for a cross-modal, multiple indicator approach is that verbal, physiological, and physical responses are concurrent multiple behaviors, all functionally sensitive to individual-environmental interactions but not necessarily interrelated in simple linear fashion. Therefore, a dual research strategy is proposed. First, because the sole reliance on verbal measurement of beliefs, feelings, interactions, and actions is conceptually as well as empirically untenable, there is need to incorporate, systematically, multiple measurements *across* verbal, physical, and physiological response modes. Second, multiple response measurements are required ***within*** each response mode.

Two applications of the cross-modal approach to the study of micropolitical behavior are currently underway in the Laboratory for Behavioral Research at Stony Brook. One deals with the psychophysical scaling of the strength of political attitudes. The second deals with the application of psychophysiological methods to the study of political behavior. The balance of this paper describes in more detail our approach to each of these applications.

Psychophysical Scaling as a Tool for the Study of Political Attitude

Introduction

Political scientists frequently need to determine not only what attributes people associate with such phenomena as political parties, candidates, institutions, policies, and issues but also how intensely people feel about the attributes they perceive. How strongly does a person identify with a political party? How much confidence does he have in the Supreme Court? How deeply does he feel about the use of busing to integrate the elementary schools in his area? How worried is he about an impending depression? a war? an increase in real estate taxes? How committed is he to democratic processes? When dealing with queries like these, concern rather obviously extends beyond the cognitive components of opinions and attitudes to their intensity (affective) components. In identifying and measuring the cognitive and intensity components of opinions and attitudes, we tend to rely, too heavily no doubt, on verbal response—typically, what people say (or write or check or mark) when responding to interviews, tests, and questionnaires.

Sometimes we want only to typify or characterize the opinions that people express. At other times we want to go beyond this and use these expressions of opinion as data from which to infer attitudes that predispose other behaviors, both verbal and nonverbal. In either case, whether we are merely surveying opinion or constructing models of political action, it is desirable to scale the relevant variables in as informative a way as is possible.

The technique most commonly used by behavioral scientists for scaling attitudinal items is category scaling. A variety of formats is available. Often we ask a respondent to select one of the category options associated with the garden variety of Likert scale or to position himself on a seven- or nine-point bi-polar adjectival scale of the kind associated with Osgood's Semantic Differential or on some form of feeling thermometer. If multiple observations are recorded, as is normal practice when using any of these category formats, a respondent's scale position is determined by averaging his responses.

Queries about the character of the intensity measures developed by category scaling procedures are hardly trivial as experience with decades of attitude research suggests. Studies have consistently reported that the degree of correspondence between attitude and behavior is typically low to moderate (DeFleur / Westie, 1958; Campbell, 1963; Tittle / Hill, 1967; Warner / DeFleur, 1969; Schwartz / Tessler, 1972). Crude measurement, to be sure, is not the only possible explanation for the disappointing degree of correspondence between attitude and behavior. Other possible causes range from a near-exclusive reliance on verbal self-report of beliefs, feelings, and activities (Cook / Selltiz, 1964; Wahlke / Lodge, 1972; Lodge / Tursky / Tanenhaus, 1974) to failure to control experimentally such intervening variables as modeling effects (Bandura, 1969) and coping behaviors (Lazarus, 1966). What makes the measurement problem such an important candidate for early assault is that as a consequence of poor measurement we cannot know whether our failure to find a closer correspondence between attitude and behavior results from weaknesses in our theoretical formulations or deficiencies in our measurements. There is pressing need, it follows, to scale the intensity of verbal response in a manner commensurate with the metric implicit in the judgments respondents can and do make.

Recent developments in the theory and methods of psychophysical scaling have amply demonstrated that the judgments people make about the intensity of many physical attributes, e.g., loudness, brightness, line length, weight, strength of handgrip, and severity of electric shock are discerning enough to warrant ratio scaling. Moreover, cross-modality matching procedures make it possible to determine the reliability and in a sense the validity of these scales. Preliminary investigation also indicates not only that psychophysical scaling methods can be adapted for use with social variables but also that people can in fact make proportionate judgments about the relative intensity of attributes associated with social phenomena (Stevens, 1966; Shinn, 1969). Among these social phenomena are the importance of political offices, racism, immorality, the severity of crimes, life stresses, and political dissatisfaction (Stevens, 1972).

In this section we have set three tasks: first, to highlight recent developments in psychophysical measurement; second, to illustrate through our own research the utility of psychophysical methods in scaling phenomena relevant to political science; and third, to report on a recently completed research program designed to construct and cross-modally validate psychophysical scales of support for the US Senate and US Supreme Court.

Background

Psychophysics is a branch of psychology concerned with sensory reactions to physical stimuli. For more than a century scientists have sought to determine the relationship between the actual intensities of physical stimuli, such as lights and sounds, and the impressions people have of the magnitudes of those stimuli. Until the 1950s, psychologists assumed that Fechner's law correctly stated the relationship between magnitude of sensation and intensity of stimulus. Fechner's law is a logarithmic function: when the intensity of the stimulus increases geometrically (i.e., each step is a constant multiple of the preceding step), the impression of its magnitude increases arithmetically (each additional step is of constant size). Mathematically represented, $R = k \log S$, where R is the perceived magnitude, S is the intensity of the stimulus relative to its threshold value, and k depends on the arbitrary choice of unit.

During the long period when Fechner's law ruled supreme, psychophysicists assumed that impressions of magnitude could not be measured directly. Rather they relied on an indirect unit of measurement called the JND (just noticeable difference). JND designates the change in the intensity of a stimulus that is just large enough to become noticeable. Since JND scales are derived from the imperfect discriminability of stimulus differences, they are sometimes called confusion scales.

Devotees of Fechner's law assumed that each JND represents an equal unit of sensation and hence that JNDs, when summed, produce a scale with interval properties. But, despite its honorific designation, as decades of experimental work painstakingly documented, Fechner's law did not accurately describe reality for most physical dimensions. Fechner's law simply did not state a lawful relationship between intensity of stimulus and magnitude of sensation.

In 1957, S. S. Stevens, in a classic essay, sounded the trumpet blast that brought down Fechner's law (Stevens, 1957). He asked *S*s to assign numbers proportional to the apparent magnitude of each of a series of light intensities presented in irregular order. As Stevens quickly discovered, once *S*s understand the concept of proportionality they can easily make numeric estimations of the magnitudes of their sensations. What is more, despite individual differences, the average result for each group of *S*s turns out remarkably stable and reproducible. As a result, one can generalize to the typical respondent (i.e., external validity). Experiments turned up comparable findings for impressions of loudness, line length, heaviness, and other prothetic modalities—those in which changes in stimulus intensity cause changes in sensation magnitude *without* at the same time altering the quality of the sensation.

Analysis of the data developed for a variety of prothetic continua by magnitude estimation techniques disclosed a consistent pattern: magnitude of sensation grows as a function of some power of stimulus intensity. In algebric terms,

$$R = kS^b$$

where

R is magnitude of response,
S is stimulus intensity,
b is an empirically derived exponent which determines the direction and degree of curvature in the function, and
k is an empirically derived multiplicative constant—a scale factor.

This is Stevens' psychophysical power function law.

The power function law does not, of course, imply that all modalities produce the same power function. In fact, exponents vary rather widely (Stevens, 1957). For example, the exponent for line length is 1.0; for force of handgrip, 1.7; and for sound pressure, .67. In other words, impressions of line length grow in direct proportion to actual increase in physical length of line. Impressions of loudness, in contrast, grow nonlinearly at a negatively accelerated rate with actual increase in sound pressure in the ear; strength of handgrip is judged to grow as a positively accelerated function of the additional force actually employed when the subject squeezes a hand dynamometer.

What Stevens' psychophysical law does state is a simple, lawful relationship between stimulus intensity and sensory response: *equal stimulus ratios*

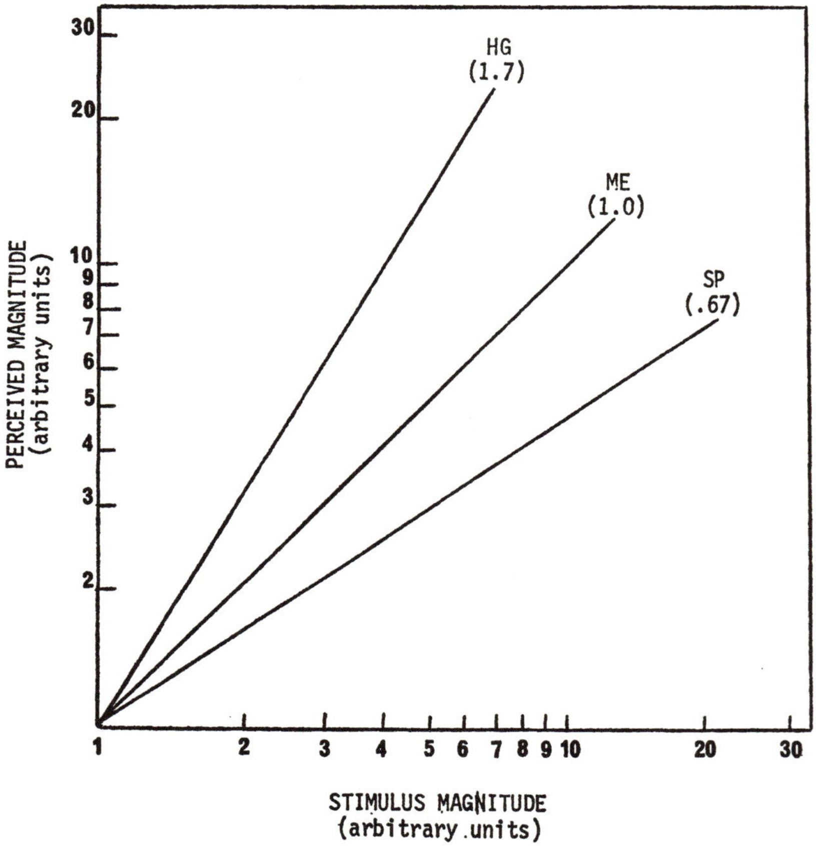

FIGURE 2. *The apparent magnitude of force of handgrip (HG), magnitude estimation (ME), and sound pressure (SP) plotted against logarithmic coordinates. When the curves of Figure 1 are plotted on logarithmic coordinates, they become straight lines. The slope of the line corresponds to the exponent of the power function governing the growth of sensation* (from Stevens, 1961)

produce equal subjective ratios. Hence, the power law ($R = kS^b$) can be restated in logarithmic terms:

$$\log R = b \log S + \log k$$

As a result, when magnitude estimation of intensities on a prothetic continuum are plotted on log-log coordinates, the graph describes a straight line whose slope provides a direct measure of the power function exponent. Figure 2 illustrates the straight lines that result when force of handgrip, magnitude estimation, and sound pressure estimates are plotted against physical continua in ratio-ruled coordinates.

Striking as the accumulation of experimental data supporting the power law seems to be, uneasiness continues about relying on numerical estimates as measures of the magnitude of sensation. After all, making proportional judgments with numbers assumes some sophistication in use of the number system. In addition, depending solely on numerical estimates entails the hazards endemic in relying on a single indicator of verbal response (see Lodge / Tursky / Tanenhaus, 1974). In an effort to dispell these uncertainties, Stevens and his associates developed a validation procedure known as cross-modality matching (J. C. Stevens / Mack, 1959). This procedure, which Stevens considered, "perhaps the most reassuring development in psychophysics," (Stevens, 1966) calls upon a subject to match the apparent magnitude in one modality (e.g., the loudness of sound in his ear) to the intensity in another modality (e.g., force of handgrip). As the investigator alters the intensity of the sound pressure, the subject squeezes the hand dynamometer to match the apparent alteration in loudness.

Cross-modality matching provides a vehicle for verifying predictors implied by the power function law. According to Stevens' law, the relationship between stimulus intensity and impressions of magnitude for two different modalities can be represented as

first modality $R_1 = S_1^{b_1}$

second modality $R_2 = S_2^{b_2}$

When the subjective value of R_1 is set equal to the subjective value of R_2, the relationship between S_1 and S_2 becomes:

$S_2 = S_1^{b_1/b_2}$

Two predictions follow:

(1) Where a power function relationship had been previously established for each modality, the matched data should also produce a power function; and

(2) The exponent for the new power function should be the ratio between the established exponents for the two modalities.

By way of illustration, let us return to the example of matching handgrip to sound pressure. Since the exponent for loudness established by means of numerical magnitude estimation is .67 and force of handgrip is 1.7, the matched data should produce a new power function with an exponent of .67/1.7 or .39. If the predicted power function is attained by cross-modality matching, original exponents based on numerical estimates of magnitude are validated. Experiments have confirmed the predicted power functions and their exponents for numerous cross-modality matches of physical stimuli.

It followed almost inevitably, therefore, that Stevens' demonstration of a human capacity to make proportionate judgments of the magnitude of sensation about physical phenomena would stimulate efforts to establish an analogue—a human capacity to make proportionate judgments about the magnitude of impressions about social (nonmetric) phenomena. To date, investigators have used magnitude estimation procedures to scale more than a dozen dimensions of social consensus. A partial listing of such magnitude estimation scales includes the following: prestige of occupations (Künnapas / Wikström, 1963); social status (Hamblin, 1971; Shinn, 1969a; Rainwater, 1971); the importance of Swedish monarchs (Ekman / Künnapas, 1963); the importance of American political offices (Shinn, 1969a); racism (Dawson / Brinker, 1971); strength of religious attitudes (Finnie / Luce, 1960); moral judgment (Ekman, 1962); the seriousness of crimes and severity of punishment (Sellin / Wolfgang, 1964); political dissatisfaction

(Welch, 1972); frustration and aggression (Hamblin / *et al.*, 1963); life stresses (Cochrane / Robertson, 1973); national power (Shinn, 1969b); national conflict and cooperation (Corson, 1970); as well as numerous studies of preferences for people, places, and things (Stevens, 1975).

Testing the validity of magnitude estimation scales of social variables is especially problematical because there is no physical continuum (criterion) whose intensities can be plotted against magnitude estimations. Lacking a relevant physical measure of the stimulus, one cannot determine directly a power function for a social variable. Hence initial efforts to validate magnitude scales of social stimuli relied on imprecise tests and on one or another form of predictive validity. But recently, following Stevens' lead, Dawson / Brinker (1971) employed a straightforward technique for cross-modality matching to validate magnitude estimation scales of social variables. Subjects are asked to indicate the intensity of their opinions about a sociopolitical variable by adjusting the strength of physical stimuli on each of two different modalities in such a way that the strength of the physical stimuli on each modality matches the intensity of their opinions. When the physical stimuli for the first modality are plotted against the corresponding matches for the second modality, a power function should result whose exponent is the ratio of the characteristic exponents of the modalities. For example, Dawson and Brinker presented their subjects with a series of thirteen statements about race and asked *S*s to match both strength of handgrip and sound pressure to their impressions of the degree of racism reflected in each statement. When handgrip matches were plotted against corresponding loudness matches, item by item, the resulting relationship was well approximated by a power function with an exponent equal to 0.39 which agrees with the predicted ratio of handgrip to loudness exponents (.67/1.7 = .39).

At this juncture, then, it appears possible to scale social variables in much the same way that sensory continua are scaled. Magnitude estimation scaling of social variables has two major advantages over category and JND scaling. In the first place, ME places fewer constraints on Rs than other types of scaling. As a result, the likelihood is enhanced that the scale will accurately reflect the judgments people can and do make of the intensity component of a social variable. In the second place, an investigator can more effectively test the validity of an ME scale because he can resort to cross-modality matches in a variety of response modes with known, reliable interrelationships.

A Review of Selected Psychophysical Experiments in Political Scaling

Recently, a number of studies have been carried out in Stony Brook's Laboratory for Behavioral Research which apply psychophysical methods to the scaling of political opinion. (The studies are described in detail in Lodge / Tursky / Tanenhaus / Cross, 1974.) The methods and results are summarized here to illustrate the power of psychophysics as a cross-modal method in the construction, evaluation, and validation of political opinion scales.

Experiment 1: The Importance of Political Offices
To test basic magnitude estimation techniques and multiple response measures in the construction of accurate and reliable political opinion scales, the first psychophysical study carried out in the Laboratory was a replication and extension of Shinn's (1969a) magnitude estimation scale on the Importance of American Political Offices. Shinn's study was selected because it scales a clear-cut dimension, importance, and because the scale items were simple and "concrete"—one-to-five-word titles of political offices ranging from "Justice of the Peace" to "President" of the United States—and the scaling dimension, importance, is a straightforward, well-understood evaluative dimension.

Shinn had *S*s estimate the importance of political offices by drawing line lengths for each office relative to the standard office of "State Senator." In our replication, *S*s made two psychophysical responses, one, numerical magnitude estimation (ME), the other, force of handgrip (HG, squeezing a hand dynamometer). Comparisons between Shinn's data and ours on the Importance of Political Offices show both scales to be remarkably close. Product-moment correlations between Shinn's line-length data and our ME and HG data are .94 and .97 respectively.

The most direct comparison of Shinn's results to ours is a plot of Shinn's line length data against the scale values derived in the Laboratory for magnitude estimations and force of handgrip, each raised to its respective power. When these points are plotted in log-log coordinates, the theoretically expected function should be linear, with a slope of 1.0. The empirically derived slope is 1.13, not significantly different (t-test, $p < .05$) from the theoretical. Given the close fit between theoretic and empirical exponents, a between-scale correlation of .954, and other comparisons between the two scales, the replication of Shinn's scale may be judged successful. What the data indicate is the "President" is perceived to be about twice as important a political office as "US Senator;" "US Senator," twice as important as "County Judge;" and "County Judge," twice as important as "Councilman of a Small Town." More to the point, the study suggests that basic psychophysical methods developed for the evaluation of physical stimuli can be adapted for use in measuring the perceived intensity of social stimuli.

Experiment 2: The Importance of Conventional Political Activities
The results of the first experiment demonstrated that relatively simple political stimuli can be scaled psychophysically using non-verbal response measures. Employing standard psychophysical procedures, the derived scale values produce a power function close to that achieved from the scaling of sensory continua. However, if these procedures were limited to proportional estimates of simple, "concrete" items along such clear-cut dimensions as importance, the utility of such psychophysical scales would be sorely limited. This experiment was designed to extend, albeit cautiously, the scope of the procedures by: *a)* employing somewhat more complex political stimuli (conventional political activities in the place of titles of office); *b)* extending somewhat the dimension "importance" by having *S*s judge the activities in terms of their "importance to the political

system;" and finally *c)* employing high school students as *S*s rather than college students which represents an initial step in the continuing process of establishing the generalizability of psychophysical methods and measures in the development of political opinion scales.

Twenty high school students were instructed in magnitude estimation and the making of proportional hand dynamometer responses. Seventeen political activities culled from Milbrath's (1965) and Verba / Nie's (1972) participation index were judged relative to the standard "Voting in a local election."

The range of this political participation scale—with items ranging from "Hold public office" to "Watch the evening TV news"—is 10:1, a narrow range compared to the Importance of Political Office scale and other scales tested in the Laboratory. Despite the bunching of items in the middle of the scale, the correlation between ME and HG was nonetheless high, .951, although the lowest correlation yet uncovered in the Laboratory between two psychophysical response modes. The HG × ME regression line fitted to these data produce a slope of 1.62 which is not significantly different from the theoretical exponent of 1.7 for force of handgrip.

These experimental findings demonstrate that *S*s (even high school students) can discriminate subtle differences within a tight cluster of stimuli. While psychophysical techniques have proved to be successful in discriminating "noise" from "signal" among physical stimuli, these data indicate that the same type of discrimination appears to be possible in the evaluation of social stimuli.

Experiment 3: Pretests of a Political Support Scale

The results of the first two experiments demonstrate the feasibility of psychophysical scaling methods in the development of relatively simple intensity scales of political opinion. In both experiments the dimension being scaled (importance) has a clearly perceived intensity, and the scale items are relatively short and simple titles of offices or statements of activities. Given this success in scaling these simple items in terms of importance, the attempt is made in Experiment 3 to utilize psychophysical methods in the scaling of complex statements typical of many political opinion items on a leading variable of political science, the dimension "support."

The steps followed in the development of the political support scale follow essentially the procedures and tests developed by Thurstone for the selection of statements expressing intensity of opinion. A large pool of statements evaluating the US Senate and Supreme Court was collected from the literature and written afresh. These statements were edited so that they were free of specific reference to the institution and free of specific evaluative content, then rank ordered by colleagues, students, secretaries, and housewives according to how much confidence, support, or trust each statement expressed. The selected items were then tested in several undergraduate classes at different universities by magnitude estimation. From this pool of statements, twenty-five items were selected on the basis of low variability and assigned to either a "Support for the US Senate" or a "Support for the US Supreme Court" scale, such that the statements roughly approximated equal log intervals and covered the full range of opinion toward the institutions.

High school volunteers were recruited and magnitude estimation and handgrip responses were elicited to both the Senate and Court scales. Product-moment correlations across modalities were highly significant, ranging from .96 to .98, while the HG × ME exponents ranged from 2.06 (statistically compatible with the established exponent of 1.7) to 2.31 (a value significantly different from the expected exponent of 1.7). The results confirmed the utility of psychophysics in the scaling of complex statements along a complex dimension—support—but highlighted the necessity for direct, cross-modal validation procedures.

Experiment 4: The Cross-Modal Validation of a Support for Political Institution Scale

Psychophysical procedures are established for the cross-modal validation of opinion scales (Dawson / Brinker, 1971; Stevens, 1972), with important refinements continually coming to the fore (Cross, 1974). Essentially, cross-modal validation involves an extension of the basic experimental design employed for the pretests in the Laboratory. Subjects are instructed to indicate the intensity of their opinions by adjusting the strength of *two different physical modalities* for which there are well-established power functions so that the strength of response on each modality matches their estimate of the intensity of the attitudinal stimulus. When the matches for one physical modality are plotted against the corresponding matches of the second modality, a power function results whose exponent should be the ratio of the exponent of the two physical modalities.

The goal of this experiment was to apply basic cross-modal matching procedures in the construction of a political support scale. Subjects were instructed to make three responses—magnitude estimations, hand dynamometer squeezes, and sound pressure adjustments—to each of the statements in the Senate and Supreme Court scales. To the extent the ratio of the exponents derived from the matches of SP and HG to the social stimuli approximate the theoretical ratio derived from matches against sensory continua, the support scales are judged to be psychophysically valid.

The magnitude estimates (ME) for the Senate support scale are plotted on log-log coordinates in Figure 3 as a function of both force of handgrip (HG) and sound pressure (SP). The product-moment correlation between log ME and log HG is .988, that between log ME and log SP is .971, and that between the two physical responses is .990. The slope of the regression line for log HG is 2.43, for log SP, 0.96. Since the regression bias in the number matches can be presumed the same in both functions, the ratio of these two exponents is of specific interest because the bias in the number scale cancels itself out. The test for psychophysical cross-modal validation is then served by the question: *does the ratio of the established power function exponents match the experimentally derived ratio?* The theoretical ratio is .392 (.67/1.7 = .392). The ratio between exponents for the physical modalities derived in this experiment is .96/2.43 = .396; essentially identical with that theoretically expected. Given the strong cross-modal correlations and the nearly exact empirical fit of .396 to the theoretical ratio

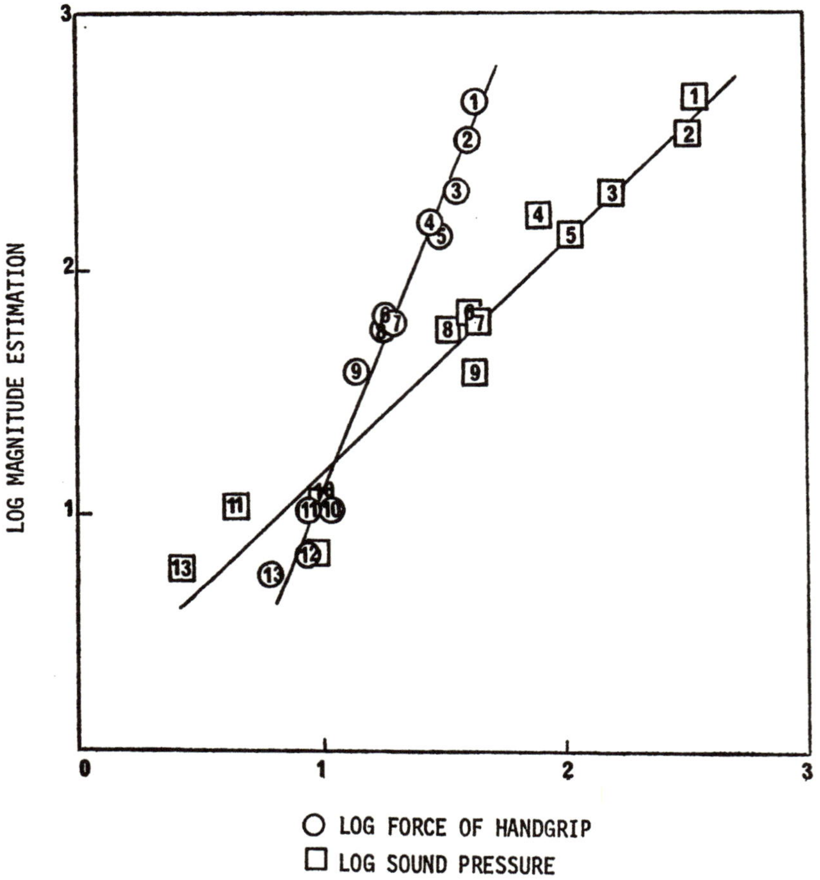

FIGURE 3. *Support for the US Senate: Magnitude Estimation plotted against Force of Handgrip and Sound Presure*

of .392, this Senate support scale meets the criterion of psychophysical validation.

The Senate support scale, with cross-modally validated scale values, is presented in Table 1.

Analyses of all three psychophysical responses for the US Supreme Court scale show similar results. Product-moment correlations across modalities range from .977-.990, and the 95% confidence limits surrounding the experimental estimate of the ratio of true exponents include the theoretical ratio of .39. A comparison of the Court v. Senate support scales shows them to be highly correlated, .986, but more importantly, the slope of the regression line is 1.05, not significantly different from the expected slope of 1.0. Given the close match between scales, both can be combined into a twenty-five-item support scale which represents a reliable and accurate (psychophysically valid) scale of support.

TABLE 1. *Support scale for the US Senate with cross-modally matched scale values*

	Statements	*Scale values* units of support*
1.	The US Senate is so perfect and so good that God himself must have created it.	168
2.	The US Senate is absolutely perfect. I completely trust them and nobody else.	161
3.	The US Senate does a real good job. You really couldn't improve on the job they do.	108
4.	I take pride in the US Senate.	82
5.	The nice thing about the US Senate is its feeling for the ordinary people.	73
6.	The US Senate is sometimes all right, sometimes not so good.	45
7.	Not good, not bad, As I see it, the US Senate is so-so.	40
8.	The US Senate is right some of the time, but not nearly as often as it is wrong.	30
9.	Sometimes you can count on the US Senate, sometimes you can't. It all depends.	27
10.	The US Senate is a bunch of troublemakers.	17
11.	None of that Washington bunch can be trusted much. And the US Senate is the worst of the whole lot.	13
12.	The US Senate should be done away with. We should get rid of that bunch of bastards.	10
13.	The US Senate is a disgrace to the country.	7

* Scale values derived from the formula: ψ — $(HG^{1.7}SP^{.67})^{1/2}$.

DISCUSSION

The scaling experiments reported here provide compelling evidence of the successful extension of psychophysics from the mesasurement of sensory phenomena to the measurement of social opinion. What the experimental results of these studies demonstrate is that lawful relationships hold between the perceived intensity of selected sociopolitical stimuli and the magnitude of response. The power of the psychophysical scaling of social opinions lies in the ability of people to match sensory responses directly to the strength of their opinion. When such responses are produced using two sensory continua for which there are known power functions, the exponent of the function mediated by the social stimuli can be compared with the exponent derived from the direct cross-modal functions of the two sensory continua. If these exponents are comparable, our confidence in the scale derived from these matching functions is substantially increased. This cross-modality matching procedure constitutes *psychophysical* validation and is a most rigorous test for building scales of opinion. This cross-modal validation procedure is a critical first step in the development of a social metric.

By means of cross-modality matching, it now appears possible to build valid scales of perceived intensity on a number of social variables. In contrast to alternative scaling techniques, these psychophysical scale-building procedures provide a means to develop interval or ratio scales and assign meaningful values to the scale items. What recommends cross-modality matching, in contrast to alternative scaling procedures, is the capacity to bring multiple response measures with known, reliable, and verifiable interrelationships to numerous sensory continua to bear on the determination of scale values. For those variables for which a metric exists (as, e.g., SP, HG, and LL), it is possible to determine directly and exactly the accuracy of the sensory scale and then to compare established exponents for the sensory (metric) continua to those derived from the social (non-metric) variable.

Because the concept "validation" carries multiple meanings, let us detail the specific meaning of cross-modal psychophysical validation in the context of social scaling. Attitude is a concept; it cannot be measured directly but must be inferred from some behavioral response(s). Typically, political attitude research relies almost exclusively on verbal self-report to indicate both the existence of an attitude (the cognitive component) as well as the strength of attitude (the intensity component). Within this framework political scientists are concerned with the meaning and measurement of verbal response. Two obvious problems come to the fore. One concerns the need to validate the *meaning* of verbal responses. (What do *R*s *mean* when they say they identify with a political party, like a candidate, favor a policy, support the Supreme Court or Senate?) The second problem involves measuring the *strength* of verbal response. (How *strongly* does *R* identify with a party, like a candidate, favor a policy, support the Court or Senate?) This distinction between the cognitive and intensity components of verbal response dates back to Thurstone, if not before (to Aristotle, see McGuire, 1969), and is critical in the analysis of verbal behavior and in attempts to infer attitude from response.

Validation of the cognitive component is difficult at best. No procedures exist today which convincingly demonstrate that the meaning an R assigns to a verbal stimulus is identical to the meaning assigned it by other *R*s or by the investigator. In time, advances in learning theory or psycholinguistic studies of concept formation may provide procedures for validating meaning. At present, rigorous experimental controls over S-R relationships offer the strongest assurance that R's response is to one specific stimulus and not another.

However, through cross-modal psychophysical validation procedures, it now appears possible to validate the *intensity* component of response. If a researcher is willing to assume that his *S*s share a common understanding of such terms as political party, candidate, policy, Court, or Senate, then it is possible—given properly constructed magnitude scales—to build interval or ratio scales and assign validated weights to the scale items. The experimental results from our efforts to psychophysically validate attitude scales demonstrate that a lawful relationship between the perceived intensity of a stimulus and the magnitude of response may be established for complex political attitude scales.

Since the beginning of these psychophysical studies, still more powerful validation procedures are being developed. Whereas our validation procedures assume no regression bias on either physical measure used in the cross-modal matches, the new procedures may enable us to detect and measure the bias for each matching variable, thereby making it possible to construct bias-free scales (Cross, 1974). We are now incorporating these new procedures in experiments designed to build, test, and validate political attitude scales.

A second question arises as to the validity of these and future scales. With the evidence available from these experiments, the scale positions and values are expected to hold across all people in the same language group as our subjects. In a very real sense the limits of the scale may well be at the borders of our culture and language. The scale should hold (be externally valid) wherever, for example, the phrase "I take pride in the US Senate" (with fifty-seven units of support) reflects five and a half times or so more support than the statement, "The US Senate is a disgrace to the country" (with ten units of support). In social science the concept of validity, when applied to social psychological variables, is dependent on language and culture. Within the small sample of language users in these studies, the scales are psychophysically valid. What needs to be done is to test systematically the cultural and linguistic boundaries of these scales, for example, to test scales within and across educational, sex, age, racial, and regional groupings. This need for external validation is common to any scale which uses language and is dependent on socially conditioned responses. Two research projects are currently underway to test the utility of psychophysical scaling procedures. In one study portable equipment is employed in the field to psychophysically validate selected political opinion scales on a stratified sample of adults in local communities. A second study employs a new calibration procedure which—if successful—will allow survey interviewers untutored in psychophysical techniques to obtain scale values from *R* in a typical survey context. Should these new scales withstand psychophysical validation, plans are afoot to externally validate the scales in local communities and then in a national survey.

Contemporary studies of political behavior have failed to account for large enough portions of variance, find strong enough correlations across scales, or better predict from scale scores to political behavior. Weak correlations, low explained variance, and poor prediction could be the result of crude measurement or weak theory. If the new scales prove to be valid expressions of the intensity of political opinion and dramatically improve our predictive power, then disciplinary advances can be made relatively quickly. But if the new scales, however superior, have little impact on explained variance, then the problem may well be weak theory, an inability to extract strong explanatory relationships from contemporary theory.

Psychophysiology as a Tool for the Study of Political Behavior

The use of psychophysical scaling methods must improve our ability to assess the intensity (affective) component of political attitudes. It is of equal importance that improved methodology be developed to assess the cognitive and conative components of these attitudes. Reliance solely on verbal behavior, no matter how sophisticated, is problematical, and other measures must be found to map these elements of political concepts and attitudes. Psychophysiological responses and techniques associated with learning theory seem to provide a reasonable approach.

Psychophysiology is broadly concerned with relating social and physiological events to biological processes. The theoretical framework underlying psychophysiological research is this: neural, chemical, and physiological processes respond adaptively and instrumentally to environmental demands and exert critically important regulatory effects on the nervous system. Since it is the nervous system which mediates the relationship between stimulus and response, any theory which claims to account for behavioral relationships (political or other) must ultimately take the nervous system into account and utilize an S-O-R (Stimulus-Organism-Response) analytical framework. Why, where, and how political scientists can use psychophysiology as a tool for the study of political behavior is the theme of this section of the paper.

Physiological Reactivity to Environmental Demands

Physiological response systems are remarkably responsive to environmental stimuli and exert major regulatory effects on behavior. These two interactions—the effect of environment on physiology, and, obversely, the effect of physiological patterns on cognitive and emotional behavior—constitute the conventional distinction between physiological psychology and psychophysiology (Stern, 1964). In physiological psychology the independent variables are physiological manipulations (e.g., electrical stimulation of the brain, drugs, or biofeedback); the dependent variables, psychological (test performance, rating scales, etc.), while psychophysiology is oriented to the effects of psychological variables on physiological functioning (one example, the effects of threat on central, autonomic, and overt behaviors). More important than this traditional distinction between the two disciplines (the classification, after all, blurs the necessary interaction between internal milieu and environment) both physiological psychology and psychophysiology focus on the role of the nervous system in behavior (Moyers, 1971; Seligman / Hager, 1972, 1974).

The nervous system is anatomically subdivided along a central-peripheral dimension. The central nervous system (CNS) is comprised of the brain, brain stem, and spinal cord. The peripheral nervous system includes all the nerve fibers throughout the body which enter and leave the brain stem and spinal cord.

> The integration of bodily responses is due to the extensive network of nerve fibers and blood vessels which run to every part of the body. Nerve impulses stimulate the secretion of hormones from various endocrine glands; circulating hormones in the blood stream affect the functioning of the nervous system; the nervous system in turn regulates the flow of blood to various tissues; and within the nervous system, centers in the brain and brain stem increase or decrease the activation of peripheral fibers, while the feedback from the periphery raises or lowers the threshold of CNS centers. (Sternbach, 1966, p. 60.)

Within this framework the autonomic nervous system (ANS) is one of the peripheral systems.

For a number of reasons, both practical and theoretical, the primary focus of psychophysiology is on the recording of ANS reactivity. Functionally, the ANS, with its centers in the brain stem and spinal cord, innervates the internal organs, glands, smooth muscle, heart, and lungs, functioning primarily to maintain optimum internal conditions in response to changing demands of the environment (homeostasis). Although a person could apparently live without an autonomic nervous system, the environment would have to be constant and benign, devoid of physical and psychological stress.

On a structural basis the ANS is further divided into a sympathetic (SNS) and parasympathetic (PNS) nervous system. A schematic of the SNS and PNS branches of the autonomic nervous system appears in Figure 4. The SNS (dashed lines) is made up of those fibers which originate in the chest and saddle regions, while the PNS fibers (solid lines) emerge from the brain stem and tail areas. Generally, the sympathetic fibers stimulate the organs they serve with hormones secreted by the adrenal glands. Hence, the effects of sympathetic reactivity are activating and widespread, similar to changes produced by adrenaline. Both the SNS and PNS fibers run to many of the same autonomically modulated organs, and their effects are for the most part antagonistic: SNS activity tends to be excitatory; PNS activity, inhibitory.

Changing environmental conditions evoke compensatory CNS and ANS responses. As Cannon (1932) found, an animal, when confronted by a situation which evokes fear, pain, or rage, responds with a complex and extensive set of physiological reactions which prepare it for "fight," or "flight." Upon perception of the threat in the cortex of the brain (indicated by marked changes in brain wave activity) impulses are transmitted down the sympathetic branch of the ANS to the adrenal glands where adrenaline is secreted.

The ensuing physiological changes in response to the threat are described vividly by Cannon:

> Respiration deepens, the heart beats more rapidly; the arterial pressure rises; the blood is shifted away from the stomach and intestines to the heart and central nervous system and the muscles; the processes in the alimentary canal cease; sugar is freed from reserves in the liver; the spleen contracts and discharges its contents of concentrated corpuscles, and adrenaline is secreted from the adrenal medulla. The key to these marvelous transformations in the body is found in relating them to the natural accompaniments of fear and rage—running away to escape from danger, and attacking in order to dominate. Whichever, a life or death struggle may ensue.
>
> The emotional responses just listed may reasonably be regarded as preparatory for struggle. They are adjustments which, so far as possible, put the organism in readiness for meeting the demands which will be made on it. The secreted adrenaline cooperates with sympathetic nerve impulses in calling forth stored glycogen from the liver, thus flooding the blood with sugar for the use of laboring muscles; it helps

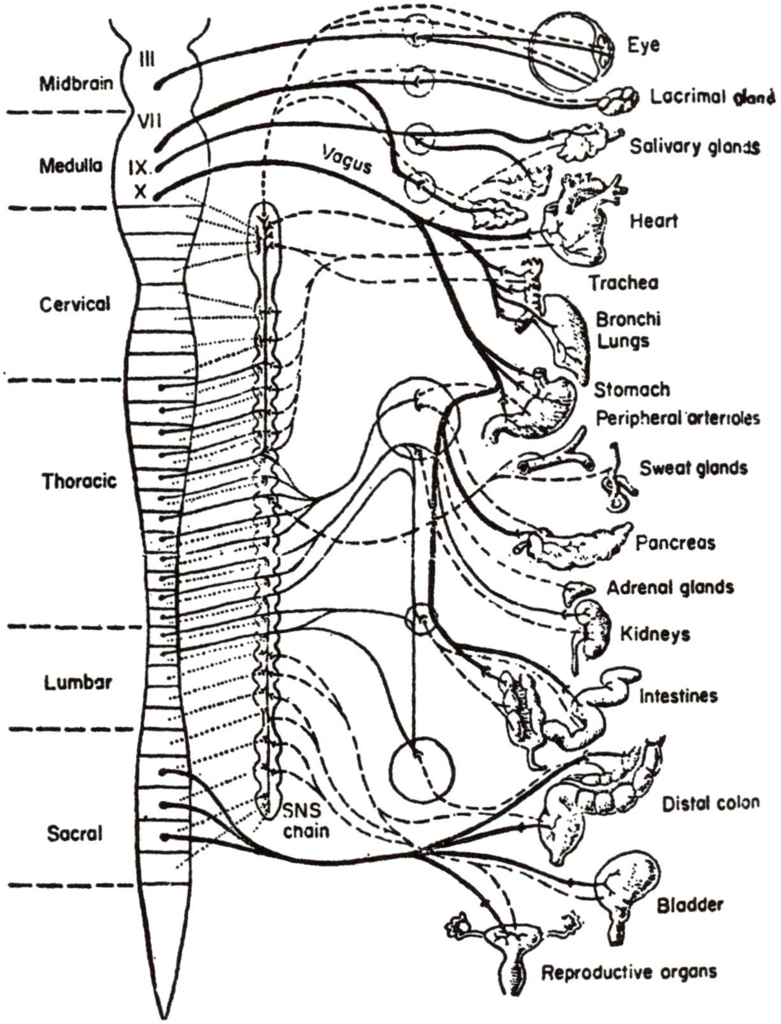

FIGURE 4. *The Sympathetic and Parasympathetic branches of the autonomic nervous system* (from Sternbach, 1966, p. 16)

in distributing blood in abundance to the heart, the brain, and the limbs (i.e., to the parts essential for intense physical effort) while taking it away from inhibited organs in the abdomen; it quickly abolishes the effects of muscular fatigue so that the organism which can muster adrenaline in the blood can restore to its tired muscles the same readiness to act which they had when fresh; and it renders the blood more rapidly coagulable. The increased respiration and redistributed blood running at high pressure and the more numerous red corpuscles set free from the spleen provide for essential oxygen and riddance of acid waste, and make a setting for instantaneous and supreme action. In short, all these changes are serviceable in rendering the organism more effective in the violent display of energy which fear and rage may involve.

This massive discharge of nervous system activity is described by Cannon as preparatory for "emergency," "fight-flight" behavior. In response to a strong threat or stress, the autonomic, somatic, and central nervous systems are immediately integrated into a "generalized activation syndrome" (Selye, 1946) which energizes the organism and mobilizes an escape or attack reaction. The experimental work of Cannon and Selye seemed to provide psychology with a physical basis for measuring the intensity or motivational dimension of approach-avoidance behavior. Given the Cannon-Selye formulation, it appeared possible to construct an activational continuum which ranged from "no response" at one end of the scale to massive SNS-type arousal at the other, and thereby to provide a continuous biological index of stimulus effects.

Psychophysiological Measures

Contemporary psychophysiological research focuses on physiological patterning—investigations of the facilitatory function of specific CNS and ANS response patterns to specific emotions and behavioral demands. The choice of physiological measures, which include all those noted by Cannon and involve virtually all of the organs illustrated in Figure 4, is determined by the physiological responses that best facilitate behavioral responses to particular environmental demands. The most popular measures, those which require only simple equipment and interfere as little as possible with the normal reactions of the subject, are *a)* galvanic skin conductance (GSC) as a summary measure of ANS activation, and *b)* heart rate (HR) and blood pressure (BP) as summary measures of cardiovascular reactivity. While most psychophysiological research deals with factors influencing autonomic functioning, the rapid development of more sensitive recording instruments and computer techniques is bringing to the fore measures of central nervous system activity, either in the form of the electroencephalogram (brain wave activity, the EEG) or in its peripheral effects on the electromyogram (muscle tension, the EMG).

For illustrative purposes a physiological record is presented in Figure 5. Channel 1 is a marker channel that delineates the segments of each trial, with each experimental event controlled by programming equipment and denoted on the paper record. Two adjacent trials are shown in Figure 5. A trial began with the onset of a red "ready" light. Five seconds later a slide appeared on a projection screen in front of the subjects. All the slides in this study contained political survey-type items. Each slide was projected for twelve seconds, enough time for a subject to read the item several times and determine the extent to which she agreed or disagreed with it. The disappearance of a slide signalled the subject to record her decision on a button box at her fingertips that offered nine choices ranging from "strongly agree" to "strongly disagree." The subject's decision was recorded on channel 8 (not shown) as a stepped voltage change, thereby allowing for the comparison of verbal and physiological responses to attitudinal stimuli.

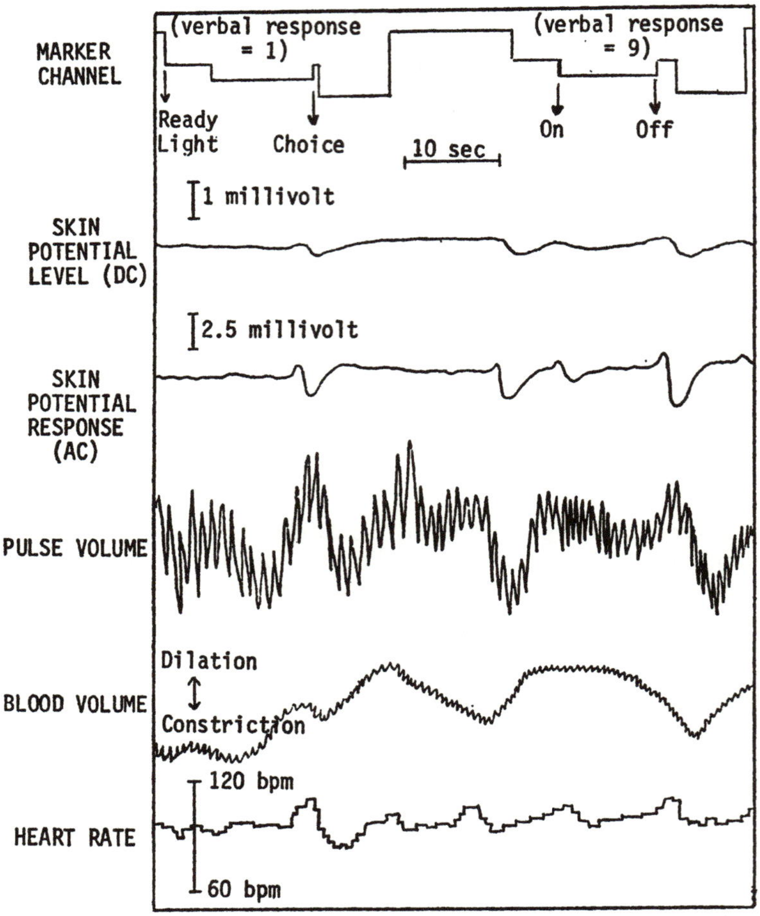

FIGURE 5. *Six channel physiological record of one subject on two adjacent trials* (from Tursky / Lodge, 1971, p. 7)

Channels 2 and 3 are measures of palm-to-forearm electrodermal activity; DC skin potentials were recorded at low sensitivity on channel 2 to provide basal level data, while on channel 3 skin potentials were recorded at higher sensitivity, through a large time constant, to detect evoked responses to the slide. Channels 4 and 5 are measures of pulse and blood volume recorded by means of a fiber optic photoplethysmograph taped to the index finger of the left hand. Heart rate, channel 6, was recorded from a pair of standard ECG electrodes placed on the left arm and right leg of the subject, with continuous beat-to-beat information processed through a cardiotachometer, calibrated, and printed out in bpm.

The figure shows two adjacent trials for one subject. On trial fourteen, the subject indicated strong agreement, button one on the choice box, and on trial fifteen indicated strong disagreement, button nine. While extreme caution is essential when dealing with such discrete responses, one may note that both AC and DC skin potential as well as pulse and blood volume are larger on trial fifteen than fourteen. These differences suggest greater reactivity to negative than to positive evaluations of the statements.

In addition to having great theoretical import in the explanation of behavior, psychophysiological techniques and measures provide an extraordinarily promising alternative to studying political behavior by purely *post facto* observational or impressionistic methods. Psychophysiological measures are remarkably sensitive to stimulus and behavioral demands on the individual, appear to be patterned in different emotional states, and vary according to the subject's coping behavior. Given adequate experimental controls, psychophysiological techniques and measures provide objective, relatively bias-free, nonverbal indices of human reactions, and thereby offer a source of relevant information about the state of the organism before, during, and after experimental manipulations. There are several major advantages of employing these psychophysiological methods in the experimental study of political behavior: *a)* the measuring devices, once properly applied, do not require experimenter intervention or direct observation, hereby eliminating a major source of bias; *b)* physiological measurements are stable—variations in responses are a function of changes in the subject, not the measuring instrument, thereby alleviating the problem of reproducibility; *c)* physiological measurements may be derived from subjects while they are engaged in other tasks, e.g., decision-making or social interaction, without interrupting the task with self-rating scales; *d)* because they may be recorded independently of the subject's verbal system, or more commonly in conjunction with other measures (e.g., subvocal speech, implicit motor acts, self-report, performance in a task, or a social behavior), physiological measures provide multiple cross-modal response data for the analysis of consistencies and incongruities in human reaction; finally, *e)* since much behavior is probably immediate and automatic, physiological measurements may be used to indicate persistent, habituated reactions of which the subject is unaware. Taken together, these advantages—all premised on the use of experimental procedures and controls—allow investigators to monitor experimental manipulations and obtain additional sources of information about experimental effects, and thereby significantly strengthen the analysis of interactions between the individual and his environment.

The Intensity Component of Responses

When individual and situational response specificity are controlled, the magnitude of physiological response increases with increasing intensities of social, psychological, or physical stimulation. The attempt to measure and account for the strength of response to social and psychological stimuli is illustrated in the long and continuing concern of psychophysiology with the study of emotion and the application of psychophysiology to social science research in studies of prejudicial attitudes and discriminatory behavior. Rankin / Campbell (1955), for example, found greater GSR reactivity among white males to a Negro than to a white experimenter by the straightforward procedure of having each of the experimenters touch the subject's arm while "adjusting" the apparatus. Westie / DeFleur (1959), using slides rather than actual people, selected prejudiced and unprejudiced groups on the basis of questionnaire data collected earlier. The subjects were then shown pictures of Negroes and whites in a variety of social settings, e.g., a Negro and white male sitting at a table in a "neutral" setting. Both GSR and pulse volume differentiated the reactions of the prejudiced from unprejudiced subjects—significantly larger GSRs were elicited from prejudiced subjects when viewing pictures of Negroes.

That greater autonomic reactivity (as measured by the GSR) as an indicator of prejudicial attitudes is also shown in a series of studies by Cooper and his associates (1956, 1959). They embedded such ethnic names as "Jews," "Swedes," "Mexicans," "Japanese" in a set of complimentary and derogatory statements, and obtained positive correlations between GSR magnitude and verbally expressed likes and dislikes. Attitudes, it appears, include a physiologically measurable affective tone. Interestingly, while both positive and negative attitudes are associated with autonomic activation, complimentary statements about disliked groups and derogatory statements about liked groups are especially arousing. Dickson / McGinnes (1966), e.g., in a study similar to Cooper's, found no overall GSR differences between pro-church, neutral, and anti-church subjects to religious statements, but once again, the kind of statement—whether pro, neutral, or anti-church—did affect the GSR. Pro-church subjects showed their largest responses to anti-church statements, while anti-church subjects responded most strongly to pro-church statements.

These relatively simple psychophysiological studies, and literally hundreds more (see the bibliographies of Bamford / *et al.*, 1968; Roessler / Kelly / Collins, 1970; Carriero / Gehringer, 1971), provide groundwork for the study of attitudes. Under properly controlled conditions they demonstrate the following: *a)* attitudes have an intensity dimension or affective component which is easily measured physiologically—intensely held attitudes are more activating physiologically than are weakly held or neutral attitudes; *b)* conflicting or ambivalent attitudes and situations are especially arousing; *c)* larger ANS responses occur when an attitude is challenged than when it is reinforced (Crider, 1970). Although lacking an extensive theoretical base, these simple psychophysiological studies suggest the utility of physiological measures in monitoring experimental effects and in complimenting subjective measures of attitudinal intensity.

Psychophysiological Response Patterning

Severe problems however, plague activational theories and the simple activational notions underlying the measurement of intensity of response. Over the past two decades psychophysiological data have accumulated which supercede the Cannon-Selye formulation of a generalized activational response and its extension into psychology as activation theory (Duffy, 1962). The assumption of activation theory that autonomic, somatic, and central nervous system reactivity is positively correlated with increasing expenditures of energy and behavioral activity has repeatedly been found wanting (Sternbach, 1966). In point of fact, the intercorrelations between measures of nervous system activity, although positive, are generally moderate. This is particularly true as individuals move away from the polar positions of coma and panic. This means that the degree of activation is variable within and between the different response systems, and that although there are some common activational characteristics in response to stimulation, the degree of commonality is generally moderate. Strong psychophysiological evidence now exists which demonstrates that within the generalized activational syndrome are patterns of response specific to individual and situational factors (Lacey, 1967; Shapiro / Crider, 1969). In short, *the kind and intensity of an individual's response to stimulation will be determined by two major factors: one, the nature of the stimulus situation as perceived by the person; the other, the individual's response hierarchy—his constitutional and acquired pattern of responding to stimulus situations.*

By way of summary, two concurrent but distinct patterns of physiological reactivity are evoked in response to stimulation—individual and situational response specificity are embedded in a nonspecific, generalized reaction. Paraphrasing Sternbach, virtually any stimulus will produce some degree of activation, some relative preponderance of SNS-like directional responses (a generalized activation syndrome) in a relaxed, resting subject. However, not all people respond in the same way to the same stimulus, nor do all stimuli produce the same responses. With continuous change possible on all the CNS innervated ANS functions it is theoretically possible (and commonsensically probable) that there are specific response patterns for each emotion and for every specific environmental and behavioral demand placed on the individual. Attempts have been made to classify patterns according to the stimulus situations which evoke them. Two examples, the effect of cognitive set and of attention on response patterning, are discussed here.

Cognitive Set

Physiological and behavioral responses can be dramatically altered by the introduction of a cognitive set. Experimenter effects (Rosenthal, 1966) and experimental demands (Orne, 1962) must be rigorously controlled or such effects can prove detrimental to the research design. Well-controlled manipulations of cognitive sets have been demonstrated to produce marked response alterations in predicted directions. Sternbach (1960), for example, had subjects swallow a pill that was in fact a sugar coated magnet that produced a record of gastric motility. One group of subjects was instructed that the pill would relax the stomach, a second group was told that the pill would stimulate contractions, and the control group was told the pill would have no effect. In all cases the pill produced an effect that was in accord with the subject's expectations.

In more comprehensive demonstration of the effect of instructional set on coping behavior, Lazarus / *et al.* (1965) recorded autonomic measures (heart rate and skin conductance) from a group of subjects while they were watching an industrial safety film designed to frighten shop workers into using safety precautions. A series of woodmill accidents are realistically presented: in one accident the finger tips of a man are lacerated; in another a worker loses a finger and an innocent bystander is killed; and in the third accident a careless worker allows a wood plank to fly loose from a circular saw and impale him. The second and third accidents occur as flashbacks, portending future danger, and allow subjects to anticipate the bloody accidents. In general, the bulk of the physiological reaction occurs during the threat periods. During these accidents scenes there is a sharp increase in both the subjective and autonomic indications of stress.

The subject's cognitive orientations to the film were also manipulated to test the hypothesis that defensive coping processes would dampen the stress reaction. Before the start of the film the Control group was told simply that some industrial accidents would be seen in a safety film, a second group (the "Denial" group) was told that the accidents were not real but dramatized realistically by actors; and a third ("Intellectualization" group) was asked to take a detached view so as to evaluate the film on a number of safety dimensions. Both intellectualization and denial effectively short circuited the stress reactions, that is, the subjective and autonomic reactions that normally result from viewing the film were obviated. Other studies (e.g., Lazarus / *et al.*, 1962) employing different stress films and procedures for manipulating cognitive orientations report similar effects as a consequence of denial or intellectualization.

Psychophysiological Measurement of Attention to Stimuli

The attention people pay to environmental stimuli is an important independent variable in explaining attitude formation, attitude change, decision-making, and other forms of social behavior. The attention variable is, for example, embedded *a)* in such related variables as awareness, beliefs, comprehension, feelings, and knowledge, *b)* in processes such as selective attention and selected perception, and *c)* in attempts to measure the saliency, valence, or interest value of stimuli.

Although attention has long been considered an important independent variable in research on attitude formation and attitude change, the vast majority of studies utilizing the variable attention do not, in fact, measure attention at all. Rather, unsatisfactory surrogates are employed, typically some self-reported measure of exposure, recall, recognition, or retention. Yet, any number of studies reveal that these surrogates are but weakly related to social attitudes and behavior. (See Lazarsfeld, 1940; Hyman / Sheatsley, 1947; Stouffer, 1955; and the review by Weiss, 1969; Bechtel / Achelpohl / Akers, 1972.)

Recent major developments in the behavioral sciences have now demonstrated the feasibility of measuring *objectively* the degree of attention paid to external stimuli. These developments, both conceptual and methodological, are largely the work of psychophysiologists. In a series of closely reasoned studies (Lacey, 1959, 1967; Lacey / *et al.*, 1963; Lacey / Lacey, 1958) Lacey and his colleagues investigated experimentally the behavioral functions of cardiovascular feedback. Lacey proposed that under certain conditions baroreceptor activity modulates cortical activity and sensorimotor integrations via projections from the medulla to higher brain centers. He reasoned that cardiovascular activity and baroreceptor firing is inhibitory in terms of reducing electro-cortical activation, thus dampening sensory inputs. Conversely, decreases in baroreceptor feedback, consequent on decreasing blood pressure and heart rate, are excitatory with respect to increasing electro-cortical activity and sensory acuity.

To demonstrate the factual basis of this theory, Lacey (1959) devised a series of experimental tasks that varied along a behavioral continuum with environmental intake at one end and environmental rejection at the other. Situations in which *S*s are required to attend to environmental stimuli are characterized by HR deceleration. Tasks involving mental effort that requires freedom from distraction, in contrast, are accompanied by marked

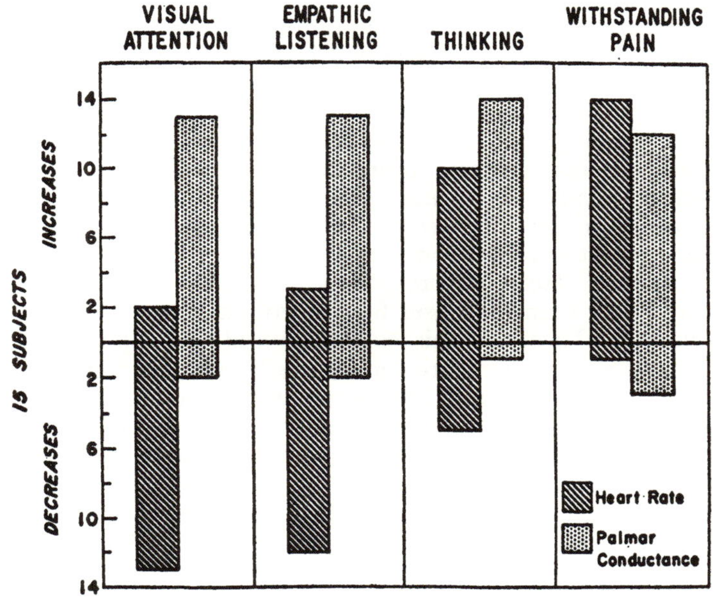

FIGURE 6. *Directionality of changes in Heart Rate and Palmar Skin Conductance as a Function of Various Experimental Tasks* (from Lacey, 1959)

HR acceleration. Figure 6 illustrates the effects of different behavioral tasks on heart rate and skin conductance. The cardiovascular pattern (blood pressure typically follows heart rate) is not a simple activational effect. Skin conductance demonstrates a clear activational effect across behavioral tasks (indicating that all four tasks were sympathetically arousing) but does not discriminate one behavioral task from the other. Heart rate, on the other hand, shows "directional fractionation of response." According to Lacey's formulation, which Black has confirmed with neurophysiological evidence (Black, 1970), cardiovascular decreases facilitate sensory receptivity, while increases in heart rate and blood pressure dampen sensory acuity, thereby reducing attention to "distracting" environmental stimuli and facilitating problem solving.

Specific to attention, HR *decreases* in situations requiring attention to environmental stimuli; and conversely, HR *increases* in situations requiring attention inward, as for example, in problem solving (Tursky / Schwartz / Crider, 1970). What is more, a number of recent studies confirm earlier findings which indicate that HR fractionation is correlated with, if not functional to, a number of discrete behavioral activities. In behavioral tasks requiring attention to environmental stimuli (e.g., carefully listening to the radio or watching television) those *S*s who produce the deepest HR deceleration to the external stimulus are the more vigilant, whereas in tasks requiring "environmental rejection" (e.g., adding numbers, remembering a list of words, problem solving in general) those *S*s who produce the largest acceleration perform these tasks the best. (See Epstein / Fenz, 1967; Fenz / Epstein, 1967; Fenz / Jones, 1972.)

The application of HR fractionation in measuring attention to complex social stimuli is illustrated by Spence / Lugo / Youdin (1972). In their study, a 17-minute passage, taken from a patient's talking in a psychoanalytic interview, was played to forty subjects, including trained therapists, therapists in training, and inexperienced undergraduates. Subjects were alerted to the organizing theme (termination of the patient's treatment) and were asked to attend to direct and indirect references to his theme. Heart rate levels averaged for thirty second periods were lower when termination clues were present on the tape than during control periods when clues were not present. Heart rate responses associated with these clues were significantly lower than responses for control passages.

The work by Lacey and his colleagues and now Spence, Lugo, and Youdin, provides the framework for the *objective* measurement of attention. In contrast to earlier work with attention, relying as it did on one or another surrogate of attention, the use of physiological measures of attention offers researchers a powerful measure for continuously tracking and differentiating inattention from close attention.

Summarily, then, two sets of problems are raised, one set methodological, the other substantive. The methodological questions concern the development of objective, reliable and valid physiological response measures of attention to complex sociopolitical stimuli. TV news programs and documentaries are examples. Moreover, it now appears possible to apply physiological response patterning to discriminating among attention, recall, recognition, and retention. Then, with objective physiological response measures of attention at hand, it becomes feasible to apply these objective measures of attention to the study of attitude formation and change.

Research on the relationships between attention and its surrogates is now underway in the Laboratory for Behavioral Research at Stony Brook. Our aim is to relate physiological patterns of attention to the *recognition, recall,* short-and-long-term *retention,* and *evaluation* of political stimuli.

Psychophysiological Measures of Concepts

Insofar as political behavior is learned behavior, the classical and operant conditioning paradigms offer a relatively simple and direct approach to the experimental study of political attitudes and other acquired behaviors. The rational for the use of the conditioning paradigm in the study of attitudes is straightforward. Attitudes—however operationalized—are mediators, the cognitive, affective, and conative components of which have been learned. The learning process is, therefore, critical to the study of attitudes. If attitudes are considered as learned responses, the assumption made here (and elsewhere: Campbell, 1963; Skinner, 1953; Staats / Staats, 1957, 1958) is that attitudes are learned in much the same way and under many of the same conditions as are other behavioral responses. If so, the established principles of classical (Pavlov, 1927) and operant (Skinner, 1938, 1953, 1971) conditioning necessarily bear on the study of political attitudes and behavior.

The characteristic feature of classical conditioning is the *association* of a novel stimulus with an established response (typically a reflexive response). The knee jerk in response to a tap on the patellar tendon is one example of respondent behavior, perspiration to heat, pupillary constriction to light, an orienting response to a sudden noise, and salivation to the sight or smell of food are others. All these responses are *un*conditional, that is, barring satiation or neurological damage, they always happen. These responses are not learned, not conditioned, but are reflexes and are defined by Pavlov as Unconditioned Responses (UCR). The stimuli, for example, the tap on the tendon, are Unconditional Stimuli (UCS).

As Pavlov demonstrated, new stimulus-reponse behaviors can be conditioned from this original UCS→UCR reflex. Reviewed simply, associational learning—the repeated pairing of a new stimulus (a CS, Conditional Stimulus) with an already established UCS-UCR relationship—will result in a new behavioral response (CS→CR). This constitutes learning, as per example, a child learning to anticipate the clap of thunder to the flash of lightning.*

* While classical conditioning is of obvious importance in the development, maintenance, and extinction of learned behaviors, the procedures do not account for a wide variety of behavioral change. The argument, first set forth by Thorndyke, then developed by Skinner, is that much learned behavior occurs without the presence of an observable UCR in evidence. Learning, it is argued, often involves more than the pairing of a novel stimulus with an established UCS-UCR reflex. It is noted that the *consequences of a behavior* (reward or punishment, most simply) determine the probability that simple behavioral responses will be repeated and shaped into complex behaviors.

This type of conditioning is termed operant or instrumental conditioning. Instrumental conditioning (learning, if you will) refers to the fact that some behavior on the part of the organism does something to (operates on) its environment. The behavior changes the environment in some way. Any activity that changes the environment in any way is an "operant" or "instrumental" act. Instrumental conditioning refers to the manipulation of the environment (the contingencies of reinforcement). Any behavior which is reinforced (rewarded) is likely to occur again under similar conditions. The more often the behavior is reinforced—other things being equal—the more likely it is to recur.

In contrast to classical conditioning, operantly conditioned behavior is weakened or strengthened by the *consequences* of the behavioral act, by the events which *follow* the response. Whereas respondent behavior is controlled by its antecedents—the association of CS and UCS to UCR—operant behavior is controlled by its consequences.

These differences between classical and operant conditioning are largely procedural (Kimble, 1961), not neurophysiologically or qualitatively different. Throughout his lifetime a human being is exposed to millions of stimuli. These stimuli produce responses which are either reinforced or punished (operant conditioning), and associations are built up between stimuli and responses (classical conditioning). This suggests that much human behavior is in large part a result of both classically and operantly conditioned responses.

The classical conditioning process would be of little value to social and behavioral scientists if its function were restricted to basic reflexive responses. It has been demonstrated, however, that psychological, physiological, behavioral, as well as reflexive responses can be conditioned by the associational procedures developed by Pavlov. What makes the conditioning process so powerful is that behavior which is reinforced or associated in one stimulus situation is likely to occur in other similar situations but not in dissimilar situations. These phenomena (Aristotle's principles of "similarity" and "contrast") are called *generalization* and *discrimination* in modern learning theory. Were it not for generalization, every response would have to be relearned for each new stimulus situation—an intolerable situation given the great variety of situations in the world. Conversely, there would be total confusion if there were complete generalization, with any behavior likely to occur in any stimulus situation. Discrimination refers to the fact that animals and men learn to respond selectively to stimuli under different situations. Discrimination, like generalization, is learned.

To further our broad objective of bringing established experimental procedures of the behavioral sciences to bear on the study of political attitudes and behavior, a series of conditioning experiments is being carried out in the Laboratory. The aim of this experimental effort is to reliably and objectively measure *S*s' *existing* concepts and cognitive sets toward things political.

The studies follow the experimental design Staats / Staats (1957, 1958, 1962) developed to measure the meaning of words and the generalization of meaning across verbal stimuli. Staats / Staats (1957) demonstrated the utility of the conditioning paradigm in a series of experiments which successfully conditioned the evaluative (good vs. bad), potency (strong vs. weak), and activity (active vs. passive) components of meaning developed by Osgood / Suci / Tannenbaum (1955) and employed in the "Semantic Differential." The Staats' work shows how the components of meaning attached to one set of words may be conditioned to another set of words by systematically associating or pairing one with the other.

The experiments conducted by Staats / Staats clearly demonstrate that it is possible to produce conditioned generalization effects across verbal stimuli. Of even greater importance is the demonstration that these conditioned effects generalize to the meaning of words which then can be used to modify an *S*'s previously existing concept. The ability of this experimental design to produce generalization across words and their meanings makes it useful for our studies on the generalization and discrimination of political stimuli.

A series of classical conditioning experiments tentatively entitled "Conceptions of the Political" are being pretested in the Laboratory. The basic aim is to develop the wherewithal to use the conditioning paradigm to determine the cognitive elements of a concept, map cognitive associations, and discover what specific political concepts generalize to and are discriminative from. Given the conventional paradigm governing political attitude research (see Figure 1), the goal of this research effort is to determine more exactly the multiple meanings which define political stimuli. If, as the results from the psychophysical scaling experiments suggest, it is possible to measure precisely the strength of attitudinal responses, the question raised in our conditioning experiments is "what are people responding to?". The same stimulus, say the "Welfare Budget," may well carry multiple connotations within and across individuals. For some, Welfare may be directly associated with "poor people," for others "fraud," and for still others the stimulus may be conceptually paired with "Black people." Until we can discriminate the specific connotation(s) of political stimuli, it will be impossible to make sense of the respondents' strength of positive or negative support.

One experiment attempts to determine the extent to which racial connotations are embedded in outwardly appearing, non-racial political stimuli. The hypothesis tested by this initial experiment is that major campaign issues of the 1972 presidential campaign were "code words" for race. (The campaign issue stimuli are derived from 1972 Gallop Poll data which asked respondents to cite their major concerns of the campaign.) Some of the test stimuli are "Welfare Budget," "Defense Budget," "The Watergate Affair," "Crime in the Streets," "School Busing," "Civil Rights," "Women's Rights."

The experiment is divided into two major parts. In the first stage (two to three weeks before the laboratory experiment), *S*s are presented with a questionnaire containing sixty-two words and phrases related to issues, problems, and slogans of the day. Some of the stimuli were obviously Black-related (Black Panther, Harlem, NAACP, Soul Food) and others were obviously not Black-related (Recognition of Red China, Suburbia, Smorgasbord, Iceland), and ten of the sixty-two stimuli were our test slides. *S*s were asked to judge on a nine-point category scale how likely it was that each stimulus related to Black people, issues, or problems.

In the experiment, *S* was brought into the laboratory where sensors to record heart rate and skin conductance were attached. Each *S* was presented with an array of thirty slides, ten of these words and phrases they had previously rated as being highly Black-related, ten were expressions they had rated as not being Black-related, and the remaining ten stimuli were the test slides—issues of the 1972 presidential campaign. The slides were arranged in a random order. The presentation of each slide stimulus was made for ten seconds. For half the *S*s the Black-related slides (CS^+) were accompanied by a one-second loud (100 db) noise five seconds into the trial, while non-Black-related slides (CS^-) were never accompanied by noise. This contingency was reversed for the other half of the subject population. All the experimental events (e.g., the presentation of slides, delivery of noise, the recording of verbal and physiological responses) were controlled and timed by electronic programming equipment.

To enhance the conditioning procedure, *S*s were instructed that noise would sometimes accompany the CS^+ trials but would never be presented on CS^- trials. The series of stimulus trials were then presented. The first five slides were paired CS^+ and unpaired CS^- stimuli. The sixth trial was a CS^+ stimulus unaccompanied by a noise to determine whether conditioning had occurred. The remaining trials were a mix of CS^+, CS^-, and test stimuli.

The learning produced by this pairing of the loud noise (UCS) and the

CS^+ stimuli is reflected in the autonomic nervous system producing highly reactive skin conductance responses in anticipation of the noise. Typically, as a result of conditioning, skin conductance responses begin as soon as *S* recognizes a stimulus as being a CS^+. With equal consistency, *S* learns not to anticipate the noxious UCS during CS^- trials and as a result produces a reduced skin conductance response on those trials. These conditioned responses are generalized to the test trials. At issue is whether the test items that *S*s consciously or unconsciously identify as Black-related produce responses similar to those produced by CS^+ stimuli in the Black-related conditioning sequence and small skin conductance responses in the non-Black conditioning process. Figure 7a shows the average skin conductance response (in micromhos) of *S*s to one of the test stimuli—"Crime in the Streets." Skin conductance shows a strong conditioned effect under the Black-related condition but a much reduced effect in the non-Black sequence. Figure 7b, depicting the skin conductance response to "Watergate," on the other hand, produces an opposite effect. These examples indicate that on average these *S*s associated "Crime in the Streets" as being Black-related, while "Watergate" is responded to as being white-related.

These data demonstrate that the use of conditioning procedures allows a researcher to discriminate different connotations associated with the cognitive component of political concepts.

Conclusions

This paper presents a case for the application of a cross-modal, multiple-indicator approach to the study of political attitudes and behavior. Under well-controlled experimental conditions, multiple cross-modal response measures provide a Stimulus-Organism-Response framework for the analysis of micro-political behavior. This framework frees the researcher from exclusive reliance on verbal self-report, provides sensitive measures to monitor an individual's complex interactions with environmental stimuli, and utilizes well-established measures from the biological and behavioral sciences in the attempt to explain political behavior.

The cross-modal approach is distinguished from the more traditional multiple-indicator approach in that the psychophysical and psychophysiological measures outlined here are theory-derived and not simply additive response measures. A second important difference is the possibility of experimentally testing these measures under controlled laboratory conditions. Empirical tests of this approach are now being carried out in the Laboratory for Behavioral Research at Stony Brook.

The psychophysical scaling experiments summarized in this paper provide compelling evidence of the successful extension of these techniques from the measurement of sensory phenomena to the measurement of social opinion. These studies clearly demonstrate that lawful relationships hold between the perceived intensity of selected social and political stimuli and the magnitude of physical and verbal responses.

Studies employing the classical conditioning paradigm suggest that psychophysiological indicators of autonomic nervous system activity can

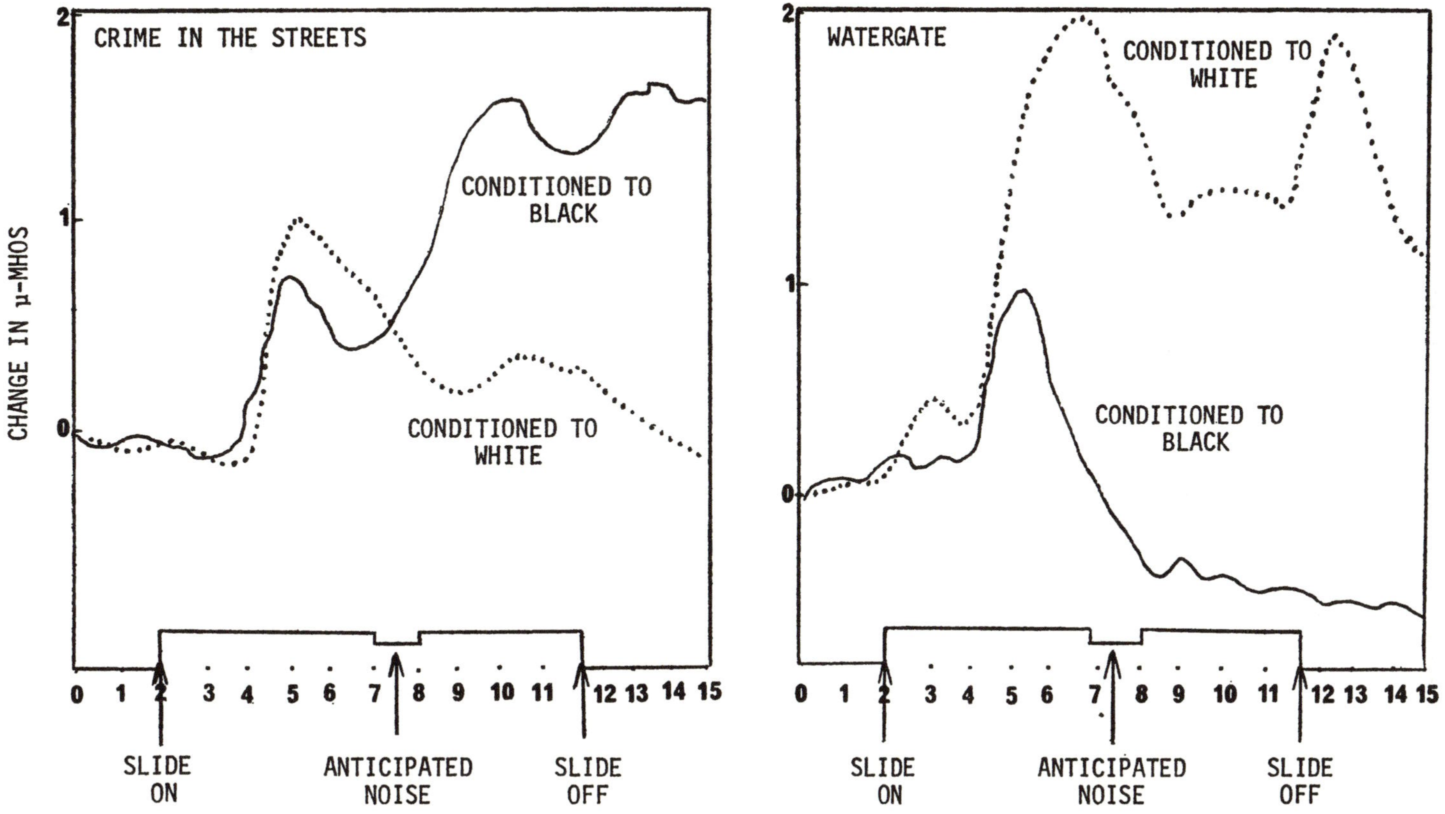

FIGURE 7a and 7b. *Computer averaged (3 samples per second) second-by-second skin conductance responses in micromho change for two representative and equated test trials. One "Crime in the Streets" was paired with Black-related CSs. The other "Watergate" was paired with white-related CSs. Seconds 7-8 (anticipated noise) is the interval during which the UCS had been presented in the CS+ trials. Both test trials show appropriate conditioned responses.*

be used to evaluate the cognitive component of social and political attitudes.

Studies are also being conducted that revolve around the variable "attention to political and social stimuli." In these studies, attention will be treated as a dependent variable to determine the effects of different beliefs and feelings on attention. Then, treating attention as an independent variable, we seek to measure the effects of varying patterns of attention on the cognitive and behavioral components of attitude.

In conclusion then, the overall aim of these proposed research projects is to bring the tools and approaches of behavioral science into political science to attack important problems of the discipline. Where, how, and why political science can (and should) move into the behavioral sciences is the theme of this paper. The overall objectives of the paper were twofold: first, to piece together from the behavioral and biological sciences an experimental approach to the analysis of micro-political behavior, and then secondly, to suggest and eventually test the worth of this approach by applications to problems of contemporary political science. Among these problems are political opinion formation, learning things political, attitude change, and decision-making, problems heretofore partially or wholly inaccessible to the more conventional methods of inquiry. The characteristics of the S-O-R approach—termed here a "bio-behavioral framework"—are: *a)* the framework is *experimental,* with primary emphasis on laboratory analyses of stimulus-organism-response behavior; *b)* the framework is *behavioral,* focusing directly on multiple behavioral responses; *c)* based on the techniques and procedures of contemporary psychophysics and psychophysiology, the framework relies on the analysis of *cross-modal* responses, on the recording of physiological, physical, behavioral as well as the verbal responses of individuals; and finally, *d)* the "bio-behavioral framework" takes its name from the proposed emphasis on two sets of interdependent explanatory variables—from the environmental side, *learning theory,* and from the *biological* perspective, the functioning of the central and autonomic nervous systems. To the extent political behavior is multifaceted, as it most assuredly is, this bio-behavioral framework seems appropriate to the conceptual and methodological tasks involved in attempts to analyze why people do what they do politically.

Bibliography

Bamford, J. L. / Ax, A. F. / Fetzner, J. / Walrad, P.
1963 *A Selected Bibliography for Psychophysiology,* Society for Psychophysiological Research, Monograph.

Bandura, A.
1969 *Principles of Behavior Modification,* New York, Holt, Rinehart and Winston.

Bechtel, R. B. / Achelpohl, C. / Akers, R.
1972 "Correlates Between Observed Behavior and Questionnaire Responses on Television Viewing," in: Rubinstein, E. / Comstock, G. / Murray, J. (eds.), *Televion and Social Behavior: A Technical Report of the Surgeon General's Scientific Advisory Committee,* vol. 4, pp. 274-344.

Black, P. (ed.)
1970 *Physiological Correlates of Emotion,* New York, Academic Press.

Campbell, D. T.
1963 "Social Attitudes and Other Acquired Behavioral Dispositions," in: Koch, S. (ed.), *Psychology: A Study of a Science*, New York, McGraw Hill, pp. 94-172.

Cannon, W. B.
1932 *The Wisdom of the Body*, New York, Appleton-Century-Crofts.

Carriero, N. J. / Gehringer, E. C.
1971 *An Annotated Bibliography of the Literature Dealing with the Physiological Correlates of Attitudes and Attitude Change*, US Army Aberdeen Research and Development Center, Human Engineering Laboratories.

Cochrane, R. / Robertson, A.
1973 "The Life Events Inventory: A Measure of the Relative Severity of Psycho-Social Stressors," *Journal of Psychosomatic Research* 17, pp. 135-139.

Cook, S. W. / Selltiz, C.
1964 "A Multiple-Indicator Approach to Attitude Measurement", *Psychological Bulletin* 62, pp. 36-55.

Cooper, J. B. / Pollock, D.
1959 "The Identification of Prejudicial Attitudes by the Galvanic Skin Response," *The Journal of Social Psychology* 50, pp. 241-245.

Cooper, J. B. / Siegel, H. E.
1956 "The Galvanic Skin Response as a Measure of Emotion in Prejudice," *The Journal of Psychology* 42, pp. 149-155.

Corson, W. H.
1970 "Conflict and Cooperation in East-West Crises: Dynamics of Crisis Interaction," Ph. D. Thesis, Harvard University.

Crespi, I.
1971 "What Kinds of Attitude Measurements are Predictive of Behavior ?," *Public Opinion Quarterly* 35, pp. 327-334.

Crider, A.
1970 "Experimental Studies of Conflict-Produced Stress," in: Levine, S. / Scotch, N. (eds.), *Social Stress*, Chicago, Ill., Aldine, pp. 165-188.

Cross, D. V.
1974 "Some Technical Notes on Psychophysical Scaling," in: Moskowitz, H. / Scharf, B. / Stevens, J. C. (eds.), *Sensation and Measurement: Papers in Honor of S. S. Stevens*, Dordrecht, Reidel, pp. 23-36.

Dawson, W. E. / Brinker, R. P.
1971 "Validation of Ratio Scales of Opinion by Multimodality Matching," *Perception and Psychophysics* 9, pp. 413-417.

DeFleur, M. E. / Westie, F. R.
1958 "Verbal Attitudes and Overt Acts," *American Sociological Review* 23, pp. 667-673.
1963 "Attitude as a Scientific Concept," *Social Forces* 42, pp. 17-31.

Dickson, H. W. / McGinnes, E.
1966 "Affectivity in the Arousal of Attitudes as Measured by Galvanic Skill Response," *American Journal of Psychology* 79, pp. 584-587.

Duffy, E.
1962 *Activation and Behavior*, New York, Wiley.

Ekman, G.
1962 "Measurement of Moral Judgement: A Comparison of Scaling Methods," *Perceptual and Motor Skills* 15, pp. 3-9.

Ekman, G. / Künnapas, T.
1963 "A Further Study of Direct and Indirect Scaling Methods," *Scandinavian Journal of Psychology* 4, pp. 77-80.

Epstein, S. / Fenz, W.
1962 "Theory and Experiment on the Measurement of Approach-Avoidance Conflict," *Journal of Abnormal and Social Psychology* 64, pp. 97-112.

Fenz, W. / Epstein, S.
1967 "Gradients of Physiological Arousal in Parachutists as a Function of an Approaching Jump," *Psychosomatic Medicine* 22, pp. 33-51.

Fenz, W. D. / Jones, G. B.
1972 "Individual Differences in Physiologic Arousal and Performance in Sport Parachutists," *Psychosomatic Medicine* 24, pp. 1-9.

Finnie, B. / Luce, R. D.
1960 "Magnitude Estimation, Pair Comparison and Successive Interval Scales of Attitude Items," University of Pennsylvania, Department of Psychology.

Hamblin, R.
1971 "Mathematical Experimentation and Sociological Theory: A Critical Analysis," *Sociometry* 34, pp. 423-452.

Hamblin, R. / Bridger, D. / Day, R. / Yancey, W.
1963 "The Interference-Aggression Law?," *Sociometry* 26, pp. 190-216.

Hyman, H. H. / Sheatsley, P. B.
1947 "Some Reasons Why Information Campaigns Fail," *Public Opinion Quarterly* 11, pp. 413-423.

Kimble, G. A.
1961 *Conditioning and Learning*, 2nd ed., New York, Appleton-Century-Crofts.

Künnapas, T. / Wikström, I.
1963 "Measurement of Occupational Preferences: A Comparison of Scaling Methods," *Perceptual and Motor Skills* 17, pp. 611-624.

Lacey, J. I.
1959 "Psychophysiological Approaches to the Evaluation of Psychotherapeutic Process and Outcome," in: Rubinstein, E. A. / Parloff, M. B. (eds.), *Research in Psychotherapy*, Washington, DC, American Psychological Association, pp. 160-208.
1967 "Somatic Response Patterning and Stress: Some Revisions of Activation Theory," in: Appley, M. H. / Thrumbull, R. (eds.), *Psychological Stress*, New York, Appleton-Century-Crofts, pp. 14-42.

Lacey, J. I. / Kagan, J. / Lacey, B. C. / Moss, H. A.
1963 "The Visceral Level: Situational Determinants and Behavioral Correlates of Autonomic Response Patterns," in: Knapp, P. J. (ed.), *Expression of the Emotions in Man*, New York, International Universities Press, pp. 161-196.

Lacey, J. I. / Lacey, B. C.
1958 "Verification and Extension of the Principle of Autonomic Response Stereotypy," *American Journal of Psychology* 71, pp. 54-73.

Lazarsfeld, P. F.
1940 *Radio and the Printed Page*, New York, Duel, Sloan and Pearce.

Lazarus, R. S.
1966 *Psychological Stress and the Coping Process*, New York, McGraw Hill.

Lazarus, R. S. / Opton, E. M. / Nomikos, M. S. / Rankin, N. O.
1965 "The Principle of Short-Circuiting of Threat: Further Evidence," *Journal of Personality* 33, pp. 622-635.

Lazarus, R. S. / Speisman, J. C. / Mordkoff, A. M. / Davison, L. A.
1962 "A Laboratory Study of Psychological Stress Produced by a Motion Picture Film," *Psychological Monographs* 76, Whole No. 553.

Lodge, M. / Tursky, B. / Tanenhaus, J.
1974 "An Experimental Cross-Modal Framework for the Analysis of Political Behavior," Paper presented at NSF Conference on Design and Measurement Standards for Research in Political Science, Delevin, Wisc., May 13-15.

Lodge, M. / Tursky, B. / Tanenhaus, J. / Cross, D.
1974 "The Development and Validation of Political Attitude Scales: A Psychological Approach," Paper presented at the NSF Conference on Design and Measurement Standards for Research in Political Science, Delevin, Wisc., May 13-15.

McGuire, W.
1969 "The Nature of Attitudes and Attitude Change," in: Lindzey, G. / Aronson, E. (eds.), *Handbook of Social Psychology*, 2nd ed., Reading, Mass., Addison-Wesley, vol. 3, pp. 136-314.

Milbrath, L. W.
1965 *Political Participation*, Chicago, Rand McNally.

Moyer, K. E.
1971 *The Physiology of Hostility,* Chicago, Ill., Markham.
1973 "The Physiology of Violence," *Psychology Today* 7, pp. 35-38.

Orne, M. T.
1962 "On the Social Psychology of the Psychological Experiment," *American Psychologist* 17, pp. 776-783.

Osgood, C. E. / Suci, G. J. / Tannenbaum, P. H.
1971 *The Measurement of Meaning,* Chicago, Ill., University of Illinois Press.

Pavlov, I. P.
1927 *Conditioned Reflexes,* transl. by G. C. Anrep, London, Oxford University Press.

Rainwater, L.
1971 "The Measurement of Social Status," Harvard University, Department of Sociology.

Rankin, R. E. / Campbell, D. T.
1955 "Galvanic Skin-Response to Negro and White Experimenters," *Journal of Abnormal and Social Psychology* 51, pp. 30-33.

Roessler, R. / Kelly, C. / Collins, F.
Personality Physiology, and Human Performance: A Bibliography, Houston, Texas, Baylor College of Medicine, Department of Psychiatry, Psychophysiology Laboratory.

Rosenthal, R. P. / Kohn, P. M. / Greenfield, P. M. / Carota, N.
1966 "Data Desirability, Experimenter Expectancy, and the Results of Psychological Research," *Journal of Personality and Social Psychology* 3, pp. 20-27.

Schachter, S.
1964 "The Interaction of Cognitive and Physiological Determinants of Emotional State," in: Berkowitz, L. (ed.), *Advances in Experimental Social Psychology,* New York, Academic Press, vol. 1, pp. 49-80.
1971 *Emotion, Obesity, and Crime,* New York, Academic Press.

Schwartz, S. / Tessler, R.
1972 "Model for Reducing Measured Attitude-Behavior Discrepancies, " *Journal of Personality and Social Psychology* 24, pp. 225-236.

Seligman, M. E. P. / Hagar, J.
1972 "Biological Boundaries of Learning: The Sauce Bearnaise Syndrome," *Psychology Today* 6, pp. 59-61, 86-87.

Seligman, M. E. P. / Hagar, J. (eds.)
1974 *The Biological Boundaries of Learning,* New York, Appleton-Century-Crofts.

Sellin, J. T. / Wolfgang, M. E.
1964 *The Measurement of Delinquency,* New York, J. Wiley.

Selye, H.
1946 "The General Adaptation Syndrome and Diseases of Adaptation," *Journal of Endocrinology* 6, pp. 217-230.

Shapiro, D. / Crider, A.
1969 "Psychophysiological Approaches in Social Psychology," in: Lindzey, G. / Aronson, E. (eds.), *Handbook of Social Psychology,* 2nd ed., Reading, Mass., Addison-Wesley, vol. 3, pp. 1-49.

Shinn, A. M., Jr.
1969a *The Application of Psychophysical Scaling Techniques to Measurement of Political Variables,* Chapel Hill, NC, University of North Carolina Press.
1969b "An Application of Psychophysical Scaling Techniques to the Measurement of National Power," *Journal of Politics* 31, pp. 932-951.

Skinner, B. F.
1938 *The Behavior of Organisms: An Experimental Analysis,* New York, Appleton-Century-Crofts.
1953 *Science and Human Behavior,* New York, Macmillan.
1971 *Beyond Freedom and Dignity,* New York, A. A. Knopf.

Spence, D. P. / Lugo, M. / Youdin, R.
1972 "Cardiac Change as a Function of Attention to and Awareness of Continuous Verbal Text," *Science* 176, pp. 1344-1346.

Staats, A. W. / Staats, C. K.
1958 "Attitudes Established by Classical Conditioning," *Journal of Abnormal and Social Psychology* 57, pp. 37-40.

Staats, A. W. / Staats, C. K. / Crawford, H. L.
1962 "First-Order Conditioning of Meaning and Paralleled Conditioning of the GSR," *Journal of General Psychology* 67, pp. 159-167.

Staats, C. K. / Staats, A. W.
1957 "Meaning Established by Classical Conditioning," *Journal of Experimental Psychology* 54, pp. 74-80.

Stern, J. A.
1964 "Toward a Definition of Psychophysiology," *Psychophysiology* 1, pp. 90-91.

Sternbach, R. A.
1960 "A Comparative Analysis of Autonomic Responses in Startle," *Psychosomatic Medicine* 22, pp. 204-210.
1966 *Principles of Psychophysiology: An Introductory Text and Readings*, New York, Academic Press.

Stevens, J. C. / Mack, J. D.
1959 "Scales of Apparent Force," *Journal of Experimental Psychology* 58, pp. 405-413.

Stevens, S. S.
1957 "On the Psychophysical Law," *Psychological Review* 64, pp. 153-181.
1966 "A Metric for the Social Consensus," *Science* 151, pp. 530-541.
1972 "Psychophysics and Social Scaling," *General Learning Press*, Module 4033V00, Morristown, NJ, General Learning Press.
1975 *Psychophysics: Introduction to its Perceptual, Neural and Social Prospects*, New York, Wiley.

Stouffer, S. A.
1955 *Communism, Conformity, and Civil Liberties*, New York, Doubleday.

Title, C. / Hill, R. J.
1967 "Attitude Measurement and Prediction of Behavior: An Evaluation of Conditions and Measurement of Techniques," *Sociometry* 30, pp. 199-213.

Tursky, B. / Schwartz, G. E. / Crider, A.
1970 "Differential Patterns of Heart Rate and Skin Resistance During a Digit-Transformation Task," *Journal of Experimental Psychology* 83, pp. 451-457.

Verba, S. / Nie, N.
1972 *Participation in America*, New York, Harper and Row.

Wahlke, J. / Lodge, M.
1972 "Psychophysiological Measures of Political Attitudes and Behavior," *Midwest Journal of Political Science* 16 (4), pp. 505-537.

Warner, L. G. / DeFleur, M. L.
1969 "Attitude as an Interactional Concept: Social Constraint and Social Distance as Intervening Variables Between Attitudes and Action," *American Sociological Review* 34 (2), pp. 153-169.

Weiss, W.
1969 "Effects of the Mass Media on Communication," in: Lindzey, G. / Aronson, E. (eds.), *The Handbook of Social Psychology*, 2nd ed., Reading, Mass., Addison-Wesley, vol. 3, pp. 77-195.

Welch, R. E., Jr.
1971 "The Use of Magnitude Estimation in Attitude Scaling: Constructing a Measure of Political Dissatisfaction," in: Nimmo, D. D. / Bonjean, C. M. (eds.), *Political Attitudes and Public Opinion*, New York, D. McKay, pp. 112-123.

Wincher, A.
1969 "Attitudes Versus Action: The Relationship of Verbal and Overt Behavioral Responses to Attitude Objects," *Journal of Social Issues* 25, pp. 41-78.

JAMES CHOWNING DAVIES

Ions of Emotion and Political Behavior: A Prototheory

THIS PROTOTHEORY SUGGESTS what may be the neurochemical substrates of the human organism's response to the politically pertinent frustration of its natural, basic needs. Individual development is the result of the interaction between the internally generating demands of a gradually maturing organism and broad environmental (including intrafamilial, socioeconomic, and political) conditions. Individual action becomes joint political action when individuals collectively blame the polity for the frustration of either old or newly emerging demands.

A critical component of individual organic development is the memory of the environment's prior responses to the organism's efforts to seek satisfaction of its sequentially emerging needs. When there has been a large accumulation of these needs within the organism and this accumulation is conjoined with broad environmental inhibitors and releasers, minute amounts of energy that have been stored for even decades in special-function particles (complex molecules) in the neocortex are released. When released from the neocortex, they help to activate higher-energy circuits in the paleocortex, midbrain, hypothalamus, and reticular activating system and also to activate endocrines. This sequence of neurochemical events produces a total but highly specific response of the organism. This response may include violent action against the political system, including the system of distributing valued objects (from material goods to affection, dignity, and self-fulfillment).

In this paper* I propose to indicate the broad social phenomena that most crucially call for biological and, more exactly, for neurophysiological analysis and to mention the social and psychological theorists who have led us to the point where we now need neurophysiological analysis most critically. Then I will state a very preliminary hypothesis about the events inside the brain and endocrine glands that relate to these broad social phenomena. Then I

* In my long-term memory neurons is permanently stored my strong appreciation for the encouragement and/or criticism of the following people who took the trouble to read with care and react to an earlier version of this paper: José M. R. Delgado, Daniel P. Kimble, David Krech, Paul D. MacLean, David C. Schwartz, Roger W. Sperry, William V. Street, Jr., Philip Teitelbaum, and Alberto Zanchetti. In an adjacent neuron or three is likewise stored my strong appreciation for the careful retyping job of Anne Parnaby.

will briefly review some neurophysiological research of the sorts that are most promising and most pertinent to the broad social phenomena, ending this paper with an indication of what I think political science most needs to know physiologically that physiologists have only begun to investigate.

Broad Social Phenomena and Some Great Theorizing About Them

There is a wave-like quality about broad social events. Human beings in their particular societies and cultures do not progress and retrogress suddenly and evenly, like the flip-flop pattern of a square wave on an oscilloscope. People in society go through prolonged periods of relatively little development or change and then, rather suddenly and very surprisingly, enter periods in which development is gross, intense, and often enduring. Then there is a return to change that is detailed, low key, and also often enduring. Change of either sort never quite restores people to the previous patterns of social and political interaction. Those profound forces which are *within the natural and more particulary the cultural environment* are relatively well understood. It is the set of forces *within human beings* and which help produce these profound changes, in varying rhythms, that are little and badly understood. It is these latter forces, massive in their consequences, that demand biological analysis.

There is consensus in social science on the facts of change in patterns of social interaction, values, systems, and polities. There is less consensus on what these facts mean. Many people dispute whether so-called primitive cultures are inferior to so-called advanced cultures. Many people dispute whether life in the Middle Ages in Europe was worse or better than life since the Renaissance, Reformation, and Industrial Revolution. But few people will argue that there are no noteworthy and durable differences between primitive and advanced or between feudal and industrial societies.

A more basic problem is one that I think is a fact, but one whose existence is very little analyzed. This is the apparent change in *periodicity* and *intensity* of change, both natural and cultural. There is a problem because such change violates common sense, that is, in this case everyday observation and extant scholarly observation. Day and night occur every twenty-four hours; there are four seasons every year. The fact that days vary in length and that seasons vary in length can be accepted because the variance is short-range and only moderately irregular. But it is hard to conceive that long-range changes—from those in the periodicity of prolonged and widespread ice ages to changes in the periodicity of the evolution of species of life and changes in the periodicity of social development—are quite irregular. When the dimensions of duration and oscillation become long-range, very intense, and occur together, then the conceptual process gets really complicated and is often lost in the observation of everyday epiphenomena. This is a major reason that it is very hard to conceptualize such things as species development and human development within culture. Some of the first people to propose theories about such matters became immediately and enormously controversial.

Indeed, Charles Darwin (1809-1882) was one of the first scientists to analyse long-cyclical change and development. His *Origin of Species* (1859) posited a theory with many implications, *most* of which were not immediately acceptable. Among the other things that people even now overlook is his argument that species-development is the product of the interaction of both genetic change (mutation) and the pressures of environmental change. The continuous interaction between genetic and environmental change selects out those individuals in all species that are best able to *both* adapt to *and* use *both* kinds of change. For decades many intellectual mastodons found it impossible to adapt to the environmental change in the intellectual climate that Darwin amounted to. As one writer put it, moribund ideas do not expire until those who have generated them do.

A second and here most relevant theorist of change is Darwin's contemporary and admirer, Karl Marx (1818-1883). The *Communist Manifesto* (1848) founded the ideology of Communism, and *Capital* (1867-1879) became a theoretical base for understanding change in the social world. In the academic world it became the most widely influential way of looking at social change. In the loose dialogue between Social Darwinism and Marxism, the latter became the dominant voice. Social Darwinists emphasized the socioeconomic *conflict* (as distinct from cooperation) that produces individual and species survival. Marx argued that the conflict was between classes and emphasized that, when the working class becomes as conscious of its common interests as the capitalist class is of its, progressive change results. Marx's principles of change were derivations of socioeconomic systems analysis, and he did not—perhaps in his time could not—elaborate on the forces within the human organism that demand change. However, Marx insisted that change is neither random nor mere oscillation but is developmental and inevitable. His argument has provided perhaps the strongest strands in the warp of the fabric of social theorizing. But he was largely silent on the woof—the forces internal to man.

Sigmund Freud (1856-1939) turned to analysis of the unconscious determinants of behavior within man (initially in *The Interpretation of Dreams*, 1900, and numerous later writings like *The Ego and the Id*, 1923). He made ineradicably clear as to individual human development from infancy to maturity what Marx had clarified as to societies at large. Freud argued that there are forces within the human organism that help determine the course of individual human development. In the form not only of nearly irresistible instinctual demands (particularly the sexual urge) but also of ineradicable memory traces of infancy and childhood, these forces interact continually with the individual's current environment to produce the orderly or the disordered pattern of his development. Freud in his 70s, in *Civilization and Its Discontents* (1930), began to apply his basic approach to broad social influences on individual behavior. In doing so, he began to cross paths with Marx.

Another psychologist, vaguely in the Freudian tradition, made developmental analysis more explicit. Abraham Maslow (1908-1970), in a 1943 paper entitled "A Theory of Human Motivation," argued that there is a hierarchy of human needs, running from the physical and affectional all the way to the need for self-actualization, and that they appeared pretty much in order as an individual moves from infancy to maturity.

Applying and modifying Maslow's fundamental idea, I have argued first that this hierarchy of needs works in an epigenetic manner whereby the first needs, the physical ones, predominate at birth and the last of them, for self-development, predominates during an individual's prime of life; and second that there cannot normally be much fulfillment of any higher need until the steady provision for the lower needs is established. I have further hypothesized that there is a synchrony between the development of the generality of people in a particular society and the development of that society, arguing essentially that in a primitive society most people remain preoccupied throughout their lives with staying alive and well, whereas the criteria by which a society may be deemed to be advanced include the extent that it optimizes the opportunities for the maximum number of individuals within it to realize their fullest unique potential (Davies, 1963, 1976).

There are linkages between the theorizing of Darwin, Marx, Freud, and Maslow. In a paper of this length it is utterly impossible to present an adequate statement showing how the work of each of these four relates to neurophysiological explanations of political behavior. But let me indicate briefly the most central relevance of each of these.

Darwin's contribution was crucial in that he established the idea of interspecies relationships and of changes that take place over long periods of time. The idea of systematic interspecies kinship and comparability became a mode of thought and of coherent analysis of man in nature. Darwin conceived of time—long periods of time—as a dimension and a yardstick to explain life as it is on earth at any particular time. His mode of analysis gradually infused the social sciences, which hitherto had largely ignored change and therefore time.

Marx's contribution in the present context is to have seen the irregular, long-cycled but progressive development of civilizations from primitive social agglomerates to highly integrated industrial societies. He saw the process of moving from feudalism to capitalism and industrialization as a relatively long one, and he saw the process of moving out of private-capital industrialization to the successive stages as being relatively briefer. He advanced, in other words, the idea of change and development on an aperiodic but cyclical basis in which each successive stage is different from its predecessor and generally better. He posited uneven cyclical change and he posited social progress.

Freud and Maslow, concerned most of their lives with the individual as the unit of analysis, studied unconscious processes functioning within the human psyche. Freud concentrated on the disturbances in the early (even infantile) stages of (sexual) development that impede the achievement of maturity. Maslow elaborated on a more complex set of drives and in a sense defined maturity as the achievement of the latest one of them to become active in all individuals: the drive of self-actualization.

One thing that is readily apparent in these contributions is that they lead to analysis of the human organism as a changing, developing entity and as the part of the equation of broad social change that is least known. Darwin saw the individual member of a species as the carrier of a kind of history of evolution of that species and of the evolution of all species. Marx, in some kind of homogenizing and blenderizing of individual humans, saw them almost exclusively in terms of socioeconomic class and the resulting political power structure. These gross events were operant on the individual as a member of the working or the middle class, each individual being totally determined in his actions by his respective class position. With the exception of his *German Ideology* (written in 1845-1846 but not published until 1932), Marx largely ignored the independent causal force of anything internal to human beings. He regarded them as altogether passive products of their circumstances. Individuals, he said, change not in accordance with any inner dynamic but as a consequence of social change.

Freud indeed made the transition from that abstraction that is called society to that abstraction that is called the individual human organism, thereby reminding his readers to recognize individual existence. But he lacked Marx's broad social matrices for judging those things that happen to individuals and which *are* caused by society. Maslow made the gigantic contribution of seeing individual development—in a process that is behavioral ontogenesis—from the human infant, an animal that in its overt behavior is not very different from other animals, to the adult human being, an animal whose nature compels it not to forget food but also compels it to become concerned with finding and fulfilling his or her individual self.

Thus, during the past hundred years biological, sociological and psychological theorists like Darwin, Marx, Freud, and Maslow have made possible the analysis of fundamental developmental processes. But these gigantic processes have thus far been tractable only to the grossest sociological and psychological analysis, and these analyses are either inadequate or wrong. People in Vietman, Montgomery, Tokyo, Berkeley, Paris, at Kent State, and in Boston have demanded things in ways that do not fit the theories, the research, or even the commonsense notions we have. The inadequacy correlates positively with ignorance of the workings of the brain and its control systems.

It is not parsimonious to assume that there is no relationship between developmental processes as they occur in the environment and as they occur in human minds. In presenting the following *proto*theory, I am trying to do no more than suggest some ideas about some neurochemical events which are already somewhat understood and which I believe do pertain to a deeper and more total explanation of processes of socioeconomic and political development. I know I cannot be very right, because I am well ahead of confirmed or even suggested interpretations of the relevant neurochemical processes. But perhaps the statement of a not-very-right prototheory can encourage a little interest among natural scientists (notably neurophysiologists and microbiologists), so that they can somewhat more efficiently come up, not with another prototheory but with a theory or theories—and with some research that will more efficiently move beyond the extant beginnings. It is hard at times to avoid the feeling that if natural and social scientists continue to go their seperate ways, without even conflictful dialogue, then we all—scientists and nonscientists alike—may become integrated, icily and thermodynamically, in some primordial and well-nigh energyless void.

A Prototheory of Ions of Emotion and Political Behavior

The ties between such contributors as Darwin, Marx, Freud, and Maslow and those who have studied human beings more directly—that is, physiologically—need not and cannot be stated here. But the ties between a prototheory of ions of emotion and the findings of relevant neurophysiology should and can be stated here. Following presentation of the prototheory as it helps explain social and political behavior, these findings will be briefly discussed. The prototheory is as follows :

> When there has been a long-term accumulation of demands within the human organism and this accumulation is conjoined with environmental inputs of two kinds (first inhibitors and then releasers), there is a rather sudden, intense, and rapid release (discharge) of minute amounts of energy stored in special-function particles in memory neurons in the neocortex, where the relatively long-term accumulation of energy has taken place. This accumulation of ionized particles (probably complex protein molecules and not organelles) is thus located in the neural substrate of complex, emotion-laden memory. The discharge of these ionized particles within memory neurons is a crucial link in the affective response to the frustration of basic drives. This sudden, intense, and rapid discharge involves activation of and feedback from higher-energy circuits in the paleocortex and the midbrain and the hypothalamus. This discharge is an essential component of both small and large changes in the environment, including the political system.

Two corollaries of the prototheory are as follows:

> When the individual is socially relatively isolated during the long-term accumulation of these internally, organically generated demands, over months, years, or decades, to which the environment's response is at first inhibition and then release, his action within his isolation is likely to be an *individual* act of verbal or violent protest—possibly a conventional crime against person or property.
>
> When the individual *shares with many others* in his society a long-term accumulation of any organic demands to which the environment's response is at first inhibition and then release, individual action becomes collective action and the protest becomes a *social* act. The releaser is the (social) *sharing* of individual frustration that appears in similar form among hundreds, thousands, or millions of individuals. In the moderate case it becomes a relatively orderly and peaceful demand for change. In the extreme case it becomes war, in which peoples (nations) engage in violent conflict with each other; or revolution, in which groups of people (usually but not necessarily socioeconomic classes) engage in violent conflict with each other.

Let me explicate the prototheory a little. The charging and discharging of the ionized particles (complex molecules carrying an electrical charge) in those neurons that are involved in social and political behavior takes place most critically for political behavior in the *information-processing* (particularly, the information-storing) parts of the neocortex, that neural part of the organism's central contral system (CCS) which is most highly developed in Homo sapiens. The CCS includes the entire central nervous system (CNS) and the endocrine system (ES), but the *charging*—the ionizing—function as I see it takes place "ultimately" and *dis*charging function "initially" in the neocortex. The parts of the neocortex most terminally involved as the receivers and transmitters of memories that are critically emotion-laden and politically portentous may be the *frontal* lobes (see, e.g., H. F. Harlow / A. J. Blomquist / *et al.,* 1968, p. 80) and their interneurons, but on that there can only be a guess until more research is done.

The accumulation or charging, as I said, may take place over months, years, and even decades. It includes the storing of memories of *high* affective content (both pleasant and unpleasant) that become permanently locked into the neocortical structure, partly in the form of ionized particles containing the emotional component of memories. These memories make available for release minute quanta of energy, with electrical potential in the order of probably no more than microvolts. When release of these electrons in the neocortex is triggered by newly received information from both the external environment (again, both inhibiting and releasing) and the internal environment (in the form of the appearance of a new and higher need like dignity, or the activation of a more basic one, like hunger), the electrons dislodged from the neocortex travel down into the limbic system (which corresponds roughly to the primitive or paleocortex).

The paleocortex, activated by signals from the neocortex, in turn activates circuits in the midbrain and in the hypothalamus, which in turn activates other neurons and endocrines (both neurons and endocrines functioning as inhibitors and releasers) that are more immediately involved in making the whole organism become alert, tense, and active. This includes feeding signals back to the neocortex, which, by producing and interpreting additional information from storage and from the environment, intensifies the entire complex process of responding to frustration of basic, organic needs.

When the intensification, the amplification, reaches certain thresholds at various parts of the body and when these thresholds have been reached in a substantial number of individuals of like social circumstances and like responses to these circumstances, then all or enough of the musculature (smooth and striated) is activated in enough individual human beings to lead in the mild case to peaceful anti-regime protest and in the extreme case to concerted acts of violence, against the domestic regime in revolution or against other governments in war.

Three further observations are pertinent to the emotive ion concept. The first of these is that the storage of memories in the still-undefinable parts of the neocortex is quite certainly not an encapsulated, sealed event. By a process that molecular biologists in recent years have very fruitfully studied, memories become an evidently permanent part of the organic structure of the brain and therefore are "permanently" available. The process is apparently akin to and indeed a case of that by which RNA carries, encodes, and programs the production of protein as living brain tissue. This long-term storage of what we call memories is not dead storage. Memories that are even decades old evidently become, in both their cognitive and affective aspects, available for directing the behavior of people in such ways as to initiate the amplification of energy in other cells, which trigger other neurons and endocrines to the final point of even violent action.

These memory neurons, in short, function as the storers of both information and affect. In at least their affective if not also their cognitive role, they function as storers of energy. In chemical terms, they contain ionized particles, and at least the affective function of memory is a special function ion. Emotion in physiological terms is here seen as a particular kind of electrically charged particle within memory neurons. The ionized particles within these cells may remain long dormant and, in every instance, they presumably do not fire spontaneously but require triggering from outside the CCS to be released. It is awesome to contemplate the effect in at fifty-year old revolutionary of memories stored in his teens, and it is increasingly possible to understand the process neurophysiologically. As a very young man, Lenin experienced the death of his father, his brother, and his sister, for reasons attributable to the Tsarist government. Decades later, these experiences became part of the fuel of a revolutionary.

A second observation is that these minute quanta of energy are highly specific in their very complex function. It is mainly for this reason that I suppose these quanta to be stored in specific kinds of molecules in memory neurons in the neocortex. Despite the big conceptual gap between two closely linked kinds of act—private and public—there occurs, at certain junctures in history, collective violence that involves hundreds, thousands, even millions of individuals. And the events taking place within the individual central control system of each human actor so involved in such collective action, from childhood on to adulthood, are a necessary component of vast "social" processes studied by social scientists. They are also a necessary component of the "individual" processes studied by psychologists concerned with the mutual roots of violence. *Concerted* violence of the *very complex and enduring sort* that sometimes occurs in human societies is absent in perhaps all other animal societies. This complexity and this duration correlate with the degree of development of the neocortex in Homo sapiens.

The very specificity of the collective response in times of individual commitment to war and revolution is both puzzling and awe inspiring. We know—and many of us can remember—the coherent, orderly, goal-oriented manner in which individuals behave in times of intense involvement in collective action. This hardly conspiratorial collaboration of individuals was apparent in the widespread cooperation of publics during the Second World War, in the particular student rebellions, worldwide in the 1960s, and in the Black Rebellion in America in the 1960s.

Underneath the seeming chaos of combat in war and on the streets, orderly neuroendocrine processes are at work in the CCS. The big question is how so many thousands and millions of individuals—more abstractly, so many CCSs—function in something like unison. They do so far more systematically than any analysis or prediction of such action would suggest—even a week before the actual outbreak of hostilities between group and group. It seems most reasonable to suppose that the system of ionized particles within memory neurons must be located in about the same areas in each of the thousands or millions of brains involved. And it further seems reasonable to suppose that the circuits that become activated between these trigger ions in the neocortex and neurons in the paleocortexes and hypothalamuses must be, within *less* precise limits than in the case of the initiating trigger ions, functionally the same circuits in all the individuals involved.

A third observation is that within each individual human organism there are substantially the same *sets* of frustratable needs, even though the particular need frustrated in one individual may be the need for food and in another the need for dignity. There is a wide variation in the particular reasons that people give for engaging in collective public action (to get away from stodgy old parents, to get away from one's pregnant girlfriend, to get away from one's creditors). There is, in addition to these seemingly altogether unpublic reasons, in some people an expressed set of widely shared "public" frustrations. People see these shared frustrations as resulting from the actions or inactions of government. Jefferson's assertion in the 1776 Declaration of Independence that "all men are created equal" resounded through the colonies and within the minds of individual colonists. They were collectively tired of being pushed around and were ready to rule themselves. They blamed the British government for degrading them and irresponsibly ruling them, and they proposed removing that government.

Jefferson's appeal to the need for equal recognition has served as a unifying motivational force also for Frenchmen in 1789, Greeks in the 1820s, Indonesians in the 1940s. And after the French control of Indochina was lost by Japan's conquest during the Second World War, Ho Chi Minh quoted that equalitarian part of the Declaration of Independence verbatim, in a manifesto that was a basis for the movement toward independence of Vietnamese people from Japanese, French, and finally American hegemony. Not all the bayonets or napalm or defoliant or explosives in the world have been able to destroy the popular, worldwide resonance of such utterances as Jefferson's—whenever a collectivity (that is, a group of like-minded people) *has reached the stage* at which they can begin to consider whether or not they are indeed created equal. An appeal during the 18th century by what would then have been a premature Ho Chi Minh to a then primitive Vietnamese people would have been an anachronism. By the mid-twentieth century they were no longer primitive and had become a remarkably solidary nation.

In sum, the extraordinary amplifier effect, the specificity of the objects in the environment that a frustrated people will violently attack, and the lesser but still significant commonality of widely shared (and thereby socially reinforced) needs—all these are phenomena which, it seems to me, can now be more directly investigated by neurophysiological means.

Let me indicate a few things which this modest proposal of a prototheory of emotive ions does *not* say, whether physiologically or psychologically. It does not say confidently where these ions are. The probability of the pertinent memories being stored in the neocortex is indeed very high. There is evidence that higher animals store memories relating to complexly integrated behaviors in parts of the brain that are not highly developed in such lower animals as reptiles and birds. Reptiles do not attack wealthy or landowning reptiles. Birds do not attack conspecifics because their grandparents made a raid on their territory. There is some intergenerational transfer of behavior patterns relating to specific enemies among such animals as elephants or wolves, but these animals have highly developed neocortexes. Complex memory is not a well-developed function in such lower animals as birds and snakes and does seem to require highly developed neocortex. But memory traces may be stored in other places, like the hippocampus (Kimble, 1968) in the limbic system, and these noncortical memory traces may relate to political behavior. So on specificity of storage sites we cannot be very certain.

This prototheory also does not say that the release of memories, notably those which are highly charged, ionized, loaded with affect and which therefore can lead to violence, is in any way *sufficient* to produce violence. Not only are extraorganic environmental events (including the kind Marx wrote about) necessary to trigger their release; so also are intraorganic events. Not only are triggers, external and internal, necessary, but also a *whole chain of many, many loops must be both activated and linked* for at least violent political action to take place. Among these loops in the chain are those in the limbic system, that border portion of the brain between the neocortex and the midbrain (notably the hypothalamus). Also in the system of activation must be included the appropriate functioning of various endocrines secreted within the nervous system and—when triggered by the CNS—secreted into the bloodstream.

This prototheory also does not say that there is no spillover from private memories to politically violent action or, correlatively, that memories of a directly political or social origin (like the memory of invasion by an ancient foreign enemy or of legislation that deprives one group of people of property and gives it to another group) do not have effects in the ways individuals privately interact. A person angry at his wife may displace that anger by joining other angry men on the street barricade set up to stop the police. A person angry at losing his job because the company has closed the plant where he worked may go home and beat his wife. There is much (but not random) crossover along the pathways from ionized, affect-laden memories to action: those with private or public potential may be interchanged en route from neocortex to overt action.

Nevertheless, there is such remarkable specificity of joint political action in times of major change (nonviolent or violent) that, despite the spillover of public and private events into each other, there must be remarkable circuit specificity among the billions of neurons that extend through the major segments of the central control systems of the thousands and millions of individuals who collaborate in such gigantic public events.

Correlatively, the chain of releasing and inhibiting neurons and of chemical releasers and inhibitors cannot be regarded as the product of the functioning of any one part of the CCS. A big mistake on the part of those who attribute highly specific political violence solely or even primarily to the limbic system is to forget one of the most obvious implications of some now classic physiological research—namely, the need for substantial parts of the neocortex to be intact for a concerted, directed attack to be made by a vertebrate. We will mention the relevant research in the following section.

And, to reiterate, this prototheory self-evidently does not argue that the function of memory neurons is solely cognitive. Not just the bare fact of prior events is stored but also an emotional evaluation. The individual's memory neurons store both the event and the affect, as both were felt at the time the event occurred. The individual remembers that he lost his job and he remembers that losing it was an uncomfortable, disconcerting event. Affect seems to be a *necessary* component of long-term memory. Millions of people remember the fact and circumstances of the assassination of John F. Kennedy, Martin Luther King, and Robert Kennedy. It is doubtful that many so specifically recall the death of even other people they knew casually but personally. The prototheory relates physiologically to the complex process which Krech has called, in an awkward but compact neologism, "perfink." By this he means that individuals do not discretely *per*ceive, *f*eel, and th*ink*: they do all three simultaneously (Krech, 1949, 1958).

One final matter merits mention. The prototheory distinguishes only quantitatively, not qualitatively, between violent and nonviolent political action in response to frustration. If an individual has memories of prior frustration that are triggered by new events in the external and internal environment, his response may vary quantitatively from grumbling about "those politicians" to joining the armed forces of either the nation or of a revolutionary organization.

Indeed differences between nonviolent and violent responses are differences in intensity and velocity more than anything else. In one nation, the environment may so flexibly respond to new demands of various segments of the public as to be early and adequate enough to reduce and maintain below a violence-critical level the individual responses to frustration. In another nation, the response may be inflexible, repressive, and so tardy as to maximize the violence of frustrated individuals.

On the side of the organism, I am assuming that generally there are such limits to the organic tolerance of frustration that most individuals at the same stage of development and sharing the same degree of frustration in a particular society will react pretty much the same. That is, if a mercantile or manufacturing bourgeoisie is well established but is denied access to political power, as was the case in France before 1789, most of its members will react negatively to the denial—even though some such people will react by writing tracts, others by organizing groups of merchants, and still others by joining such a group or a revolutionary army. Similarly, if a working class is well established but is denied access to political power, it will, like its bourgeois predecessors, react in a variety of ways to that denial; and those ways will be generally negative. The central control systems, whether bourgeois or proletarian, are reacting in substantially the same way to similar political frustration.

Some of the Relevant Neurophysiological Research

There is by no means adequate neurophysiological research as yet, to either confirm or disconfirm the general outlines of this prototheory. Enough work has nevertheless been done to suggest its fit with observed phenomena. And the prototheory itself may facilitate the process of discovering empirically the links between long-stored memories of anger and frustration among thousands or millions of people and their collective demand for change, even when it disturbs the peace.

Let me present examples of the sorts of research that bear on a physiologically based theory of political behavior. While my choice of research to mention is not arbitrary, neither is it comprehensive or conclusive.

The research and theory I will mention is divided into segments that correspond in a general way to the sequence of neural events outlined in the prototheory. These are research and theory on the neocortex, paleocortex, hypothalamus, and the endocrines; and research and theory on motivation as neurophysiologists conceptualize it. The CCS (central control system, including the brain and glands) processes the receipt and storage of perceived happenings in the environment that pertain to the individual human being's needs and his or her expectations of their fulfillment. Following Maslow, I posit that these needs emerge in some sort of sequence from the earliest need for food, warmth, dryness, and emotional sustenance to needs that develop after these physical and affectional needs are sufficiently satisfied so that the growing infant, child, and adolescent no longer is continuously concerned with satisfaction of the physical and affectional needs. As we shall see, this concept of motivation is not shared by neurophysiologists, but without it I find it impossible to explain political behavior in any depth, psychologically or physiologically. There is ample evidence that man is first of all concerned with individual and species survival. There is ample evidence that man is also deeply concerned with fulfilling other needs.

1. *The Neocortex*

The receipt, processing, storage, and retransmittal of information relating to need satisfaction takes place mostly in the neocortex. The here-relevant research is thus that which relates to these processes that start with inputs from the environment external to the organism or to its CCS and end with outputs to the same environment.

Only the most general kind of findings has appeared thus far on the input process. Before even that could be done, the very basic research by John Eccles (1968) and others had to be done, on the ways that electrical impulses are transmitted from nerve to nerve across the synaptic cleft. During signal transmission, ions of sodium, potassium, and chlorine pass through a presynaptic nerve membrane into the junction and then move from the cleft, through the postsynaptic membrane, and into the nerve, carrying electrons across the synapse in the process. The choice between sodium and potassium is critical, passage of the former into the postsynaptic cell producing excitation and of the latter inhibition. Acetylcholine is the neurotransmitter in this process. Acetylcholinesterase is the enzyme that metabolizes acetylcholine and thereby quite precisely permits just the "right" amount of electrical signal to move from nerve to nerve. (Nerve gases used in combat interfere with this enzyme; electrons pass freely across synapses; muscle spasms caused by unlimited electrical signals put muscles out of control; the gassed person dies.)

A group of psychologists and physiologists at Berkeley undertook in the early 1950s a program of research that, once dismissed as crude and of no possible utility, has now been accepted as demonstrating that different kinds of environment produce significantly different kinds of brains in laboratory mice. Mice whose laboratory environment was relatively exciting, with toys and gymnasium equipment to play with in brightly lighted surroundings, turned out in post-mortem examination to have heavier cortexes and larger amounts of acetylcholinesterase than mice raised with minimal handling, few things to play with, in darkened rooms (research summarized in Quay / *et al.*, 1969, and Krech, 1968). Additional research with mice that were provided with even more stimulation—that is, with an environment much like that of mice in a "wild," non-laboratory environment—indicated even larger cortical and neurotransmitter development (Wallace, 1974).

This research tells little thus far about the socially relevant *kinds* of inputs, but it does suggest that an environment that is lacking in *amounts* of stimulation leads to underdevelopment of the brain, including the cortex where information is stored. From this research we do not yet know, physiologically speaking, whether stimulation which the organism "likes" is differently treated from stimulation which it "dislikes," but we have some basis for saying that stimulation is necessary for normal development of physical structures of the CNS.

There is growing evidence that the effects of external environmental inputs are in many ways permanent. The start of such evidence came with the findings of ethologists, most notably Konrad Lorenz, as to the virtual ineradicability of attachments to either adult ducks or to human beings which were established at critical periods in the duckling's first days out of the shell (E. H. Hess, 1973).

It is believed that macromolecules, giant molecules, in a not yet well understood process involving at least RNA, store information (Hydén, 1967). While the process by which information is locked into the brain's information storage molecules remains rather obscure, there is already some information on what can interfere with the long-range process. Agranoff and others have argued that an antibiotic, puromycin, inhibits the synthesis of protein that is a basic part of the process of permanent memory storage. The investigators used a classically Pavlovian experimental situation, in which goldfish were trained to avoid electrical shock by swimming out of the shocking part of the tank when a light went on. By injecting puromycin into the skulls of goldfish immediately before or immediately after the training period, they were able to stop the long-term retention of the avoidance response. The fish were unable to remember to swim out of the charged part of the tank because the puromycin, infusing into their brains, evidently prevented the normal storage of information that involves protein synthesis (Agranoff, 1967a, 1967b). In ways also not clear, the hippocampus, a major part of the limbic system, is involved in establishing and recalling memory (Hydén / Lange, 1968; Kimble, 1968).

There is no present political significance in the work of the Berkeley group, of ethologists, and of Hydén, Agranoff, and others working on memory. We do not yet know whether deprived or otherwise manipulated people are going to remain dull, apathetic individuals attached to whomever in their formative years they found to be their leaders—or that they will become dull, occasionally violent individuals who will attack their leaders and their environment. But the work of these people does indicate that a process of a clearly neurochemical sort does occur in response to inputs from the environment and that the stored information remains a major determinant of subsequent behavior. It is logical to assume that the political consequences of naturally or artificially frustrating environments have neurochemical substrates, and fortunately little research on humans has been possible.

We do know from psychologists who have studied infant humans and monkeys that there can be catastrophic consequences (presumably involving long-term memory storage) of early experiences that frustrate the basic need for affection. René Spitz (1949, 1965) in his study of the deprivation of infants of adequate care by adults in a foundling home, found that emotional deprivation produced a death rate of a third of the infants before they reached two years of age. He also noted an alternation in their behavior between virtually total withdrawal and brief periods of diffuse violent action, in which the object of the violence was usually the infant himself but sometimes other humans. Since the 1940s, John Bowlby (1969, 1973) has independently, and meticulously, studied the effects of maternal and other affectional deprivation in children.

Harry Harlow and associates (1959, 1962, 1970), studying maternal deprivation in monkeys, have been able to spell out the sequence more systematically than Spitz and Bowlby could in their studies of non-laboratory isolation of human infants. Harlow and associates found that monkeys deprived of normal mothers would nevertheless cling to their substitute terry-cloth mothers even when surprise jets of forced air or spikes suddenly emerging from the substitute mother made it harder for the infant monkey to cling to such support as they did get from these mechanical monstrosities. They further found that when such deprived female monkeys reached maturity, they not only lacked interest in sex but also were less likely to achieve successful pregnancy. They found that when these mother-deprived females were impregnated and bore offspring, they were brutal to their offspring. And they further found that the new generation monkeys, whose mothers had come back to their own unsuccessfully repellent artificial mothers, similarly clung to their natural but brutalized mothers—and eventually restored their brutalized mothers to rather normal affection.

The gap persists between what physiologists have found and what Spitz, Bowlby, Harlow, and other psychologists have found by observing overt behaviors. But the gap is narrowing, partly because of the beginnings of post-mortem examination of brain tissues of mother-deprived monkeys. If the beginnings of direct histological research that includes examining brain weights, amounts of neurotransmitters, and storage of memories are not dealing with the same phenomena that psychologists have dealt with, then indeed there is cause for scientific despair.

The processing of information, the events taking place presumably in the cortex between the receipt and storage on the one hand and re-transmission on the other is something about which, as far as I can tell, not much is known. It is usually accepted that the processing, especially decision-making, takes place in the cortex, and it is supposed that highly complex decision-making takes place in the frontal lobes, in front of the sensory and motor parts of the neocortex. But very little appears to be known about this circuitry that can help us here.

The ability to process information and to make effective decisions may depend partly on the postnatal covering of critical neurons by the fatty sheath called myelin and on the multiplying of glia cells, the fatty substances that surround and nourish neurons in the brain (Altman, 1967, pp. 730-731). In any event, an interesting study of that trembling genetic mutant of mouse called Quaking Mice suggests that the very low ability of that mutant to withstand stress in the form of stimulation correlates with a subnormal amount of myelin (Kanfer, 1970). It may be that human infants with less than normal ability to withstand stress, like Spitz's foundlings, have suffered deficient development of myelin and glia cells.

From psychological investigation it is apparent that under stress humans' decision-making processes are altered. These differences surely have neurochemical substrates, whether because of tissue damage, changing patterns of firing in nerves and neurotransmitters, or changes in endocrine secretions. Perhaps there are changes in all three.

As to the re-transmission of information that has initially been accepted, stored, and then processed, evidently even less is known. All of these processes, involving the activity of cortical tissue, are crucial parts of the events taking place between the input of information into the brain (from both the external and internal environment) and the output, in the form of a command to the muscles to act.

We know at least that *the demand for change* in both natural and social environments (including socioeconomic and political system changes)—even at the price of violence undertaken to get the change—is behavior characteristic of human beings and not much of lower animals. And we know that the human cortex is much more highly developed that that of other species. But we know very little about events taking place between the distincts parts of the cortex where complicated information is received, stored, processed, and re-transmitted. What little we can guess from research on lower parts of the brain is of high negative inferential value. Since the lower parts of the brain in man are like those in lower animals and since lower animals almost exclusively adapt to rather than change their environment, the demands by humans for change must involve the highly developed human cortex, in the neural events that lead to demand (even *violent* demand) for change and this neocortical involvement must be both cognitive and affective.

The data are by no means yet adequate to confirm the theory. If it were known that memory cells did or did not contain emotive ions as storers of affect, there would be no need for a theory. But, as an Italian physiologist has written, in noting that "emotion" travels up from the reticular activating system: "the stimulus may well start from forebrain mechanisms and project downward to arouse the activating reticular system and the hypothalamus, with consequent re-excitation of limbic and neocortical areas" (Zanchetti, 1967, p. 614).

2. *The Paleocortex, the Limbic System: Is It the Emotional Brain?*

It is to those lower parts of the cortex, the ones which are about as well developed in lower vertebrates as they are in Homo sapiens, that we can now turn. Here more research and more definite physiological (that is, metapsychological) knowledge is available. We are mainly concerned with the old cortex, the paleocortex, and its subcortical connections, roughly an equivalent term for the limbic system, which is roughly equivalent to the emotional brain.

It is somewhat easier to make generalizations about the paleocortex than about the neocortex. One reason is that the paleocortex has been so extensively studied partly because it is not necessary to reduce humans to laboratory animals to find out how it works. The paleocortex seems to function pretty much the same in man and at least all higher vertebrates. Another reason is that emotional ("limbic") disturbances interfere more with good human relationships than rational ("neocortical") disagreements. Another is that a growing number of humans with severe emotional disorders have recently had parts of their limbic systems successfully surgically treated. In contrast, earlier gross surgery (frontal lobotomies) on the neocortex had proven rather futile.

Before mentioning some of the very fruitful research in the paleocortex, it should be noted (for the first but not last time in this paper) that there is an enormous knowledge gap as to the nerve tracts connecting the various parts of the limbic system with various parts of the neocortex. Not only are the sites of memory storage in the neocortex not well mapped and the circuitry within the neocortex not well mapped. Even more frustrating is our lack of knowledge about the neural pathways between the paleo- and neocortexes, between the excessively dichotomized emotional and intellectual parts of the brain. Nerve tracts have in general been clearly mapped (MacLean, 1958, 1970, p. 341; Livingston, 1967), but there is not yet much actual tracing of signals that pass back and forth from the old and new cortexes.

However, since nineteenth-century beginnings in Europe and continuations in America and Europe in the twentieth century, neurologists using ablation and cutting techniques have severed the neural connections between the diencephalon or upper brain stem and the paleocortex and have made disconnections at various points within the upper brain stem, which contains the thalamus and hypothalamus. To generalize as to the findings, it appears that when the cut is made high enough up toward the cortex, a highly specific rage response can be readily produced, in which the laboratory animal (cat or dog) will attack the experimenter; and when the cut is made lower, nearer that is to the brain stem, the rage reaction becomes more generalized (Zanchetti, 1967, p. 604). Experimentation with electrodes implanted in the hypothalamus has indicated that stimulating parts of it in cats produces a rage response. Unfortunately it is not clear (because electrical impulses do not necessarily travel only downard or only upward from the stimulated region) what if any parts of the brain above the stimulated part are involved (Von Holst / Von St. Paul, 1962). Prior research had, however, made clear that the neocortex is the inhibitor of immediate reactions to threat (W. Hess, 1964, esp. p. 58).

The significance of this research by ablation and electrode implantation must be seen in its time context. An American psychologist who had trained in Germany, William James, argued that the seat of the emotions is in the (sensory and motor) neocortex (James, 1890, chap. 25, esp. p. 473). Walter Cannon and others since (like MacLean) have argued that the site is in the very primitive parts of the brain, parts which Homo sapiens structurally shares with animals at least as far down the phylogenetic scale as reptiles.

My prototheory is generally consistent with James's, the difference being that I am arguing only that the very complex political responses to frustration and threat involve storage of affect-laden information in the neocortex. The response to environmental threat quite certainly involves the limbic system, where it seems that *generalized* positive and negative affects are triggered. But the stored affects—the stored good and bad memories—that relate to past and present political objects are highly specific and it seems very doubtful that these specifically political responses could be triggered by the upper brain stem without any affect from the neocortical memories being involved in the response circuit. My prototheory argues that the neocortex, by an amplifier process that probably involves feedback to the neocortex from midbrain and limbic system, activates these lower structures. These in turn alert the neocortex to scan itself for other relevant information and to plan action against the change-resistant parts of the environment.

The history of the idea of an emotional part of the brain is an interesting one. The French brain surgeon and anthropologist, Pierre Paul Broca (1824-1880), was perhaps the first in a sequence of students of the brain to delimit a part of the brain, notably the rhinencephalon, as the site of the emotions, in an 1878 statement. In 1937 an American brain surgeon, James Papez, more precisely delimited this emotional brain which in Broca's term is the limbic (that is, the border) system that overlies the thalamus and hypothalamus and forms the phylogenetically primitive part of the cortex.

Papez not only designated the parts of the brain to be included as the site of the emotions but also very generally indicated how these various parts are linked to the neocortex on the one side of the vast circuitry and to the upper brain stem on the other. Papez's statement was a bold integration of the theory and the research of others, and it advanced theoretically the issue of just where and how emotions are to be studied physiologically.

Paul MacLean, a tiny fraction of whose work is cited above (*supra*, p. 112), intensified and elaborated the systematic physiological study of the emotions, concentrating on what others and he have called the self-preservation and species-preservation functions of the limbic system. He and such others as José Delgado (1969) have provided large increments of knowledge as to the effects of electrical and chemical stimulation of such parts of the limbic system as the amygdala, fornix, pyriform cortex, hippocampus, and cingulate gyrus. The work of Papez, MacLean, and Delgado, mainly with rats, cats, and monkeys, has been extended to human beings by such brain surgeons as Vernon Mark (Mark / Ervin, 1970).

It has now been well established that there are specific sites in the limbic system where the display of different emotions can be produced by electrical and chemical stimulation (Delgado, 1969, 1974a, 1974b). Fear, anger, and rage have been elicited by direct limbic stimulation in the amygdaloid region and have become manifest in such overt behaviors as arched back, hair standing on end, snarling, baring of teeth, and either fight or flight. In very general ways, the sequence of sites at which such behaviors are triggered has been mapped, making it possible to say that there are neural links between successive nuclei in the limbic system, between the limbic system and the neocortex, and between the very basic part of the brain known as the reticular activating system and the limbic system and the neocortex. But, as I have already said once, there is a lack of detailed mapping of the nerve tracts between the neocortex and paleocortex and midbrain. This lack of maps is one of the major roadblocks to understanding how human beings specifically respond to specific threats. There is an unfortunate tendency of researchers not to state but to imply that an electrode that triggers an emotional response at one point in the brain indicates that *that* point is *the* site of that emotion—and thus to imply that no other site or circuit is involved. Brain circuitry is complicated enough without muddling the complexity by oversimplification.

3. *The Hypothalamus*

Perhaps the most intensive study of any part of the brain that relates to the emotions has been done on the hypothalamus, that central switchboard in the upper brain stem through which perhaps nearly all pleasant and unpleasant signals are channeled as they enter and leave the brain. The hypothalamus is very precisely structured, so precisely that specific and opposite emotional responses can be triggered by electrical or chemical stimulation at locations immediately adjacent to each other.

Parts of the hypothalamus have been studied which are major links in the long neurochemical chain whose output is behavior such as eating, drinking, and sex. Among others, Teitelbaum (1967, pp. 98-100) has reported that the insertion of a strong saline solution into the "thirst" center of the hypothalamus can cause a goat to drink till it is bloated almost to the bursting point, because the "thirst" center tells the brain that there is too much salt in the system. Extensive research, both with electrode and cannula, has produced quite precise mapping of those areas of the hypothalamus that are involved in sensing and responding to such basic physical needs as those for food, water, sex, and proper body temperature.

The now classic research of James Olds and associates put together both technologies and findings of such prior pioneers as the Swiss W. R. Hess, the Canadian Hebb, and the American MacLean. In the early 1950s Olds began to report research demonstrating that there are "pleasure" areas in the forebrain and upper brain stem, notably in parts of the hypothalamus and the medial forebrain bundle. When a rat's brain is so wired that it can stimulate itself in those areas, it continues to press the bar that produces the electrical charge. Some rats have pressed up to seven thousand times per hour (roughly two times per second) and some as long as twenty four hours, stopping only when physically exhausted. The sensation is so pleasant (or at least compelling) that rats produce it not only in preference to rest but also in preference to food. Correlatively, Olds found centers very close to these pleasure centers which, when stimulated, cause such an evidently unpleasant sensation that the rats very speedily stop pressing the bar that produced the stimulus (Olds / *et al.*, 1956, 1958, 1963).

4. *The Endocrines and Chemical Transmitters*

Chemical investigations of the hypothalamus have spelled out in detail the places where injection of various steroids, including testosterone, estrogen, and progesterone, produce behavior that contradicts the normal sex-linked behavior. That is, male rats can be caused to behave sexually like females and females can be caused to behave like males. And there is much research, some of it already decades old, that has established the interactions between the hypothalamus and the hypophysis or pituitary, the peasized endocrine attached to the underside of the hypothalamus and controlled by it, with some feedback of it to the hypothalamus.

As the master regulatory gland which is controlled by the brain through the hypothalamus and which controls many other endocrines and in certain ways the brain itself, the pituitary secretes several substances that directly relate to overt behavior and indirectly relate to any theory of social and political development that includes organic determinants. It secretes growth hormone, which helps regulate the growth and maturation of the organism. At certain stages of the organism's maturation, the pituitary secretes gonadotropic hormones. Carried through the blood stream to the sex glands, these hormones cause the sex organs to mature and in turn to secrete sex hormones: androgens in the male and estrogens in the female. Endocrines from the hypophysis also cause the secretion of both androgens and estrogens from secondary sources in the cortex (outer layer) of the adrenal glands.

But the hormones in the hypophysis that are probably most pertinent to political behavior are the catecholamines, notably nonadrenaline (norepinephrine) and the very similar chemical which is a metabolite of it, adrenaline (epinephrine). These hormones are produced not only in the hypophysis but also at the sympathetic nerve endings neurally connected to the hypophysis and in the medulla (the core) of the adrenal glands. These two catecholamines relate directly to stress—to tension of perhaps every sort, both assertive (positive) interactions (as in love relationships) and aggressive (negative) ones (as in hostile interactions). The catecholamines function in parallel with the hypothalamus in preparing the entire body (including the brain) to respond to stress—from the somatic stress of illness to the psychic stress of interpersonal, more broadly social, and of political frustrations.

When triggered by the hypothalamus, norepinephrine and epinephrine are almost instantaneously released at the sympathetic nerve endings, thereby activating those organs and tissues with which the sympathetic nervous system is linked. And the adrenal medulla, neurally linked to the hypothalamus, secretes stored norepinephrine and epinephrine into the blood stream, by which these endocrines are carried throughout the system, including into the brain. Catecholamines secreted at sympathetic nerve endings immediately prepare the body generally to respond to nerve signals which control the specific responses to threat. Catecholamines secreted into the bloodstream from the adrenal medulla sustain that readiness to respond to nerve signals.

Norepinephrine causes constriction of the blood vessels in both striated and smooth musculature, accelerates the heartbeat, relaxes the intestines, and releases sugar into the blood stream. Epinephrine does all of these things but has less tendency to constrict blood vessels in muscle tissue, greater tendency to make the heart beat faster, and many times greater effect in generally increasing the metabolic rate of the whole body, making it more ready to respond as the instrument of the brain's commands.

These catecholamines are thus very vital links in the chain of events that starts with the sensory cortex's receipt of stimulation from either outside or inside the organism. They are critically involved in a long but often very rapid sequence of neural and endocrine events that *functionally* involve decision-making and action. And they are critically involved in a sequence that *structurally* includes the *entire* central control system, from the sensory cortex and limbic system to the decision-making areas (presumably in the frontal cortex), to the motor cortex and limbic system and thence to the hypothalamus, pituitary, and to the sympathetic and peripheral motor nerve systems. They are thus centrally involved at all stages of the response of the CCS to social and political stress.

And so stress, coming from the cortical processing of information that indicates threat to need satisfaction, produces secretion of catecholamines into the bloodstream. But this does not tell us the sequence by which the catecholamines are manufactured in the first place. The sequence, chemically speaking, is from tyrosine, an amino acid and one of the basic building blocks of protein, to dopamine to norepinephrine to epinephrine. The conversion of tyrosine to dopamine is controlled by the enzyme tyrosine hydroxylase and of dopamine to norepinephrine by dopamine-beta-hydroxylase. The conversion of norepinephrine to epinephrine is catalyzed by phenylethanolamine-N-methyltransferase, an enzyme whose pretentious name belies its task, which is merely to methylate (add a methyl (CH_3) group to) norepinephrine.

When we look back into the chain of causation in the production of the catecholamines and their antecedent enzymes, again we get back to stress, which triggers the process of manufacture as well as the previously discussed process of releasing these catecholamine neurotransmitters. This is to say that the process by which the organism deals with tension involves an intimate and usually orderly interaction between the firing of nerves and the secretion of hormones. The response to stress indeed emphasizes the neurochemical interdependence and interaction of the nerve centers and endocrine glands.

Although hardly yet well understood, the interaction process is already dramatically demonstrable in the effects of various stimuli on neurochemical states and the effects of these neurochemical states on the general mental state. There is an upper limit of stimulation, beyond which the CCS is unable to respond effectively. Stimuli which are neither too intense nor long-lasting tend to trigger the manufacture and use of the catecholamines, whose function in generally preparing the body to respond to stress is so central. Norepinephrine concentrates at receptor sites in the brain as well as the body. The whole person is ready to respond, physically as well as mentally. Excessive stress does not produce this effective response. When the stress has resulted from object loss (as in the loss of a beloved one, loss of limb, or even a career) and produces bereavement and depression, the hormonal concomitants of the behavior include a *reduction* in norepinephrine and epinephrine levels. A similar effect has been produced by injecting into monkeys' brains 6-hydroxydopamine, which (contrary to dopamine-beta-hydroxylase) does not produce norepinephrine but destroys it. The injected monkeys became withdrawn and affectless, like people in a state of depression (Redmond / *et al.*, 1973).

As is increasingly well known, catecholamine deficiencies and imbalances that are associated with depression and its opposite, mania, can be chemically treated with remarkable success. But, as suggested above, there are so many links in a long chain back to the processing, by the neocortex, of stimuli producing the depressed state that the effect of drug treatment restoring norepinephrine and epinephrine to normal levels is about as durable as the effect of various psychotherapies. The drugs treat a consequence (endocrine imbalance) rather than the cause of depression. It helps the individual in the short run, but only resolution of the stress that has resulted from the failure to meet the demands of the organism (both somatic and psychic) can "finally" restore the person to good physical and mental health. If this restoration is not possible, the person either becomes a relatively mindless body or dies: he become a vegetable or a corpse.

Eventually it should be possible to explain the political apathy in our time of perhaps the vast majority of human beings partly in physiological terms. This political apathy may be one consequence of the chronic inability of the CCS to respond to the chronic stress of unfulfillment of basic somatic and psychic needs and the resultant chronic depression.

5. *The Physiological View of Motivation*

The progress of neurophysiologists toward understanding the basic neurochemical processes has been outstandingly and increasingly rapid in recent decades. But, as seems inevitable, research and theorizing have been done where the payoffs are most substantial and verifiable and applicable. And, because of a variety of factors that include the moral and the economic limitations of research, it is merely the vegetative functions—those causes of stress that Homo sapiens so obviously shares with other primates and vertebrates—that have been studied. These are the functions, in MacLean's words, which relate to the innate desire of life to survive as individuals and species. Self-preservation and species-preservation are so natural and universal that they are not only the first law of life but also the first and almost the only object of physiological research. They remain the only human or animal motives that physiologists have examined. Fitting nicely the feral mood, the wild and near-total preoccupation of the 1970s with the natural environment and population growth, some social scientists have sought to explain and evaluate political behavior only as it relates to the basic desire that man shares with perhaps all other species of life to survive as individuals and as species. These social scientists, too, ignore all other motivations that self- and species-preservation.

This brings us to the problem of conceptualization of other demands that are present to some degree in many vertebrates and certainly in man as an organism. In the psychological terms of Maslow, the needs of humans that neurophysiologists have considered in their continuing preoccupation with self- and species-preservation are the physical and affectional ones (Valenstein, 1973; Kimble 1973). Neurophysiologists since at least Walter Cannon have long been studying stress that results from physical deprivation in the form of bodily discomfort (as from electrical shock, hunger, and thirst). More recently, they have also been studying stress that results from affectional deprivation (in the form of being denied sex contact or of bereavement coming from loss of maternal care).

Unfortunately, from the standpoint of understanding the realities of human interaction, these are not the only origins of stress among human beings, particulary the very real stress that sometimes explodes violently in politics. People become tense not only for lack of food and love: they also become tense when others betray their trust, ignore them, and depreciate or degrade them. They become tense when the "others" are close to them (parent, spouse, or friend) and not so close (bosses, policemen, soldiers, politicians, and even chiefs of state). And people become tense over their own loss of ability to work efficiently—quite apart from the relevance of their work, their job, to being appreciated or making a living. The suicide of Ernest Hemingway was a dramatic case in point. It has been easy—too easy—to continue doing research based on the false premise that the total spectrum of needs of higher vertebrates (cats, dogs, rhesus and other macaques; the great apes) and even lower vertebrates (birds, snakes, etc.) is coextensive with the total spectrum of human needs.

From the research of such people as Teitelbaum and others who have studied the physiology of hunger, thirst, and lack of warmth, we know physiologically rather well the neurochemical responses to the stress produced by physical deprivation and we know a little about emotional deprivation. From the research of those like René Spitz, John Bowlby, and Harry Harlow, who have studied the effects of emotional deprivation, we know generally the behavioral responses to that kind of stress. And from the physiological research of those who have begun to work on the neurochemical effects of an environment that is emotionally or otherwise sensually depriving, we are beginning to learn about the sequence of events within the CCS of that kind of stress. Some of the events as observed in high and low vertebrates are surely comparable to man, particularly those taking place at the level of the hypothalamus, the pituitary, and the sympathetic nervous system.

As to specific neural and chemical events that take place in response to the kinds of stress that relate to socioeconomic or political change including violence, we know next to nothing. It is almost a truism, a banality, to assert that in politically related stress, norepinephrine and epinephrine levels rise, at both the sympathetic nerve endings and in the bloodstream. A banality because that is perhaps the penultimate stage of the response of human beings *to stress of every sort. We need to know the circuitry, from*

sensory cortex, frontal cortex, limbic system, hypothalamus, to reticular activating system to understand political stress neurophysiologically.

It is a reasonable presumption that there are indeed large differences in degree (though not necessarily in kind) between the innate potential demands of human beings and of other primates and that these differences involve different circuitry in the neocortex for particular kinds of stress. If we *are* to understand the neurophysiology of political phenomena, it can be understood only by using as the stimulus-inputs in research those which have, directly or analogically, social and political content.

My prototheory of particles within memory neurons (probably macromolecules) that are affectively ionized may give some direction to the search for biological explanations of political behavior. The prototheory posits a neurochemical substrate for the affective content involved in the processing, storage, and re-transmission of information that the neocortex has received. Are the "known" facts of brain function so totally inconclusive as to make *any* other prototheory better than one of ionized particles that are part of what we call long-term memory? And is *no* prototheory better than *any* prototheory? If the determinants of political behavior that are within the human organism were not so evidently critical to understanding political behavior, it would be easy to say "yes," categorically, to these questions. If there were not extant such a substantial body of physiological research that is iterative and reiterative of work that was done even decades ago, it would also be easier to say categorically "yes" to these questions.

We do "know" a few things. We know that emotional states are remembered. This affective component of stored information and of currently received information involves activity (in at least the limbic system and endocrine system) signaling the neocortex that certain basic demands of the whole organism are not being met. The neocortexes of thousands and millions of people may blame their individual and collective stress on specific responses or failures of the environment to respond to demands that large collectivities of people make on their environment. The pain and the frustration call for release by interaction with the environment (in the political case, the government) to make it responsive to these ultimately innate demands.

As we have noted, the presence of ionized—that is of electrically charged—particles of potassium, sodium, and chlorine, is perhaps common to all transsynaptic events in the CCS. Reasoning by analogy, it seems appropriate to suppose that the storage of information relating to political action has not just its cognitive but also its *affective* content carried by special-function complex protein molecules that in some way are electrically charged. As past memories of frustration accumulate over time—a few days, weeks, months, and years—they are in some way connected to present, precipitating stimuli (as, for example, a sudden rise in food costs or an abrupt insult to one's integrity when he is shoved or beaten by a police officer or a soldier). When these old and new stimuli are neurochemically conjoined, they function as an amplifier system, interacting with the lower centers of the brain to produce political action.

It would, I think, be absurd to suppose that there is any such thing as a broadly social, nationally communal, or political area of the neocortex.

Absurd because there is such a close and continuing interaction of forces within and between individuals (including stored memories and ongoing expectations) and broad social and political forces, as these become parts of the neurochemical events in the brain. It would be enormously hard to predict, by electrical stimulation of particular parts of the neocortex, that some particular events would quite surely lead to political interaction, because the contexts in time and space of these events are so very variable.

On the other hand, it should be possible to focus both theorizing and research on new modes of thought that will indicate far more precisely than the present modes just what is the complex of environmental *and* neurochemical events which, when they combine, produce demands for political change. It seems most likely that the pertinent intracranial events include the affect-laden, the charged state of large *sets* of memory neurons and that these acts are hooked up in a complicated but orderly way to those multifunction pathways of response to stress that have been observed in the limbic system and the hypothalamus.

What Political Science Needs to Know about the Central Control System

Two large things are needed before there can be much advance in such politically relevant physiological research. Both are implicit in what I have already indicated. One is an investigation of the physiology of motivational states beyond those that have to do with individual and species survival and rudimentary affiliative and affectional needs. Some people who have implied that man wants only to survive and procreate may deny that there is a neurochemical substrate for those other intense motivations that are so manifest in human behavior.

It would on this matter be interesting to look at the physical and metaphysical arguments *against* the existence of interactive segments of the central control system that structurally serve the functions of wanting and demanding—from family, society, and polity—such goods as equality, dignity, and freedom. To date there has been no argument: the issue has been ignored by those natural scientists who have not become philosophers.

I doubt that many neurophysiologists would argue, metaphysically or otherwise, against the existence of nerve and endocrine tissues whose functioning relates to what some psychologists call innate, organic needs of human beings. But neurophysiologists, unnecessarily I believe, continue to regard both man and apes and cats and rats as only individual and species survival mechanisms. While physiologists readily use the terms insult and trauma to tissue, they have yet to consider physiologically insults and traumata to the psyche, by which I mean neither more nor less than the central control system as it has been elaborated by the permanent storage of memories with affective content.

The second large thing is a reexamination of the controversy between William James and Walter Cannon, in which James argued that the cortex is the seat of the emotions and Cannon that the upper brain stem is. As I see it, the process of affective interaction with the environment is too neurochemically integrated to allow for saying that either the neocortex *or* the paleocortex *or* the hypothalamus *or* the pituitary is *the* locus of emotion. The sensing, processing, and response to events that produce emotional states is a total and totally integrated one involving perhaps *all* major parts of the CCS. Rather than pick the neocortex or the paleocortex, it makes more sense to hypothesize that all neural and endocrine structures are *sequentially involved* in establishing ionized particles *in the neocortex* where memories are stored, and that these ions serve as the affect-laden, the charged component of memories relating to political (as well as other) action.

If the locus or loci of emotive events are not well established, then the research that would be most helpful is a mapping of the circuitry between the limbic system and the neocortex, mapping which will make it possible to locate particular areas of the neocortex where response-sets are established that relate to needs other than to survive and procreate. If it should turn out that there are no definable areas, then this will be valuable information. But the specificity of response to stimuli in various other areas where the emotional content is manifest, as in the pleasure-center research of James Olds, should encourage investigators to look for comparably specific areas in the neocortex. Unfortunately, only the most preliminary steps have thus far been taken, in indicating generally the large number of areas in the cortex to which tracts in the medial forebrain bundle lead. This kind of mapping is very difficult and almost impossibly precise work. And of course, the kind of research that W. R. Hess in the 1920s pioneered with implanted electrodes was impossible before he did it. And the fruit of that technology has been very abundant after Hess made it possible.

A third large area of research, growing out of the kind that Hydén and Agranoff have separately done, would be to ascertain whether the affective content of memories does indeed have—as seems reasonable to suppose—a neurochemical substrate in the form of special-function ionized macromolecules in memory neurons. This line of investigation takes the problem into the research technology familiar to molecular biologists.

It may be too early and too presumptuous for social scientists to ask questions of neurophysiologists. But certainly it is better now to ask such questions than for political philosophers to continue to talk in a mystical way about human nature and for physiologists either to ignore the political implications of their findings or to assume that social scientists or natural scientists turned philosophers are quite able, unaided, to provide adequate explanations for profound, earth-shaking processes of social and political change—or to assume that social science is and always will be softly and solemnly futile and banal. To at least this political scientist, it is laughable that any explanation for political phenomena can be adequate that omits the human organism as a partial but integral determinant of socioeconomic and political development. It is very difficult for neurophysiological research to contribute to the better understanding of politics. The most pertinent phenomena remain largely unknown. But the need for this contribution is not only large but also critical.

Darwin, Marx, Freud, and Maslow faced and turned our attention to the big questions: how does epigenetic development from zygote to fetus to neonate and to child, adolescent, and adult proceed? How does this epigenesis affect socioeconomic and political processes? And how do stages of socioeconomic and political development affect individual development? Manifestly it is possible only conceptually and abstractly to separate the influences of the organism from those of the environment, because each of these "abstractions," these processes, develops pari passu with the other.

But this is no reason, as Schubert in this volume emphasizes, for arguing or assuming that there is no structure, no order, no continuity or sequence, in the emergence of internal forces from within the organism. There is no reason for prematurely asserting that neurophysiology can have no relevance to social science. As a continuing pioneer in such research, Roger Sperry, has said: the frontiers are between the ears. And, as Paul MacLean (1969, p. 17) has said: "The story of the tree-of-the-knowledge-of-good-and-evil would appear to represent a paranoid delusion of early man, delusion in which the weapon of procreation was mistaken for the weapon of destruction." To put it in other words, the only thing we have to fear is fear of our unknown selves.

Bibliography

In this list of a few of the more pertinent readings, I have sometimes used semi-popular rather than specialized-journal sources, where a very notable contributor has written clearly for an unspecialized readership. While some may object to my preference for not citing the most technical source, written for the most cloistered communicants of neurophysiology, I maintain that the general subject and the material listed are so important that even a proper academic pride in not being able to communicate with everyone should here be put aside.

Agranoff, B. W.

1967a "Agents That Block Memory," in: Quarton, G. C. / Melnechuk, T. / Schmitt, F. O. (eds.), *The Neurosciences*, New York, Rockefeller University Press, pp. 756-764. — A clear historical summary of past and recent research on memory.

1967b "Memory and Protein Synthesis," *Scientific American* 115-122, June. — An extensive report of the goldfish experiments.

Altman, J.

1967 "Postnatal Growth and Differentiation in the Mammalian Brain, with Implications for a Morphological Theory of Memory," in: Quarton, G. C. / Melnechuk, T. / Schmitt, F. O. (eds.), *The Neurosciences*, New York, Rockefeller University Press, pp. 723-743.

Bowlby, J.

1969 *Attachment*, New York, Basic Books.

1973 *Loss*, New York, Basic Books. — This and the 1969 volume report the research commenced during the Second World War on children separated by war exigencies from their parents.

Davies, J. C.

1963 *Human Nature in Politics*, New York, John Wiley. — A non-mystical approach to analysis of the organic roots of political behavior.

1976 "The Priority of Human Needs and the Stages of Political Development," in: Pennock, J. R. (ed.), *Human Nature and Politics*, in press.

Delgado, J. M. R.

1969 *Physical Control of the Mind*, New York, Harper and Row. — A provocative report on brain research, with speculation on social implications; tends to ignore endocrine processes.

1974a "L'agressivité: pulsion ou réponse à l'environnement?," in: *Les entretiens de Rueil*, 13 May, pp. 85-96. (Available from the author.)

1974b "Brain and Behavior," paper as part of exhibit at 26th International Congress of Physiological Sciences, New Delhi, India, October. (Available from the author.)

Eccles, J. C.
1968 *The Physiology of Nerve Cells,* Baltimore, Md., Johns Hopkins Press. (First ed. 1957.) — A very comprehensive summary of basic research on the synapse.

Ganong, W. F.
1969 *Review of Medical Physiology,* Los Altos, Calif., Lange Medical Publications. — The most useful of four general texts on physiology that I have found.

Harlow, H. F.
1959 "Love in Infant Monkeys," *Scientific American* 200 (6), June, pp. 68-74.

Harlow, H. F. / Harlow M. K.
1962 "Social Deprivation in Monkeys," *Scientific American* 207 (5), November, pp. 136-146.

Harlow, H. F. / Blomquist, A. J. / Thompson, C. I. / Schlitz, K. A. / Harlow M. K.
1968 "Effects of Induction Age and Size of Frontal Lobe Lesions on Learning in Rhesus Monkeys," in: Isaacson, R. L. (ed.), *The Neurophysiology of Development,* New York, John Wiley, pp. 79-120.

Harlow, H. F. / Suomi S. J.
1970 "Induced Psychopathology in Monkeys," *Engineering and Science* (Pasadena, Calif., California Institute of Technology), April, pp. 8-14. — A report of second-generation effects of maternal deprivation, when mother-deprived monkeys become mothers.

Hess, E. H.
1973 *Imprinting: Early Experience in the Developmental Psychology of Attachment,* New York, Van Nostrand Reinhold. — A thorough summary of research that "began" with Konrad Lorenz.

Hess, W. R.
1964 *The Biology of Mind,* Chicago, Ill., University of Chicago Press, esp. ch. 2. (First ed. 1962.) — A many-sided, nonparochial statement, very well balanced in discussing both neural and endocrine processes.

Hydén, H.
1967 "Biochemical Changes Accompanying Learning," in: Quarton, G. C. / Melnechuk, T. / Schmitt, F. O. (eds.), *The Neurosciences,* New York, Rockefeller University Press, pp. 765-771. — A statement by one of the most innovative pioneers.

Hydén, H. / Lange, P.
1968 "Protein Synthesis in the Hippocampal Pyramidal Cells of Rats during a Behavioral Test," *Science* 159, March 22, pp. 1370-1373.

James, W.
1950 *The Principles of Psychology,* New York, Dover Publications. (First ed. 1890.)

Kanfer, J. N.
1970 "Cerebral Sphingolipids in the Quaking Mouse," in: Bowman, R. E. / Datta, S. P. (eds.), *Biochemistry of Brain and Behavior,* New York, Plenum Press, pp. 65-90.

Kimble, D. P.
1968 "Hippocampus and Internal Inhibition," *Psychological Bulletin* 70 (5), pp. 285-295. November. — An historically based report of research relating to the function of the hippocampus in inhibiting perceptual and memory processes.
1973 *Psychology as a Biological Science,* Pacific Palisades, Calif., Goodyear Publishing. — An elegantly clear introduction, which, like Valenstein 1973, shows that physiologists regard human and other vertebrate motivations to be self- and species-preservation only.

Krech, D.
1949 "Notes toward a Psychological Theory," in: Bruner, J. S. / Krech, D. (eds.), *Perception and Personality: A Symposium,* Durham, NC, Duke University Press, pp. 66-87, at pp. 79-82.
1968 "The Chemistry of Learning," *Saturday Review,* January 20, pp. 48 ff.

Krech, D. / Crutchfield, R. S.
1958 *Elements of Psychology,* New York, A. A. Knopf.

Livingston, R. B.
1967 "Brain Circuitry Relating to Complex Behavior," in: Quarton, G. C. / Melnechuk, T. / Schmitt, F. O. (eds.), *The Neurosciences,* New York, Rockefeller University Press, pp. 499-515. — A very clear summary.

MacLean, P. D.
1958 "The Limbic System with Respect to Self-Preservation and Preservation of the Species," *Journal of Nervous and Mental Disease* 127 (1), July, pp. 1-11. — A basic article by the man who has perhaps done more research on "the emotional brain" than anyone else.
1969 "The Paranoid Streak in Man," in: Koestler, A. / Smythies, J. R. (eds.), *Beyond Reductionism: New Perspectives in the Life Sciences,* New York, Macmillan, pp. 258-278, at p. 274.
1970 "The Triune Brain, Emotion, and Scientific Bias," in: Quarton, G. C. / Melnechuk, T. / Adelman, G. (eds.), *The Neurosciences,* New York, Rockefeller University Press, pp. 336-349.

Mark, V. H. / Ervin F. R.
1970 *Violence and the Brain,* New York, Harper and Row. — A report on research in use of electrical stimulation to find and treat sites in the limbic system that are in the violence-producing circuit.

Moyer, K. E.
1971 *The Physiology of Hostility,* Chicago, Ill., Markham. — A comprehensive and balanced summary of neural and endocrine factors.

Olds, J.
1956 "Pleasure Centers in the Brain," *Scientific American* reprint. — An extensive summary of his research.
1958 "Self-stimulation of the Brain," *Science* 127, February 14, pp. 315-324. — Includes a report on the effect of chemicals on rats' barpressing to produce reward of electrical stimulation of the pleasure centers. Very complete summary of pleasure-center research.

Olds, M. E. / Olds, J.
1963 "Approach-Avoidance Analysis of Rat Diencephalon," *Journal of Comparative Neurology* 120, April, pp. 259-293. — Contains a long bibliography on intercerebral stmulation.

Papez, J. W.
1937 "A Proposed Mechanism of Emotion," *Archives of Neurology and Psychiatry* 38 (4), October, pp. 725-743. — A classic statement and a great theoretical leap forward, before MacLean's research.

Quay, W. B. / Bennett, E. L. / Rosenzweig, M. R. / Krech, D.
1969 "Effects of Isolation and Environmental Complexity on Brain and Pineal Organ," *Physiology and Behavior* 4, pp. 489-494.

Redmond, D. E. / Hinrichs, R. L. / Maas, J. W. / Kling, A.
1973 "Behavior of Free-ranging Macaques after Intraventricular 6-Hydroxydopamine," *Science* 181, September 28, pp. 1256-1258.

Spitz, R. A.
1949 "The Role of Ecological Factors in Emotional Development in Infancy," *Child Development* 20, pp. 145-155.
1965 *The First Year of Life: A Psychoanalytic Study of Normal and Deviant Development of Object Relations,* New York, International Universities Press, ch. 14. — A very complete report on his research in the foundling home.

Schubert, G.
1975 "Politics as a Life Science: How and Why the Impact of Modern Biology Will Revolutionize the Study of Political Behavior," in this volume. — An expression of criticisms and hopes that I share.

Teitelbaum, P.
1967a *Physiological Psychology: Fundamental Principles,* Englewood Cliffs, NJ, Prentice-Hall. — A clear and comprehensive intoduction.
1967b "The Biology of Drive," in: Quarton, G. C. / Melnechuk, T. / Schmitt, F. O. (eds.), *The Neurosciences,* New York, Rockefeller University Press, pp. 557-567. — Indicates the self- and species-preservation view of physiologists toward motivation.

Valenstein, E. S. (ed.)
1973 *Brain Stimulation and Motivation.* Glenview, Ill., Scott Foresman. — A convincing demonstration that physiologists continue to regard physical (including sex) drives as the only basic, organic ones. Cf. also Teitelbaum 1967b and Kimble 1973 for the same artificial limitation on neurophysiological investigations.

Von Holst, E. / Von St. Paul, U.
1962 "Electrically Controlled Behavior," *Scientific American*, reprint March. — A report on the induction of attack behavior by electrical stimulation of primitive parts of the brain of a primitive vertebrate (chickens).

Wallace, P.
1974 "Complex Environments: Effects on Brain Development," *Science* 185, September 20, pp. 1035-1037. — An excellent summary of research on the effects on the brain of different cognitive and affective environments, including a summary of the Berkeley enriched and empoverished environment studies of rats.

Zanchetti, A.
1967 "Subcortical and Cortical Mechanisms in Arousal and Emotional Behavior," in: Quarton, G. C. / Melnechuk, T. / Schmitt, F. O. (eds.), *The Neurosciences*, New York, Rockefeller University Press, pp. 602-614. — An excellent integration of historical and contemporary research.

PETER A. CORNING

Toward A Survival Oriented Policy Science

I think we are delicately poised right now. I genuinely think that the next decade could either be a period that in retrospect will look like one of the great periods of human creativity, or it could be the beginning of extraordinary disarray.

Henry Kissinger (1974)

Madam Beverly Harrell complains that some 25% fewer customers have been making their way north to her nearly Cottontail Ranch. "Keep in mind that the bordello is at the bottom of the lift," she says philosophically. "Food and rent come first."

Time (1974)

The future's not what it used to be.

Kenneth Watt (1974)

NEW ANALYTICAL TOOLS are required to cope with the kind of political problems that have emerged in the past few years. There is an urgent need for policy analyses that are grounded in the fundamental imperatives of the human survival enterprise and informed by a sophisticated understanding of man as a biological species. This paper is addressed to both the normative and empirical implications of this challenge. A number of propositions are put forward in order to establish the "cornerstakes" of a biological paradigm, and five general implications are discussed. The normative implications of this paradigm involve a "survivalist" or "stewardship" ethic, which is advanced as a value-base for policy analyses within a biological framework. Ten concrete decision-rules are then proposed as potentially applicable to survival-oriented decision-making. Three specific modes of analysis are discussed:* a) *a form of* social indicators *oriented to measuring how well basic survival needs are being met (or "survival indicators");* b) forecasting, *with particular emphasis on basic, survival-related processes; and* c) *whole* systems analyses *focussed on the* interactions *between human populations, their natural and cultural environments, and the technological infrastructures that mediate these relationships.*

* The author wishes to thank W. J. M. Mackenzie, Kenneth Shepsle and Nannerl Keohane for their suggestions and insightful criticisms. If, in the end, I did not take all of their good advice, the fault is mine.

The era of abundance is over. Almost without realizing it, we have passed into an era of scarcity and find ourselves caught up in what can fairly be called the politics of survival. If we are to negotiate this unfamilar terrain successfully, we will need policy decisions which are both firmly grounded in the biological realities and boldly addressed to the kind of life and death problems we confront—and probably will confront for the foreseeable future. Since it is no longer possible to sustain the illusion that our survival can be taken for granted, we must learn how to subordinate our behavior to present and future survival needs. We must organize ourselves if we want to save ourselves.

As Maurice Strong, Secretary General of the UN Conference on the Human Environment, has expressed it: "Doomsday is possible—even probable—if we continue on our present course, but it is not inevitable. It is possible to opt for a future of unparalleled promise and opportunity for the human species. But this future can come about only if we make a radical change in our present course."[1]

In order to do this, however, we must develop a much clearer picture of where we stand with respect to the on-going survival problem and, more important, what course we should be steering to ensure our posterity. This is obviously not a mission for any one discipline or area of expertise; multi-disciplinary efforts will be essential. However, political scientists can make important contributions. Specifically, there is an urgent need for new analytical tools—both of a normative and an empirical nature—to assist policy-makers who are seeking to come to grips with our present predicament.

One could take a narrow view, of course, and seek to limit the scope of "biopolitics" to the sort of policy issues that impinge directly upon the work of life scientists or medial practitioners. There is, for instance, the problem of determining for legal purposes when death has occurred, or the problem of what social constraints should be imposed on the technical means that are now available for manipulating behavior, or the problem of where the burden of proof should lie for demonstrating the safety of a new medicine. These and other "biosocial" issues have arisen in large measure from the increasing technical virtuosity of modern science, and they are all worthy of sustained evaluation from various points of view, including ethical, economic and political. (Indeed, such evaluations are already being made under the auspices of the Institute of Society, Ethics, and the Life Sciences.) I would certainly not rule out consideration of such issues, but my purpose here is to advance a more expansive vision of what a biologically-oriented policy science might contribute to the policy-making process.

The Biological Paradigm and the Social Sciences

If there is any single, dominant theme in the biological realm, it is the problem of earning a living in the environment—the problem of survival and reproduction through time. Though different species adapt to their environments in vastly different ways, nonetheless the survival problem is the common denominator; it is the paramount problem shared by all species (and by all human societies for that matter) and it is utterly inescapable. It is a concommitant on life itself.

In the life sciences, the survival problem provides the basis for a unifying theoretical and analytical framework. Natural selection, the "prime mover" and master "cause" of biological evolution, refers to the outcomes of those organism-environment interactions which differentially affect, no matter how minutely, an organism's chances of survival and reproductive success. Natural selection selects preeminently for transgenerational survivability. Accordingly, every extant species has been shaped—and repeatedly tested—for adaptiveness (or functionality) in relation to the biological imperatives; the entire edifice of evolution represents, in essence, a history of contingent and only intermittently successful solutions to the survival problem.

To be sure, the pattern of adaptation for any given species may involve many interrelated aspects and many situation-specific factors. For most species, in fact, survival entails much more than merely avoiding catastrophes. The problem subsumes an array of biologically-based "needs" which must be satisfied more or less continuously (or in some cases seriatim) if survival is to be assured. "Survival" thus implies a complex enterprise. Furthermore, the survival of any "higher" organism requires successful adaptation at many different "levels" of biological organization, all of which are intensely interdependent. Processes at the molecular and cellular levels exert causal or feedback influences at the behavioral and social levels, and vice versa. Nevertheless, the problem set, or "paradigm" (to use Thomas Kuhn's term[2]) of modern biology is focused upon various dimensions of the survival problem—and the ways in which living organisms attempt to cope with it.

As recently as five years ago, it was still possible for social scientists to deny that the biological paradigm is equally applicable to humans and to human societies. For example, one prominent social scientist back in that smug and, by comparison, innocent era wrote: "Now that the problem of survival has been solved, the great unsolved problem of our time is self-actualization." Such myopia is hardly conceivable today. Though some scholars still question "the tightness of the evolutionary vice" for man,[3] there is now much more readiness to accept the notion that the biological paradigm is also an appropriate theoretical scaffolding for the social sciences.[4]

Some Basic Propositions

This emergent viewpoint has numerous theoretical and methodological ramifications,[5] but for the present purpose, perhaps it will suffice simply to adumbrate an abbreviated set of basic proportions (and a few amplifying statements) by way of establishing the corner stakes, so to speak, for a biologically-oriented policy science. These propositions include the following:

— Biological evolution, of which man is a part, is a unique and irreversible historical process, and social analysts must be prepared to come to terms with this aspect of their subject-matter.

— Though evolution is historical in nature, it is also cumulative: It is characterized by both continuities and changes through time, and the process cannot properly be described or understood without reference to both of these aspects.
— Behavior as well as morphology is survival-relevant. In the biological paradigm, there is no sharp boundary between the organismic and behavioral levels; the two levels interact.[6]
— Cultural evolution is a dimension of the overall evolutionary process and can also be accommodated within the biological paradigm; any observed trends in human evolution, furthermore, are strictly contingent and potentially maladaptive.
— As with all other species, the basic problem for Homo sapiens is survival and reproduction. The problem is also multi-faceted; many simultaneous needs exist.
— Homo sapiens, like many other higher mammals, is an interdependent social species: Man copes with his survival problems by and large through a division of labor and collaborative behaviors within a collectivity. Thus, a human society may be characterized as a "collective survival enterprise."
— Accordingly, the survival problem for man is at once an individual and a collective affair; the needs of individuals and of the collectivity overlap, but they are not identical with one another.
— Social processes may have quite different consequences for individuals and for groups (an extreme case is the sacrifice of an individual for the benefit of the group). There are, in fact, several possible "levels of analysis" within this paradigm including: *a)* the individual; *b)* various social collectivities (from proximate groups to the international "system"); *c)* the future needs of these collectivities. An analyst working within the biological paradigm may therefore evaluate social phenomena in terms of any one of these levels of analysis, or all of them (see below).
— Causation in the biological realm involves interactions between various levels of organization, and between organisms and their environments. The processes involved are goal-oriented, multivariate, interactive and nonlinear in nature.
— The organizational "architecture" of the biological realm consists of hierarchies of cybernetic systems—that is, goal-oriented, feedback controlled systems that interact with other systems (or sub-systems) at higher or lower organizational levels, as well as with their environments. The cybernetic model, furthermore, is equally appropriate at the molecular level and at the political level.[7]
— Human cultures and, by extension, human politics and policy processes are functionally (or dysfunctionally, as the case may be) related to the survival problem; social and political "systems" represent, at bottom, levels of biological organization. ("Functions" here refer to "designs for survival"—designs in the special sense of means-ends relationships that are produced by natural selection or by evolved adaptive mechanisms, including the mechanisms of learning.)

— The interplay between human "purposes" (or subjective "utilities") and survival functions is complex and often indirect. An adaptive process or artifact is frequently not consciously "intended" to serve survival needs. Conversely, human purposes, even when they are intended to satisfy survival needs, may lead to maladaptive outcomes.
— Politics, in the terms stated above, is the process by which human societies go about cybernating their behavior—the process by which man attempts to cope with "public" problems, make authoritative decisions for the group and organize and coordinate the behavior of his fellows. Another way of putting it is that politics involves the authoritative selection and implementation of society's collective survival strategies.[8]
— Because of the historical and political nature of human societies, both the past and the future (past experience and anticipated outcomes) have causal efficacy in social life. Indeed, anticipations of future outcomes are central to political decision-making (see below).
— Finally, the principle of "complementarity," which Heraclitus aptly characterized as the "harmony of opposites," is a fundamental property of the biological realm and an important rubric for policy science analysis. Reformulated for modern scientists by physicist Niels Bohr, the principle of complementarity concerns the paradox that many optimal relationships in nature involve a balancing of seeming opposites—innovation and replication, freedom and order, cooperation and competition, determinism and autonomy, individual and group, the universal and the context-specific. (There are both theoretical and empirical grounds for believing that such balancing acts often represent adaptive solutions to the problem of optimizing simultaneously for two or more survival values.)

Needless to say, almost every one of the propositions above represents an intellectual thicket; some are controversial and all could certainly benefit from being elaborated upon at greater length.[9] However, this is not the place to do so, except as necessary to develop further the policy science implications.

Some Implications

To summarize the above section, then, politics is viewed as a process that is vitally linked to the problem of survival. A biologically-oriented policy science must therefore be grounded in the survival imperatives. But, at the same time, different ecological, historical, and cultural contexts pose different configurations of problems for a human population. (Energy needs, for example, are partly a function of climatic conditions, and partly also a function of the cultural and economic milieu.) Indeed, survival is preeminently a problem that must be defined in terms of the *relationship* between organisms and their environments. It follows that a prime requisite for policy analyses is that they be context-sensitive. There can be no universally applicable solutions; a policy can be successful only under a more or less limited set of circumstances.[10]

By the same token, it is not possible to do just one thing (as Garrett Hardin puts it). Because humans inhabit an intricately interconnected biosocial system, there is no such thing as an "isolated" act. All actions by humans consume resources and impact in some manner on the biosphere, not to mention the social environment. Thus, one must be conscious of systemic effects.

A further implication is that policy analyses must be responsive to the configural, interactive and multi-leveled nature of social causation. To take one example, although "perceived relative deprivations" have been identified as a significant factor in determining the magnitude of civil strife, only a small portion of the variance can be explained with this variable.[11] What remains unaccounted for is why so few relatively deprived individuals and almost none of the absolutely deprived actually revolt. Why is it that most political violence is perpetrated by young upper-class or middle-class males? And why do so few women, regardless of culture, engage in political violence?[12] It is quite clear that other social variables, and very probably some biological variables, are also involved, and that these must be taken into account in making public policy.

The propositions above also direct our attention to human "needs." But this is obviously not a simple matter. Men have been debating the subject for centuries. Among modern social scientists, there are many competing points of view and very little consensus. One of many unresolved issues involves the question of how we distinguish between what any given individual wants and what he needs. And between what he says he needs and what solicitous outsiders—or authorities—think he needs. There may in fact be a substantial gap between perceived and actual needs, and between culturally-defined needs or "dependencies" and bedrock biological needs. In addition, there are the truly awesome problems of differentiating between individual and collective needs and between present and future needs. There will be no ready-made or certain answers to these problems, but it is precisely in this domain that a rigorously developed biological perspective may be able to make a substantial contribution (see below).

Finally, the "levels of analysis problem" and "complementarity principle" imply that a critically important set of analytical problems concerns value tradeoffs and multiple value optimizations. How does one go about choosing between individual needs and collective needs (or present and future needs) if a choice becomes necessary? And how does one decide which of several competing values to favor in a given situation? For instance, what environmental tradeoffs are permissible to obtain energy? And how much energy do we really need? There are in fact no *a priori* answers or all-purpose commandments. However, I shall use the biological perspective to suggest some guidelines—some decision-rules and some criteria with which it might be possible to commensurate seeming incommensurables.

In sum, biologically-oriented public policy must be: *a)* context sensitive; *b)* attentive to systemic effects; *c)* alert to the interactional nature of social causation; *d)* oriented as far as possible to objectively determinable survival "needs;" and *e)* focused on attempting to optimize for many survival-relevant values simultaneously.

THE NORMATIVE BASES OF A BIOLOGICALLY-ORIENTED POLICY SCIENCE

At the outset I suggested two overarching missions for a biologically-oriented policy science: *a)* clarification of the normative implications of a survival-oriented public policy; and *b)* development of analytical modalities—concrete techniques of policy-relevant analyses within a biological framework. No definitive resolution of these complex matters is possible here, of course, but the task I have set for myself in the present context is to outline some possible approaches to these problems.

The key to a survivalist ethic, in my view, involves a reformulation of the venerable and much-debated concept of the "General Welfare." To Plato, the general welfare was more or less equivalent to an harmonious and well-ordered community, one in which individual talents were fitted to necessary social tasks. Aristotle went a step further and added to the definition an aspiration for the "good life"—for the full realization of human potentialities. For both theorists, though, interdependence and synergy in terms of meeting the needs of individual members was the basic *raison d'être* of society. Yet, if it is true that they attached value to the community as a transgenerational collectivity, at no time did they seriously doubt the long-term viability of the human survival enterprise.

Modern capitalist societies, by contrast, derive their public values, and their view of the general welfare, from a nexus of political and economic ideas tracing back to Hobbes, Locke and the classical economists. Locke argued that civil society exists and can be justified only in terms of its contributions to the satisfaction of private utilities. "Property," he asserted, is "the great and chief end of men uniting into commonwealths."[13] The "social contract" which binds men together, then, is grounded in the reciprocal benefits it provides to individuals; individual "interests" therefore have moral priority.

To this normative perspective, the classical economists added the reassuring thesis of an "invisible hand" which was presumed to underlie the supply-demand dynamics of the economic market place. The presumption was that the free, competitive pursuit of self-interest by individuals would lead to the efficient satisfaction of "needs" and to a self-equilibrating society. Thus, the general welfare would emerge from the aggregate satisfaction of individual interests.

Edmund Burke thought otherwise:

> Men are qualified for civil liberty in exact proportion to their disposition to put moral chains upon their own appetites. Society cannot exist unless a controlling power upon will and appetite be placed somewhere, and the less of it there is within, the more there must be without. It is ordained in the eternal constitution of things, that men of intemperate minds cannot be free. Their passions forge their fetters.

Thomas Hobbes was even more skeptical of human nature. He argued that men "in the state of nature" (that is, unfettered by any external constraints on their behavior) would lust for "power after power" restrained only by their equally strong fear of death. Therefore, only a strong curb on individual appetites, in the form of absolute political authority, could be expected to maintain public order for very long.

Karl Marx, however, had more confidence in human nature. With the anticipated demise of Capitalism, the social and communal nature of man would be liberated, he believed. Socialist man would become community-regarding, and the economic life of society would be reconstituted so as to serve every individual's needs on an equitable basis. Genuine needs and not artificially stimulated wants would thus become the engine driving economic life. However, Marx was unclear about the details. There were also in Marx's writings implicit assumptions of unending economic abundance and ultimate socio-economic homeostasis, so that he foresaw no scarcities. Indeed, he did not even entertain the possibility of conflict between the needs of the present generation and of future generations.

The liberal economist's response to Marxism was to downplay the processes of Capitalism (and even to tolerate some modifications) and to concentrate on manipulating aggregate measures of the *outcomes* of economic activity, on the assumption that individual need satisfactions could be fairly well represented by Gross National Product, National Income and related measures. Presumably, then, growth in these measures of economic activity would mean improvements in the satisfaction of human needs.

The political concommitant of neo-classical and Keynesian economics is the political liberal's thesis of a self-regulating bargaining system among political interest groups. Theodore Lowi (who disapproves) calls it "interest group liberalism"[14] and Robert Dahl (who approves) call its "polyarchy."[15] To interest group liberals, the Public Interest or General Welfare evolves out of the interplay among organized interest groups as they freely pursue their seperate goals. Like Marx, liberals recognize the reality of organized interests, but they hold these interests to be inevitable—and inevitably multiplied by the complexities of modern economic systems. However, if the "countervailing power" of these groups is given free rein, a form of political homeostasis will result.[16] A network of bargains and compromises will be struck by the various interests in society. The maintenance of this self-regulating network, then, constitutes the public interest.

Attacks on economic and political liberalism have come from three sources—the political left, the environmentalists, and theoreticians in welfare economics and political economy. Political criticisms center on the unequal distribution of power among contending interests, the lack of organization (and thus representation in the bargaining system) for vast numbers of citizens, and the manifestly inequitable social and economic outcomes of interest-group liberalism; while the "wants" of the wealthy few may be satisfied, the clearcut "needs" of many others remain unmet. Nor do economic measures adequately account for human values. In the name of GNP, many other social values have been sacrificed, they argue.

An equally damaging line of attack comes from ecologists (and neo-Malthusians), who argue that many economic "goods" result in ecological "bads"—pollution of the environment, disruption and wanton destruction of delicately balanced eco-systems, and wasteful use of scarce resources. Furthermore, in a world of ultimate scarcities, the satisfaction of non-essential wants for some individuals can only occur at the expense of the genuine needs of others, or of future generations. As Barry Commoner points out, it may not be enough simply to "internalize" into the economy such "externalities" as pollution. The externalities may be so massive that internalization would make many economic activities totally unprofitable. But more to the point, there is the question of "whether the operational requirements of the private enterprise economic system are compatible with ecological imperatives."[17] In all too many cases, a high rate of profit is directly attributable to practices that damage the environment or use resources in a profligate manner. These practices cannot simply be "cleaned up." Yet if they are restricted, profits decline accordingly.

A third line of attack on liberalism comes from political economists. There are, first of all, "public goods," the benefits of which may be indivisible within the community, whereas the costs to provide the goods may be divisible. The fire department is an obvious example. As Mancur Olson has shown in his classic study, *The Logic of Collective Action,*[18] government is a necessary concommitant of the provision of public goods because of the "free rider" problem—the problem of coercing people to share the costs of indivisible benefits when self-interest alone won't motivate them to do so. The converse problem involves the prevention of collective "bads," a problem elucidated by Garrett Hardin in "The Tragedy of the Commons."[19] In this case, individuals may temporarily benefit from a practice that will have cumulatively deleterious effects on everyone. The frequent overgrazing of common pasture lands during the Middle Ages furnished the title for Hardin's essay, but the main case in point is overpopulation. In this situation, Hardin argues, mutual restraint is necessary to prevent a Malthusian disaster. And since short-term self-interest most often dictates a lack of restraint (current US birth rates are presumably an exception to the rule), only through mutual coercion can mutual restraint be achieved. In this class of situations, the free rider is not the one who avoids the costs, but the one who reaps benefits from activities that in the aggregate produce a collective malpractice.

Although these and other "economic" critiques have restored to the political dialogue some sense of a generalized "Public Interest"—and a corollary role for government as an instrument of the public interest—the resulting "political vision" (to use Sheldon Wolin's term), is still quite limited and circumscribed. It is still essentially a Hobbesian view of government as policeman and referee for certain rules of the game which reasonable men can see would advance their "enlightened self-interest."[20] This line of argument does not really address itself the to difficult but all-important question of how scarcity is to be allocated between generations. Should the substantive needs of future generations have priority over the wants, or perhaps even the needs, of the present generation in situations where the human survival enterprise may become, in effect, a transgenerational zero-sum game? The dilemma must be faced squarely. If the exigencies of transgenerational survival are not totally compatible with even the most far-seeing ethic of enlightened self-interest, then we must entertain the theoretical possibility of a transcendent general welfare—even if we cannot be certain in advance of its precise contours. We must admit, in other words, that there may be biologically-adaptive norms which cannot be derived from an individualistic frame of reference and which transcend the sum of all the individual welfares of the living members of a given society, or even of all contemporary societies in the aggregate.

Toward a Stewardship Model of Social Ethics

In case it needs to be reiterated, this is not the same as saying that individual self-interests or liberal values will *necessarily* conflict with the general welfare. Nor do I assert that most of us are motivated by anything beyond a more or less enlightened and socialized self-interest. As I conceive it, the general welfare may not even be an "operative norm," in the sense of actually influencing behavior in any significant way. If the needs of the species as a whole or of future generations conflict with operative norms, we may not be aware of it or care about it. I am not, in other words, asserting a theory of behavior but rather a theory of feedbacks and outcomes.

If we accept the argument that there may in principle be autonomous "species needs" which for want of a better term we can call the general welfare, the next question might be, so what? The existence of a general welfare does not *a fortiori* create an ethical imperative—as David Hume and whole legions of political theorists ever since have been at pains to point out.

Robert Heilbroner has stated the issue baldly:

> Will mankind survive? . . . Who cares? It is clear that most of us today do not care—or at least do not care enough. How many of us would be willing to give up some minor convenience—say, the use of aerosols—in the hope that this might extend the life of man on earth by a hundred years? Suppose that we knew with a high degree of certainty that humankind could not survive a thousand years unless we gave up our wasteful diet of meat, abandoned all pleasure driving, cut back on every use of energy that was not essential to the maintenance of a bare minimum. Would we care enough for posterity to pay the price of its survival? . . . I doubt it.[21]

Of course, many of us exhibit concern for at least the near future when we invest in our children and grandchildren, buy life insurance, endow university chairs and so forth. A few of us make far greater sacrifices for an even more distant future, but this is not the rule.

In short, we cannot depend upon enlightened self-interest or a spontaneous concern for the future to ensure the long term survival of the species. Appeals grounded in "reason" are not likely to be persuasive. If we are to establish a basis for policy-making in the interest of posterity, we must base our appeal on some other ground than the logic of liberalism and "rationality."

The only alternative, in my view, is to try to awaken in both policy-makers and the public a sense of *responsibility* for the future of the species—for the general welfare. This might be done in two ways. One is to recapture an awareness of our debt to the past. Edmund Burke probably overstated it when he declared that "people will not look forward to posterity who never look backward to their ancestors," but we might do well to revive Burke's view of society as "a partnership, not only between those who are living, but between those who are dead, those who are living, and those who are to be born."

Whether we like it or not, each of us is in effect a steward of the human species—a bearer of both biological and cultural legacies that have been passed down to us against long odds by ancestors who trace back beyond our capacity to imagine. It is humbling to think that none of us would be here were we not the beneficiaries of an unbroken chain of life, including ancestral species who were not at all like Homo sapiens but who, nonetheless, fought the good fight and prevailed. If we can recover a sense of the continuity (and precariousness) of life, and a sense of our role as a link between the past and the future, then perhaps we might be more ready to accept responsibility for the future. We might then take to heart the dictum of John Buchan (Lord Tweedsmuir) that "we can only pay our debt to the past by putting the future in debt to ourselves."

The second way of trying to awaken a sense of responsibility for the future is to document in a convincing manner the linkages between our choices (either as individuals or as polities) and the fate of the species. If it can in fact be shown "with a high degree of certainty," as Heilbroner puts it, that our actions will measurably affect the survival chances of future generations, then it will be far more difficult to avoid a *subjective feeling* of obligation. This is precisely the contribution that can be made by the proper use of such analytical tools as future forecasting, societal modeling and social indicators, about which more will be said below.

Will such efforts succeed in eliciting the appropriate social and political responses? Nobody knows. The more relevant question is "do those who care about the future of the species consider it worth while to try to persuade others?"

As I indicated above, an ethical framework that is concerned with maximizing the long-term survival chances of the human species as a corporate entity—what might be called the "stewardship model" of social ethics—need not exclude liberal values, but it would not be limited to them, bound by them. It was noted that survival is a multi-faceted and multi-leveled process, and that the survival needs of individuals, societies and the species as a whole are partially independent but not mutually exclusive. One way of illustrating this relationship is through the use of the familiar Venn Diagram, in which the needs of individuals, social groups or polities, the human species in its entirety, and future generations can be circumscribed as overlapping sets.

As Figure 1 suggests, many of the needs (or wants) of any given individual might be consonant with the needs of more inclusive social entities. In some instances, though, individual needs might be conflict with "higher level" needs. By the same token, the needs of any given society might or might not conflict with those of other societies—depending.

The survivalist ideal, then, would consist of a set of social and political arrangements in which there is a perfect mesh between each of these sets—a situation in which the "interests" of individuals, societies, the global community, and future generations were in total harmony. In such a biological Utopia, the satisfaction of individual needs would invariably

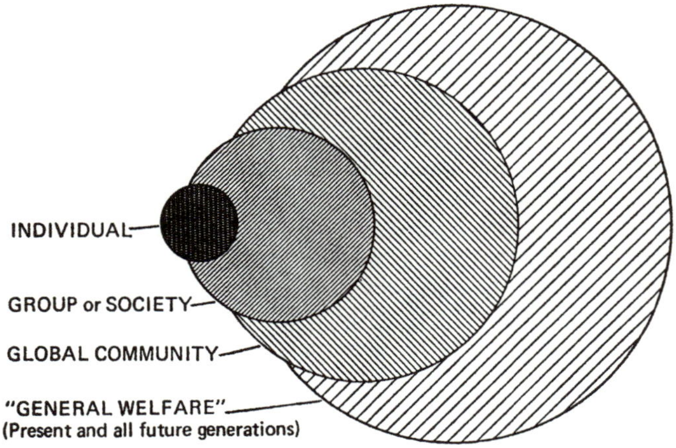

FIG. 1. *Relationship between needs at different "levels of analysis"*

also serve the needs of posterity—or the general welfare. Since this Platonic situation is not likely ever to exist, we shall have to consider some ethical directives for the second best of all possible worlds. We shall have to play Aristotle and suggest some possible ways of trying to approximate the general welfare in the real world, when hard choices are necessary and must be made under conditions of extreme uncertainty.

Some Normative Implications

What, then, are the normative implications of a stewardship ethic? Among the more important are the following:

- The primary value, or social goal, should be the survival of the human species for as long as possible and, only secondarily, as well as possible. A focus on long-term survival need not be antithetical to human aspirations and cultural achievements, but survival is viewed as the *sine qua non* for such other attainments. Individual and societal self-fulfillment may or may not be instrumental to that objective, but fulfillment is certainly not viewed as an end in itself. Freedom, and even democracy, justify themselves within this value system only insofar as they are adaptive. (However, the free flow of ideas and freedom to pollute drinking water are quite obviously different matters.)
- "Progress" as conventionally defined is not necessarily desirable; indeed, some conventional forms of progress may even be antithetical to long-term survival. Progress in *this* framework refers to improvements in the long-range survival chances of the species (or any of its members, insofar as the improvements are Pareto-optimizing—that is, the improvements should not be made at the *expense* of others).

— Survival *needs,* and not "wants", are the basic concern of a stewardship ethic.
— The complementarity principle and the context-specific nature of the survival problem suggests extreme caution about absolutism of any kind; neither absolute egalitarianism nor unfettered individualism are necessarily adaptive. (To take one obvious example, a completely equal distribution of the Earth's known energy resources among all of the Earth's inhabitants would in fact be inequitable in terms of the survival needs of peoples in different latitudes.) Similarly, competition, struggle and change may be beneficial in some circumstances but are not always and everywhere adaptive. (The "tooth and claw" struggle of Social Darwinism was a gross and self-serving caricature.) Stability, security, and altruism also have their uses.
— Contrary to the views of Social Darwinists and Marxists, political systems are not ephemeral artifacts of recent historical vintage that will in due course wither away. It may be that the Capitalist state will wither away, but the biocybernetic model of politics views political organization and collective decision-making as an integral part of the human survival enterprise, with behavioral precursors that in fact predate the emergence of man. Indeed, government at its best can serve as probably the most potent instrumentality for positive action on behalf of the general welfare.
— A stewardship ethic does not take an *a priori* position in favor of either the status quo or radical change—or anything in between. The survival implications of continuity and change depend upon the context. Nor are particular institutional arrangements—for example, government ownership of the means of production—necessarily called for, or ruled out. Institutional arrangements in all cases should be judged by their performance in meeting human needs. We are not discussing here, in other words, a concrete program for action but a broad set of guidelines for decision-making and a program of empirical research.
— Finally, a stewardship ethic should not be identified with the "interests" of any particular political unit, or stage of "development." At one extreme, many small band-level societies are essentially autonomous units in terms of meeting the survival needs of their members. At the other extreme, many modern nation-states are highly interdependent, despite their nominally "sovereign" status. Neither political unit is necessarily more adaptive, or in a survivalist sense "morally" superior.

The General Welfare Versus National Socialism, '1984', etc.

It is understandable that those who remember the atrocities of the 1930s or who have lived under a totalitarian government may be disturbed at the assertion of any "higher good" which should be allowed to take priority over all other social values. Isn't that what Hitler did with such destructive effect? And isn't that how Big Brother was justified in *1984*? It is, after all, deeply offensive to the Western liberal tradition to have collectivist values stated so baldly. Nor does it help matters to point out that, despite their professed values, even the most freedom-loving and democratic societies have always hedged liberty with actions in the "public interest." Perhaps there is a fear that, to concede openly a superordinate public interest, or general welfare, may be tantamount to condoning governmental abuses of power; even if we don't live up to liberal ideals, shouldn't we continue to assert them vigorously, so as to deter Big Brotherism? Some may argue that it is perhaps better not to expose and make us own up to those adaptive little hypocrisies we have worked out over time. Maybe it would be better to leave well enough alone.

Unfortunately, I cannot agree. The very murkiness of the relationship between our professed values and our actions is part of the problem, in my judgment. Clarifying and asserting our true values is an important prerequisite to more adaptive behavior. Furthermore, the danger of encouraging tyranny in the name of the general welfare is less real than apparent, if one reads carefully what was written above. No one could seriously construe my argument as a warrant to ride roughshod over individual values. The key, again, lies in an understanding of the implications of the complementarity principle and what it suggests about the complex nature of the survival problem. The idea of a dynamic balance between seemingly contradictory values is not easily reconciled with absolutism.

Perhaps the most important counter-argument, though, is that nothing said above could be interpreted as an attempt to seduce us into believing that any particular political leader is acting in accordance with the general welfare just because he says he is. Vigilance to prevent abuses of political power and a healthy skepticism regarding the behavior of men in authority are if anything encouraged.

Some Algorithms and Decision-Rules

Keeping in mind the caveats expressed above (and recognizing how limited human capacities are for understanding complex biosocial process and predicting future outcomes), nonetheless, it would seem desirable to make a start on trying to incorporate into the policy-making process calculations relating to a larger compass and a longer time-frame. Systematic attention to future needs, future costs and future benefits beyond the limits of conventional economic and social planning—a sort of "General Welfare Impact Analysis"—might well be attempted in a more sustained way, particularly now that multi-variate, interactive computer models are available for analyzing complex dynamic processes over long periods of time.

We shall have more to say on this subject below, but first I should like to consider whether it might be possible to formulate some guidelines, some algorithms or decision-rules which, if followed, would encourage decision-making for the general welfare? Though I feel strongly such an effort should be made, I am sufficiently awed by the task to disclaim in advance any hope of producing something definitive. What I suggest below, therefore, should be treated as a first trial—as a basis for discussion and, hopefully a stimulus for further research by other investigators. Herewith are some possibilities:

1. To the maximum feasible extent, seek to harmonize and satisfy the "needs" (as distinct from "wants") of all the "subsets" identified in Figure 1 above—all the members of the present and future generations. (Given the present world situation, this clearly implies a more equitable distribution of basic resources.)
2. Structure economic and social incentives as far as possible so as to encourage the widest possible satisfaction of needs and discourage non-essential consumption, consonant with the rules enumerated below.
3. Where constraints exist and tradeoffs are required, seek to optimize for the needs of the most inclusive unit (for the largest number over the longest period of time).
4. To the extent that either present or anticipated future needs remain unfulfilled, needs should take precedence over wants. More particularly, the needs of future generations should take precedence over the wants of the present generation.
5. Where critical renewable resources are concerned, avoid exceeding "sustained yield" exploitation.
6. Where critical non-renewable resources are concerned, seek to eliminate consumption unrelated to needs (or to developing replacements). Urgent priority should be given to the development of replacements for any critical resource that is consumed, so as to prevent per capita depletion of resource stocks, if at all possible.
7. As far as possible, seek to minimize future costs and risks and maximize future benefits before exploiting present opportunities. Opportunities that involve benefits only for the present but probable costs for future generations should be strongly discouraged.
8. Where there is serious uncertainty regarding deleterious future impacts, proceed on the basis of conservative, or "worst-case" assumptions, even if increased present costs must be shouldered in order to minimize future risks. (A prime case in point, in my judgment, is the current debate over nuclear power development, as against such longer-term, higher-cost options as solar power and geothermal power.)
9. There have been times in the past and there may be times in the future when absolute scarcities occur. In such "zero-sum" situations, needs-satisfaction for some individuals (or collectivities) means depriving others of their needs. Hard choices are inescapable under these circumstances. A survivalist decision-rule in such situations would involve doing whatever would stand the best chance of furthering the long-run survival enterprise—the general welfare. Such dilemmas may seem improbable in an historically affluent society, yet there are many historical precedents and contemporary examples, including some in our own society.[22] Needless to say, one hopes never to be confronted with such terrible choices,

and there are no Platonic solutions to the question of who should make such decisions. One can only hope that they will be made, if they must be, with wisdom and compassion.

10. In crisis situations where resources are insufficient to meet everybody's needs, the technique of "triage" could be applied.[23] If it is possible to differentiate between: *a)* those who cannot survive, even with outside help; *b)* those who can probably squeak through without help, even though they may be deprived and suffering; and *c)* those who can survive *only* if help is provived, aid priorities might follow in the reverse order. That is, proffer aid first to those for whom assistance would mean the difference between life and death, then to the "walking wounded" for whom aid would merely relieve their distress, and lastly to those who can only be comforted but cannot be saved.

The above rules are obviously not exhaustive and are not meant to be carved in stone. It is assumed that critical reactions and further analysis will lead to significant revisions. The purpose, rather, is to initiate a dialogue on the normative implications of a survival-oriented political ethic. If we define the general welfare as the survival of the human enterprise for as *long* as possible, and only secondarily as *well* as possible, what should be our order of priorities ? To the question "who should survive and under what qualitative conditions ?" the answer suggested here is that it will depend upon the particular context. Though we might wish for the "good life"—for the survival of everyone in optimal economic and social circumstances—this may well be an unattainable goal. We must face that possibility and cope with it intelligently if we are to avert apocalyptic outcomes.[24]

Even the most soundly drafted set of normative guidelines is of little use, however, without concrete measuring rods—some means of determining just how well we are in fact doing with respect to the on-going survival problem and, more important, whether the survival strategies we are pursuing are likely to ensure our continued viability into the foreseeable future. A necessary corollary of a stewardship ethic, therefore, must be an effort to develop appropriate analytical tools. Admittedly, this objective involves some formidable problems and, realistically, some unfathomable uncertainties. But there is every reason to believe that we can, with sustained effort, substantially improve our capabilities in this regard. Accordingly, I would like to outline here some approaches being explored by a group at Stanford.

On Measuring Survival

One of the most productive insights of modern evolutionary biology concerns the fact that information about the environment has been of fundamental importance to the evolution and perpetuation of life. Whether biologically-relevant information is coded in the genes or acquired during the lifetime of an animal (or both), information is the key to survival. Thus, what the modern social scientist dignifies with the term "social intelligence"[25] is neither of recent vintage nor unique to man. Indeed, from Stonehenge to *The Farmers' Almanac* to weather satellites, the human record has been marked by the quest for useful (i.e., predictive) knowledge. A complex society could not operate without it. As Otis Dudley Duncan notes:

> It is a banal observation that many kinds of social forecasting are routinely carried on in an implicit way. The very possibility of social organization rests on our ability to anticipate correctly, in a large proportion of instances, the behavior of others. Society, from this point of view, is an intricate reticulation of expectations and commitments concerning future actions and reactions.[26]

And yet, despite the veritable cornucopia of social data now available to policy analysts and decision-makers, it is abundantly clear that we do not have enough of the right kinds of data. The "social indicators" movement, for instance, arose out of a disatisfaction with our traditional reliance on economic measures of societal well-being, or on unsystematic social data collected primarily for bureaucratic or political purposes. By the same token, the "environmental crisis" thrust forward a whole new set of social problems, about which there was at the outset an extraordinary paucity of data.

Perhaps the single most important source of our informational malaise is the growing uncertainty about the human prospect. As long as human survival was taken for granted, and as long as growth and material abundance were taken as an article of faith, predictions about the future for the most part took the form of optimistic extrapolations of existing trends. Indeed, under such assumptions, the more resources consumed the better. But once we began to entertain the alternative proposition that there might be both limits to environmental tolerance and ultimate resource scarcities, the implications for the future became radically different. In the latter case, the more resources that are consumed the worse off a society might be. Continued growth and maximum resource exploitation might be highly *un*desirable in terms of the needs of future generations. For this reason, the need to know more about the future (or, more specifically, about the future ability of the planet to support industrial man) has become a high priority concern.

As suggested above, there are several desiderata for survival-oriented policy analyses: they must be grounded in the basic survival "needs," sensitive to the environmental context, capable of handling patterns of causation that are non-linear, interactive, configural and multi-leveled, and attuned to the systemic and multi-valued nature of human societies. To that end, three closely interlinked modes of analysis are, in our judgment, required:

1. A form of *social indicators* oriented to measuring how well basic survival needs are being met (these we are calling "survival indicators");
2. *Forecasts,* or rigorously disciplined speculations, about probable trends and future problems in basic, survival-related social processes;
3. Holistic, or whole *systems analyses* (some of which would also involve forecasting) focused on the basic interactions between human populations and their environments, resource bases and technologies.

Although none of these modes of analysis is entirely new, their full potential is very far from being realized. Not only will they require extensive efforts at further development but they will in most cases need to be better focused in terms of the survivalist—on stewardship—ethic outlined above. This is the objective of the work underway at Stanford, which I will describe briefly.

Survival Indicators

The "Survival Indicators" project at Stanford involves a variation on the "social indicators" concept.[27] An "indicator" in the context of social intelligence refers to facts about values: Data concerning behavior or social processes that affect social values or goals. Indicators are concerned with change. They provide selected and often capsule information relevant to such questions as: Are we getting healthier? Is crime increasing? Is air pollution better or worse this year than last? Etc. As the report of a government panel on social indicators expressed it:

> A social indicator, as the term is used here, may be defined to be a statistic of direct normative interest which facilitates concise, comprehensive and balanced judgments about the condition of major aspects of a society. It is in all cases a direct measure of welfare and is subject to the interpretation that, if it changes in the "right" direction, while other things remain equal, things have gotten better, or people are "better off."[28]

There is nothing new about the use of indicators, of course. We can trace far back into history such commonplace environmental "indicators" as the barometer and the miner's canary, or medical indicators like blood pressure and pulse. In the social sciences, economists were the first to make systematic use of the indicators concept, beginning in the 1930s, and until quite recently were almost alone in the indicators field. In the past several years, however, the growing discordance between economic "progress" and perceptions of societal well-being has given rise to the so-called social indicators movement. As the government's social indicators panel explained: "Just as we need to measure our income, so we need 'social indicators,' or measures of other dimensions of our welfare, to get an idea how well off we really are."[29]

Although the social indicators movement has succeeded in stimulating many new ideas, and creating a more holistic approach to measuring social welfare, in our view there are still serious shortcomings. The social indicators work has no coherent value base, and there is no substantive integration. If there is any normative focus at all, it remains conventional and middle-class. Indeed, the fundamentals are mostly taken for granted, and that is our major objection.

By contrast, the survival indicators project is focused primarily on the fundamentals. The objective is to try to measure as rigorously as possible the basic survival and reproductive needs of a human population; we are interested in establishing criteria and measuring rods for determining whether or not the basic survival minima are being satisfied.[30]

Our approach involves trying to disaggregate the survival problem into a number of concrete components (or "needs") which, at least in principle, can be differentiated from *perceived* needs, wants, or dependencies. *We use the term need in the biological sense of a requisite for the continued functioning of an organism in a given environmental context; that is, denial of the posited need would significantly reduce the organism's ability to function and/or reduce the statistical probability of its continued survival and successful reproduction.* We maintain, furthermore, that a theoretical distinction is useful for heuristic purposes between the level of needs-satisfaction required for: *a)* minimal life support; *b)* minimal ability to sustain transgenerational continuity (meaning both successful reproduction and nurturance of the young during the maturational process); and *c)* optimal life support (meaning maximally efficient functioning and the maximizing of survival and reproduction capabilities).

Our work, then, is focused upon eleven key variables which, as a first approximation, appear to encompass the basic survival needs. These variables include: *a)* food; *b)* water; *c)* energy; *d)* basic raw materials; *e)* shelter (including housing and clothing); *f)* technology (the knowledge, skills and organization required to satisfy primary biological needs); *g)* environmental quality; *h)* health; *i)* physical security; *j)* reproduction (including nurturance of the young); and *k)* certain postulated social-psychological needs. Some of these we view as "input" needs. Others are considered "output" needs. But all involve organism-environment interactions.

Although it is not possible to engage in a detailed discussion of this research here,[31] it should be noted that we find systematic variation among needs in three dimensions: biological, ecological and (depending upon how one defines "needs") cultural. Biological differences between individuals based on such factors as age, sex, and even genotypic differences represent one major source of variance. For instance, very young children and the elderly require very much less food, in general, than do young, active adults. A pregnant or nursing woman, conversely, requires twice the protein intake of a woman who is not in the reproductive mode.

Another source of variation among survival needs arises from differences in climate, topography and other ecosystem characteristics. To take one example, the problem of climatic adaptation is radically different for arctic and equatorial populations. Among the former, the basic problem is heat conservation; among the latter, the problem is heat dissipation.

Finally, there are the variations in needs arising from the cultural milieu in which an individual is embedded. We prefer to use the label "dependencies" with reference to such needs, since they are not in fact elemental and are subject always to further cultural modification. An example is a transportation system. In many primitive societies, transportation on foot is all the mobility any individual requires. In a large and highly differentiated society, on the other hand, an individual might have to be able to traverse long distances relatively rapidly in order to sustain himself. The particular mode of transportation might be highly variable, of course, and the degree of dependency on such a system might be greatly altered or even eliminated over some period of time. But at any given moment in time, such cultural dependencies might be every bit as imperative to survival as are the bedrock biological needs.

Can such needs be established empirically? Though we cannot know for certain without sustained research, it is clear even now that a surprising amount of data bearing on this question already exist. Unfortunately, the data are widely dispersed among various disciplines—from ethology to physiology. Often the data were not generated with this sort of question in mind, and doubtless we will conclude that there are many gaps and uncertainties. We don't expect to be able to produce any more than a partial synthesis or first approximation—and a shopping list of researchable questions—at least for some time.

On the other hand, we hope to be able to make a creditable case for the underlying validity of this approach. In opposition to relativist or crude behaviorist models of man, we hope to lend support to a complex bio-cybernetic model in which input needs and output needs are organized in conformance with the characteristics of a hierarchical control system.

We are hopeful, furthermore, that we will be able to validate at least some of our postulated needs well enough to provide the kind of measuring rods we seek. Once this has been accomplished, it is a fairly straightforward matter to apply these criteria to individual cases (or to aggregations of individuals) and determine whether or not a particular need is in fact being satisfied in a specific context. In point of fact, models (after a fashion) for this sort of analysis already exist with respect to nutritional requirements, health and housing. We believe the same approach can be applied to other human needs.

However, standards for evaluating whether or not basic needs are being satisfied is only one side of the coin. The other side involves analyses of the extent to which any given individual, or population of individuals, is consuming *more* resources than are needed for minimal, or even optimal, survival. What, in other words, might be the margin of profit—or the margin of overconsumption (or perhaps even overproduction, in the case of "outputs")?

In either event, the existence of such a set of survival indicators would give direction and focus to our social monitoring activities; we would be able to track the current status of the survival enterprise with far greater precision and sensitivity than is now the case.

In addition, once such criteria have been established, they can be applied to future-oriented research. Whatever assumptions we care to make about basic demographic trends (as the independent variable), certain aggregate levels of needs would become the dependent variables. We would have, then, an alternative means of evaluating the consequences of population trends apart from merely projecting forward existing per capita consumption rates, or even growth in per capita consumption rates. In other words, we would have reasonably objective standards for *survival minima* at some specified future date—what people at that time will need, as distinct from what they may want or *think* they need. Such minimal needs would, of course, have to be set against the existing pattern of cultural dependencies and consumption trends. But at least some non-subjective, non-culture-bound measuring rods would also be available to researchers.

What we could *not* do with survival indicators is answer the obvious next questions: What is the likelihood that the requisite resources and environmental conditions will be available to satisfy anticipated future needs, and what steps might have to be taken to ensure the needs of posterity—the general welfare. We would not be able to ascertain, for example, whether the stocks required to satisfy a given need are increasing or decreasing over time. Indeed, we would have no way of knowing from such indicators precisely what population levels to anticipate in the future. In short, survival indicators would not tell us very much about the dynamics of survival as an inter-temporal process.

Forecasting in a Survivalist Mode

An analytical complement to survival indicators, therefore, involves the use of forecasting techniques. Forecasting—or sometimes "futures research"—involves a process of imaginative but disciplined speculation about probable, possible, desirable, or even hypothetical future states of some variables, or set of variables.[32]

As a social act, forecasting is ubiquitous, as Duncan noted (above). Psychologists confirm that anticipated outcomes represent a major causal variable in shaping human behavior. (Indeed, behavior involves a process of reciprocal causation, in which anticipations or forecasts modify behavior, which modifies outcomes, which may in turn modify anticipations.) It is precisely because the human environment and the human future is partially susceptible to human control that we worry about it, "weigh options" and make "decisions." As Davis Bobrow observes:

> Through forecasts we seek the opportunity to make our personal or collective future different from what it might otherwise be. We seek to determine or mitigate the consequences of situations yet to come and to plan effectively to achieve wanted states of affairs.[33]

In recent years, forecasting has become an increasingly important area of research among professional social scientists and policy analysts. As Theodore Gordon explains:

> Policies steer us through a maze of interlocking possibilities. Futures research is a means of discovering and articulating the more important of the alternative futures and estimating the trajectory likely to be produced by the contemplated policies. Thus, forecasting is perceived to be an aid to decision-making in the present. . .[34]

As an emergent discipline, forecasting might be said to be guided by several rubrics, or touchstones:

— *Candor about personal value preferences*—that is, explicit aknowledgement of one's own values. This is viewed as preferable to trying to appear value-free.

— *Maintenance of the spirit, and research morality, of science*—empiricism in one's investigations, rigorous and systematic treatment of the subject-matter, honesty in gathering, using and reporting data, and openness to inspection, or replication, of research.

— *Common sense*—since so much in forecasting is ultimately judgmental, sound results require good sense as well as good methodology.

— *Intuition*—just as great scientific achievements involve intensely creative acts, so the art of anticipation depends heavily upon human imagination.

Several implicit assumptions underlie policy-oriented forecasting: *a)* that the aggregate or collective behavior of social entities is partially controllable through the political sphere; *b)* that significant social changes are required or highly desirable; *c)* that the complexities involved in bringing about controlled social change require careful analysis; *d)* that long lead-times are required to make major changes in complex social systems, even when "revolutionary" *political* changes clear the way; and *e)* that early intervention can greatly facilitate change and minimize disruptive effects.

A case in point is the current debate over the leasing of offshore oil reserves for oil company exploitation. The question is "should the US government encourage the most rapid possible development of offshore reserves?" Setting aside environmentalist objections, the intuitive response might be yes, of course, in order to ease our dependence on oil imports.

However, the Senate Commerce Committee's National Ocean Policy Study (NOPS)[35] concluded that, depending upon which forecast regarding the exhaustion of domestic reserves one chooses to accept, full-bore development of offshore oil might be a very *un*wise policy. If M. King Hubbert's low estimate were used, our reserves would be exhausted by 2008. This would probably not provide sufficient time to develop and bring on line a satisfactory substitute for fossil fuels. On the other hand, if the US Geological Survey's high estimate were used, our oil could be expected to last until 2086, which would provide several times as much lead-time.

In the latter case, all-out exploitation of offshore oil would probably not jeopardize the process of adaptation to a world without fossil fuels (assuming that suitable alternatives will be developed at all). However, if the low estimate is more accurate, then strenuous conservation measures are called for in order to stretch out our reserves. In this case, it would be wiser to proceed with offshore drilling at a much slower and more deliberate pace. (Of course, if we add to the decision calculus the still-great potential of environmental damage from offshore oil drilling, the case for slower development becomes even stronger.) But there are so many unknowns in the situation that a definitive choice is not easy to make.

Here in a nutshell, then, is an example of both the problems and the potential value to policy-makers of sophisticated forecasting.

From a survivalist perspective, though, such forecasting efforts should be focused on the kinds of normative criteria elucidated above—on the "needs" relating to survival. These needs provide minimal objectives for future-oriented planning, and standards against which to evaluate current trends.

To cite one example, a critical assumption made by the Meadows team in the notorious *Limits of Growth* volume[36] was that, at a certain threshold level of pollution, a rapid increase in the death-rate would occur; in some of the Meadow's alternative "scenarios," pollution was posited as the ultimate cause of world-wide "collapse." Apart from the obvious criticisms, such as the fact that many forms of pollution are concentrated locally and are not uniformly distributed in the biosphere, the most telling point is that there are only fragmentary and inconclusive data available regarding the actual relationship between pollution and human mortality. The issue is of critical importance, yet we do not at present have measuring rods, or indicators, relating to potentially hazardous pollutants, such that we can assess the survival significance of currently measured levels and time-series trends.

Systems Analyses

Although systems analysis probably needs no introduction, its application specifically to what we have called "general welfare impact analyses" requires some explanation. The rationale for whole systems modeling stems from the cardinal assumption that many of the phenomena with which human decision-makers must deal cannot be treated in isolation; they are in fact parts of larger systems—networks of processes which may interact with one another in complex ways. Often, both the pattern of *causation* and the *outcomes* of any "point events" are diffuse and configural in nature, with many non-linear relationships. Not only must these complex webs of cause and effect be studied systemically in order to be fully understood, but in the case of social processes the outcomes impinge upon many human values and goals simultaneously. As suggested above, a given policy decision may involve tradeoffs and problems of optimizing for several values at once.

Systems analysis techniques were developed precisely for the purpose of helping us to understand such processes. Systems analysis is obviously not a panacea. Nor is it a substitute for intelligence and good judgment. Rather, it is an extension—an amplification—of human analytical abilities.

How, then, might systems analysis be applied to survivalist concerns? The objective would be to construct models that are built around the values—and the measuring rods—related to the survival indicators discussed above.

The *Limits to Growth* study was a crude first attempt in this direction. Our preference, and a goal toward which we are working, involves a "system" whose boundaries correspond to the "level" or levels, at which political decisions are *actually* made and implemented.

To be precise, we are interested in starting with a nation-state model and then using this model as a basis for constructing a multi-leveled global system from the ground up, so to speak.

Models applicable to the nation-state level of analysis already exist, of course. There are both econometric models employing differential equations and systems dynamics models utilizing simultaneous equations. However, there are many technical problems with these models. The state of the art is still primitive; many of the underlying causal relationships are at present poorly understood; and there are many gaps in the data bases required to model the real world.

But most important, we do not at present have rigorously developed measuring rods for survivalist values. As we move forward with the survival indicators project, however, we will at the same time be attempting to develop just such criteria. Thus, the survival indicators project and whole systems modeling efforts are ineluctably linked to one another.

What could such models be used for? First, if the models are constructed more inductively than has been the practise all too often in the past, the very process of building the model, though much slower, will impel research and analyses that could shed light on causal linkages, feedback relationships, times delays, etc. Early efforts may well precipitate research efforts that would only be fedback into follow-on versions of the model several years later. This is not, clearly, a short-term venture.

Second, if threshold values for survival needs can be specified with some precision, the tradeoffs involved in policy decisions could be analyzed from a survivalist perspective. We would be in a position to model precisely those systemic effects on survival values referred to above. Systemic effects of changes either in the input quantities (or initializations), or change in the underlying relationships within the system itself could be analyzed extensively.

Thus, such models are obviously applicable to forecasting efforts. Time series analyses for whole systems, versus single sectors, could be done from a more solid base and from the perspective of the level where political decisions are actually made.

A further use for such models would be to explore systematically the relationship between the existing, real-world system and some hypothetical alternative system—such as, for example, one which would entail equal access to the world's resource stocks, as proposed by Robert C. North.

A multi-leveled version of the model would provide a tool for analyzing the interactions between any given political decision-making level and other levels. One could explore the consequences for the "international" system of decisions at the national level, or *vice versa*.

Similarly, multi-national interactions at the same "level of analysis" are highly feasible. Recent work in international conflict analysis clearly points us in this direction with respect to survivalist models.[37]

Finally, and in some respects most urgently, the analytical tools envisioned here would enable us not only to observe possible consequences of fairly obvious trends, but we could make systematic analyses of various possible "unforeseen" disasters—a change in the climate for the worse (which some meteorologists in fact predict), a pandemic, a severe blight hitting basic grain stocks, or (an event which is fairly certain but nonetheless "unforeseen"), the loss of nitrogen fertilizers when oil stocks are gone, within perhaps fifty years. Indeed, what are the implications for the food-population relationship of the loss of artificial fertilizers? Can the soil nutrients be replaced by human wastes or other alternatives, and at what cost in terms of capital investments? And what are the health risks involved? And what if presently foreseeable alternatives cannot make up the difference? How much of a food shortfall might there be in terms of population levels projected for 2025?

Again, these tools would not be able to predict outcomes, but they would stretch man's understanding of the systems in which he is embedded and the significance of the trends he observes.

Toward the General Welfare

Of course, the forms of analysis described here can only reinforce and lend greater precision to what our own internal computers are already telling us. The human species is in serious trouble, and powerful forces are corroding whatever measure of the "good life" we may have achieved in some areas of the globe. No one, it is now clear, is immune from this corrosive process. Man is threatened by his very success and is locked into unstable and even dangerous modes of adaptation. The cost of extricating ourselves from these difficulties will be high, and before we are through we may pay a terrible price in terms of human suffering. Heroic measures to deal with our problems are still possible, but political leadership on a scale that does not presently exist will be required to steer us through the narrows that lie ahead. Hard measures will be easier to take, however, if we have a clear sense of what our objectives are and some confidence that the steps we are taking are commensurate with those objectives. Though it is no small task, it is certainly worthy of our best efforts.

Afterword

This is clearly not the forum for a detailed critique of Rawls' *A Theory of Justice*. But, in view of its current prominence among political theorists and political economists, certain "biopolitical" criticisms seem warranted.

Rawls may not like to think of himself as being in the liberal, social contract tradition, but the normative focus of *A Theory of Justice* is cut from the same cloth. Rawls seeks to win assent for certain principles of fairness, or "justice" which enlightened men might, if they all started from the same hypothetical "original position" specified by Rawls, accept as being advantageous to themselves as rules of the game for economic and social life. Rawls does not ask us to overthrow self-interest as a basis for social ethics, but only to start from some unrealistic premises which have the effect of rigging the outcome in advance. We are asked to imagine that all parties to the "social contract" start from a position of equality and rationality (in the sense of being able to fit means to ends) but behind a "veil of ignorance" about what their precise interests and stations in life will be. Since they do not know what their exact interests will be, they will presumably choose rules that would maximize their chances of realizing whatever interests they may in the end turn out to have. This leads Rawls to postulate two fundamental rules: *a) the equal liberty rule,* which guarantees every individual the maximum of freedom compatible with similar liberty for others; *b) the just distribution,* which would arrange social and economic inequalities so that it would be even to the benefit of the least advantaged to participate and would provide everyone with a fair and equal opportunity to strive for various offices, goals and benefits. At the risk of oversimplifying a complex argument, Rawls asserts that rational man in his "original position" will naturally see it as being in his self-interest to agree to a system which permits free competition but which gives some special dispensations to the disadvantaged and buffers the losers with a "social minimum," so that their basic needs will at least be met. To the discerning eye, Rawls' position may resemble the contemporary American value system: "life, liberty and the pursuit of happiness," as modified by the values of the welfare state.

Setting aside the questionable premises behind his logic, the relationship between Rawls' construct and the urgent problems of the real world is not satisfactory. Rawls in effect takes "life" for granted, gives priority to "liberty" and anticipates no serious conflicts between life, liberty and the pursuit of happiness. But, in point of fact, that is the nub of our present conundrum. The message of the ecologists and population experts is that liberty and the pursuit of happiness may, in many respects, be jeopardizing life. The problem is how to constrain liberty and the pursuit of happiness in the interest of life, both for the present and for future generations. The choice we may face is between less liberty and less life. Indeed, all societies set limits on liberty; the public debate generally centers around whether or not there is, in a given context, either too much or too little. (Rawls himself defines the common interest as being the maintenance of public order, which presumably would include maintenance of his principles of justice.)

The basic problem with *A Theory of Justice* is that is does not take account of the very real survival consequences of the principles Rawls espouses. As is the case with most major works in political theory, the entire edifice rests on the prior assumption of survival and further, on the assumption of surpluses (if not also progress). Inequalities are defensible because everyone's minimal needs can be met, and the acquiescence of the poor can be bought off with the surpluses of progress.

Although Rawls does make an attempt to extend his principles to future generations, his approach does not really solve the problem. He attempts to establish a linkage with the future through the device of suggesting that, because his hypothetical men in their hypothetical original positions might not know which generation they belonged to, they would not opt to limit the principles of fair play only to one generation. Their self-interest would dictate transgenerational rules of fair play. But on the face of it, the argument is specious, because hypothetical man is presumably alive, at least, while future generations have not yet been born—and may not be. Therefore, why should hypothetical man care what the rules of the game will be beyond his finite lifetime?

But more to the point, how are scarcities to be allocated between generations when human survival becomes a transgenerational zero-sum game? In fact, liberalism is based on the twin assumptions of surpluses (beyond minimal survival needs) and progress—or at very least continuance of existing surpluses into the foreseeable future. In the end, liberalism depends upon an act of faith—the extrapolation into an uncertain future of the historical experience of a segment (but not by any means the majority) of the human species. It is precisely this assumption which the "limits to growth" proponents are calling into question.

Notes

1. M. Strong, "The Case for Optimism," *Saturday Review/World*, December 14, 1974, p. 7.
2. T. S. Kuhn, *The Structure of Scientific Revolutions*, Chicago, Ill., University of Chicago Press, 1962.
3. E. B. Haas, "How Whole is the Hole ?," 1973, unpublished manuscript.
4. As David Easton's commentary for this conference demonstrates, there remains some resistance, though.
5. These ramifications are discussed in detail in P. A. Corning, *Politics and Survival*, Princeton, NJ, Princeton University Press, in press.
6. There have been a number of systematic efforts to elucidate this complex subject in recent years. In particular, see the monumental three volume compendium edited by Y. A. Cohen, *Man In Adaptation*, 2nd ed., Chicago, Ill., Aldine Publishing, 1974.
7. For more detailed treatments of this subject, see especially J. G. Miller, "Living Systems," *The Quarterly Review of Biology* 48 (2), 1973, pp. 63-276. For politically-oriented discussions, see K. W. Deutsch, *The Nerves of Government*, New York, The Free press, 1966; P. A. Corning, "Politics and the Evolutionary Process," in: T. Dobzhansky / et al. (eds.), *Evolutionary Biology*, VII, New York, Plenum Publishing, 1974; also, P. A. Corning, "Human Nature Redivivus," in: R. Pennock / J. Chapman (eds.), *Nomos*, XVII, New York, Liber-Atherton, in press.
8. The definition I have developed is close to that of Deutsch and David Easton, and is similarly focused upon collective "functions" and consequences. The differences relate to my emphasis on the biological population, rather than superordinate and/or reified social or political "systems" and to my explicit emphasis on the relationship between politics and various aspects of the human survival problem.
9. The author has undertaken to do so elsewhere; see the references above.
10. By "successful", I mean policies which are functional, or adaptive, in the sense of satisfying biologically-based survival and reproductive needs, inclusive of certain instrumentally important "psychological" needs.
11. T. R. Gurr, *Why Men Rebel*, Princeton, NJ, Princeton University Press, 1970.
12. A study of women and violence has recently been completed by Constance H. Corning.
13. Locke, *Second Treatise*, 124.
14. T. J. Lowi, *The End of Liberalism*, New York, W. W. Norton, 1969.
15. R. A. Dahl, *Polyarchy*, New Haven, Conn., Yale University Press, 1971.
16. J. K. Galbraith, *American Capitalism: The Concept of Countervailing Power*.
17. B. Commoner, *The Closing Circle*, New York, A. A. Knopf, 1971, p. 254.
18. M. Olson, Jr., *The Logic of Collective Action*, Cambridge, Mass., Harvard University Press, 1965.
19. G. Hardin, "The Tragedy of the Commons," *Science* 162, 1968, pp. 1243-1248.
20. An important variant of the political economists' perspective is John Rawls' *A Theory of Justice*, Cambridge, Mass., Harvard University Press, 1971. A brief "Afterword" concerning Rawls' arguments follows the text. Among the critiques of Rawls, see in particular K. J. Arrow, "Some Ordinalist-Utilitarian Notes on Rawls' Theory of Justice," *The Journal of Philosophy*, May 10, 1973; B. R. Barber, "Problems of Psychology, Politics and Measurement in John Rawls' A Theory of Justice," 1973, unpublished; and "Symposium on Rawls' A Theory of Justice," *The Quarterly Journal of Economics* 88 (4), 1974.
21. R. L. Heilbroner, "What Has Posterity Ever Done for Me ?," *The New York Times Magazine*, January 19, 1975.
22. Among the contemporary examples are the practices of the many societies with little or no subsistence surpluses, as in Amazonia. The Sirionо, for instance, practice infanticide with deformed infants, with infants born to a mother who is still nursing, or with twins. If an older person becomes disabled or too old to contribute productive labor, he is left behind when the band moves its village. (For more detail, see B. J. Meggers, *Amazonia*, Chicago, Ill., Aldine, 1971.) Another instructive example is the conduct of the Biafrans during the Nigerian civil war. When food became critically short, young children were sacrificed in favor of adults in their child-bearing years. As a means of minimizing consumption and dependency, and maintaining rapidly renewable fertility, this was in fact a highly adaptive coping strategy. In our own society, there have been occasions when a new medical technology (kidney dialysis machines, for example) could only be made available to patients in need on a selective basis. In such cases, medical decision-makers have been known to take into account such social considerations as the person's "social value." A father with children, to take a case in point, might be given preference over a single man, all other things being equal.

23. An important discussion of the concept of triage can be found in William and Paul Paddock, *Famine 1975! America's Decision: Who Will Survive?*, Boston, Mass., Little, Brown, 1967.
24. Just in case there may be any misunderstanding, I do not suggest that the aspiration for a "good life" should be abandoned, but I do maintain that some tradeoffs may have to be made if the species is to endure. We may be forced to make such choices. Indeed, if the environmentalist critics are correct, we may already be making such choices without always realizing it. "Non-decisions" to continue destructive social and economic practices may amount to decisions against what I have defined here as the general welfare.
25. See B. M. Gross, *Social Intelligence for America's Future*, Boston, Mass., Allyn and Bacon, 1969.
26. O. D. Duncan, "Social Forecasting—The State of the Art," in: *The Public Interest* (17), Fall 1969, p. 88.
27. The concept of Survival Indicators is discussed in P. A. Corning, "Evolutionary Indicators: Measures of Adaptiveness," 1971, unpublished; and *Politics and Survival, op. cit.* A volume on *Survival Indicators* is currently in preparation.
28. US Department of Health, Education and Welfare, *Toward a Social Report*, Washington, DC, US Government Printing Office, 1969, p. 97.
29. *Op. cit.*, p. xii.
30. A different minimalist approach is being pursed by O. W. Markley / et al. ("Intolerability Thresholds and Minimum Standards for Quality of Life," 1975, unpublished) who are focusing on the standards set by law, court decisions and custom.
31. These issues are addressed in a volume currently in preparation. It might be noted here, though, that the concepts underlying each need are by no means self-evident and unambiguous, nor are the relationships between needs straight-forward. To take one example, "energy" can be treated as both an "output" need (in the sense of productive labor) and as an "input" need which intersects with nutritional needs, shelter (most particularly temperature maintenance aspects) and transportation needs. Such complexities require careful elucidation and analysis.
32. Forecasting is often distinguished from prediction on the basis of the degree of certainty involved. Whereas, in this line of reasoning, predictions involve deterministic expectations of a particular event or outcome, forecasts are said to be couched in probabilistic terms. As Nazli Choucri puts it: "A forecast involves contingencies." Perhaps a more fruitful distinction might be made in terms of where the causal relationships are to be found. Forecasting may in fact involve a single projected outcome, but forecasts generally involve implicit "if-then" statements, because they assume that human intervention is an unknowable (in advance) but potentially important causal variables. We would not bother with making forecasts unless there was reason to believe that they might inform human decision-making and thus affect the outcomes. Predictions, on the other hand, generally imply that the outcomes of certain trends are effectively beyond human control. Thus, for example, M. King Hubbert's estimates of the total US oil reserves (past, present and future) are it the nature of a prediction. Either the oil is there in economically recoverable amounts, or it is not, regardless of what we do. On the other hand, the date of exhaustion of these reserves involves forecasts based on one or more sets of assumptions about the rate at which we consume our reserves.
33. D. B. Bobrow, "Criteria for Valid Forecasting," 1973, unpublished.
34. T. J. Gordon, "The Current Methods of Futures Research," in: A. Toffler (ed.), *The Futurist*, New York, Random House, p. 165. For a select bibliography on futures research see the Toffler volume and recent issues of *The Futurist*.
35. United States Senate, Committee on Commerce, "An Analysis of the Department of the Interior's Proposed Acceleration of the Development of Oil and Gas on the Outer Continental Shelf," Washington, DC, US Government Printting Office, 1975, p. 13.
36. D. H. Meadows / et al., *The Limits to Growth*, New York, Universe Books, 1972.
37. Especially noteworthy is the work of N. Choucri / R. C. North, *Nations in Conflict: National Growth and International Violence*, San Francisco, Calif., W. H. Freeman, 1975.

GLENDON SCHUBERT

Politics as a Life Science: How and Why the Impact of Modern Biology Will Revolutionize the Study of Political Behavior

A "LIFE SCIENCE" APPROACH *implies much closer interaction between the biological and social sciences than obtains now*. The impact of the revolution in modern biology was greater and came earlier to psychology and anthropology than to sociology, economics, and political science, and because of differences in their component structures and disciplinary histories. A sustained interest in biopolitics, by political scientists, became evident only during the latter sixties.*

The morphology of political science, as a set of social organizations, lags behind the thrust of contemporary research, which features behavioral (viz., social science) analysis of political attitudes, decision-making, and policies. Unlike earlier, prebehavioral political science, which emphasized political description and dealt with motivation only speculatively, political behavioralism does attempt to analyze causation; but it does so on the basis of naturalistic (ethological) observations of the externalities of individual and group activities, relying particularly upon inferences from human speech (both oral and as recorded).

Contemporaneous with the development of biopolitics, there has been a movement of "postbehavioralism" among political scientists, emphasizing radical social change in both the discipline itself and its subject polities. But the atheoretical (albeit highly ideologized) and anti-methodological bias of the postbehavioral movement—many if not most of whose supporters seek to supplant, with a rampant humanism, any type of scientific work—begs (even more than does political behavioralism) the critical questions upon which depend the future survival of our species.

A biobehavioral approach, as adumbrated in the concluding section of this chapter, explores the basic biological needs of humans acting politically, and relates those needs to the psychophysiological processes mediating the arousal of appetitive behaviors and their interaction with the natural (as well as social) environment, plus feedback from the environment to the psychophysiological processes whose homeo-

* See also my "Biopolitical Behavioral Theory," *The Political Science Reviewer* (5), 1975, pp. 402-428.

static balances are essential to life (including political life) for the organism. It is argued that the chief task confronting the political science of the present and future will be to study in depth interaction between the rational and cognitively determined political actions hypothesized by political behavioralism, and the biology of human organisms which supports and constrains the possibility of any and all political behavior.

The biological revolution came much sooner and more easily to psychology and anthropology than to sociology or economics or political science, and for very basic reasons relating to the fundamental structures of these academic disciplines. Comparative and developmental psychology have been involved in natural science research procedures and in anatomical-physiological work for over half a century; while anthropology also divides into sets of dissimilar subdisciplines, with a physical side integrated with the natural science disciplines of geology, botany, and zoology (inter alia), and which indeed is itself considered to be a branch of biological science. There never has been such a natural science component in either sociology or economics or political science; and there is none now. Such sociological subfields as demography and human ecology and aging take a social science approach; the present field structure of political science is a subject of this paper. It is hardly surprising, therefore, to find biologically oriented basic textbooks in psychology[1] and monographs in anthropology;[2] but biologically oriented works such as there are just beginning to appear in sociology,[3] and they have not yet begun to appear in either economics[4] or political science.[5]

*

It is barely a decade since Keith Caldwell heralded the advent of what since then has become a new approach to the study of political science. It was Caldwell's view that "biopolitics" suggested "political efforts to reconcile biological facts and popular values [. . .] in the formulation of public policies," a synthesis made both timely and urgent by the "population explosion" and the "concurrent explosion of biological knowledge, an accelerating geometrical expansion of knowledge, the culmination of long years of accumulating inquiry in the various bio-sciences."[6] It is, he said, "the contemporary convergence of these two explosions—of people and biology—that justifies, indeed necessitates, a focus on biopolitics."[7] Caldwell went on to discuss primarily issues of ecology and of human physiology, in relation to governmental science policy and the difficulties of translating biological knowledge (or the lack thereof) into political action;[8] and biopolitical themes explored by other political scientists during the sixties covered a wide range of questions, including ethological theory,[9] comparative mortality in relation to political-psychogenic causes,[10] the effects of malnutrition on political participation,[11] neurological and endocrinological substrates of political behavior,[12] and psychopharmacology.[13] Already by the summer of 1970, when a score of papers were presented at panels in Munich organized by Albert Somit for the Eighth World Congress of the International Political Science Association, it was apparent that so many different political scientists in divers countries were becoming involved in biopolitical research that a new dimension had been added to the scope (and competence) of the discipline.[14]

I. Behavioralism

1. *The Morphology of Political Science*

In order to specify how a greater infusion of biological knowledge is likely to change political science, it is necessary to be able to state what political science is now. We all, of course, have our own notions about that; but for present purposes a more objective basis for appraisal is required. Unfortunately, but also quite naturally, we can speak with much greater confidence about how the profession used to be organized, than about what it is now doing. The most recent authoritative and comprehensive study is the decade-old report by Somit and Tanenhaus, which discusses political science in terms of its traditional (classical) subdivision into subfields, ranked according to their importance as this was perceived then, as follows:[15]

— Comparative government.
— General politics and political processes.
— International relations.
— Public administration.
— American government and politics.
— Public law.
— Political theory.

There is a subsequent report by the Behavioral and Social Sciences Survey Committee,[16] but it is not very helpful here because of its preoccupation with administrative and budgetary matters rather than substantive problems. Political science is defined as a hodge-podge of "focuses," reflecting an eclectic array of theoretical sources as an accommodation (one infers) among the members of the political science panel, who must have voted (and logrolled, in the process) in order to have produced *this* list: power, institutions, policy processes, functions, ideologies and movements, international relations, *and* (sic: in 1969 !) "political behavior."[17] There probably is not a single political science department in the entire world that was organized along the lines of these categories, either before or since the committee wrote—whereas there remain many, and perhaps a majority, which continue to describe themselves in their catalogues pretty much in terms of the classical model employed by Somit and Tanenhaus.

One can turn to the rubrics sponsored by the American Political Science Association; but little help is forthcoming from that source, either. The organization of panels for the national conventions of the Association shifts from year to year, reflecting the interests and aspirations of the chairman and his committee for the particular meeting. The annual listing of dissertations appears, in the Association's newsletter,[18] under the headings:

— Political philosophy, theory, and methodology.
— Government and politics of the United States and its dependencies.

— Constitutional and administrative law in the United States.
— American state and local government and politics.
— Canadian government and politics.
— Public administration.
— Foreign and comparative government and politics.
— International organization, politics, and law.

And by the clever combination of the second through fourth plus the sixth categories, and the merging of the fifth with the seventh, the above set of eight is reduced to half that number for purposes of classifying the book reviews of the *American Political Science Review,* as follows:

— Political theory, history of political thought, methodology.
— American government and politics.
— Comparative government and politics.
— International politics, law, and organization.

The Association's current (1973) *Biographical Directory* does somewhat better, even though it retains a categorization that is largely traditional:

— *Foreign and cross-national political institutions and behavior.
— International law, organization and politics.
— *Methodology.
— *Political stability, instability, and change.
— Political theory.
— Public policy: formation and content.
— Public administration.
— United States political institutions, processes, and behavior.

But of the *subcategories* almost half (as listed below) are explicitly behavioral. Furthermore, over a fourth of these subcategories relate directly to the kind of studies with which biopolitics has been and will be concerned:

— Analyses of particular systems or subsystems.
— Decision-making processes.
— Elites and their oppositions.
— Mass participation and communications.
— Values, ideologies, belief systems, political culture.
— Experimental design.
— Field data collection.
— Survey design and analysis.
— Cultural modification and diffusion.
— Personality and motivation.
— Political leadership and recruitment.
— Political socialization.
— Revolution and violence.
— Science and technology.
— Natural resources and environment.
— Ethnic politics.
— Urban politics.

And I have not included half a dozen of the methodological subcategories (computer techniques, content analysis, epistemology and philosophy of science, measurement and index construction, model building, and statistical analysis) which, as it seems to me, have to be the concern of anyone who is going to do any empirical political research at all. No explicit subcategory of "biopolitics" is listed; and neither is there any indication of any other subfield with an orientation toward the natural sciences—the subcategory "Natural resources and environment" is social science in approach, in political science just as in sociology.

The major categories which I have asterisked (above) are the ones under which the *Directory* groups most of the subcategories listed above; but the crossruff in vivo applies much more broadly than that: e.g., the study "elites" and "masses" is not confined to the field of "foreign... behavior," nor is the study of "legislators" in practice limited to "United States political... behavior." Clearly, however, these subcategories do define points of articulation, in terms of which biopolitical (life-science-oriented) work can be related to previous research, both traditional (humanities-oriented) and behavioral (social-science-oriented).

It would be helpful, of course, to have a better sense of what the direction has been of political science work during the seventies, as classified in terms of at least these major categories of traditional, behavioral, and biopolitical research. But such an understanding implies a rather substantial research project, with content analysis, for the period beginning in 1970, of: at least a dozen major political science journals, some published in the United States and others abroad; both textbooks and monographs authored by political scientists; doctoral dissertations completed in political science; and a survey of catalogue descriptions of departmental curricula. Also desirable would be a field survey of a representative sample of political scientists, whose interviews would be supplemented by copies of their course syllabi. I could not get all that done and analyzed in time for the writing of the present report; and so instead I did a few things much more modest in scope that were intended, nonetheless, to bring this discussion down to earth by anchoring our consideration of political behavioralism in relevant empirical data.

2. *Political Behavioralism*

Let us consider first several recent year's output of what is purportedly the profession's leading journal. In doing so I hasten to admit to skepticism second to none in regard to the *American Political Science Review*, under its present managing editor and his immediate predecessor alike, concerning (inter alia)[19] such matters as whether the bulk of the articles published by them are qualitatively as good as what has appeared (during the same period, of course) in the *American* [Midwest] *Journal of Political Science* and the *Journal of Politics*; and the extent to which the articles published in the *Review* are representative of the research that political scientists have carried out in recent years. Consequently it is only with reservations (and less confidence than one would like to feel) that I report the summary results of my examination of eleven issues of the *Review* for the three years 1972, 1973, and 1974.

For reasons that are apparent prima facie, I could not use any of the taxonomic schema outlined above. The completely traditional categories would not have been helpful in classifying the non-traditional work with which we are primarily concerned here; and even the *Directory* subcategories combine only as hash—they overlap, leave gaps, and reflect so many differing theoretical and methodological premises that the notion of reconstructing any coherent, logical structure out of them quite staggers the imagination. Instead, I categorized the 128 articles[20] using the following induced set of behavioral concepts, with observed frequencies as indicated:

Mass political behaviors	38	Attitudes	11
		Electoral voting (causal)	09
		Participation (orthodox)	11
		Participation (protest and violence)	07
Elite political behaviors	32	Attitudes	03
		Legislative voting	05
		Party competition for legislative seats	08
		Other behavioral outputs/inputs (causal)	05
		Interaction (empirical)	11
Other behavioral	27	Policy analysis	10
		Decision-making theory (mathematical models of rationality)	14
		Methodological	03
Prebehavioral	31	Paradigms of interaction	13
		Description	06
		Traditional	12

Most of the above articles are concerned with an attempt to explain how and why people act politically. Only about a fourth rely upon philosophical or legal explanation, or eschew explanation in favor of description; these are classified as "prebehavioral." But the causative variables upon which explanation continues to rely almost exclusively, in the three behavioral groupings, are derived from sociology, cultural anthropology, and social psychology. They direct empirical inquiry, that is to say, for evidence that consists of observations of what humans as complete organisms[21] do in relation to each other. Sampling is always involved, because all human "populations" that have ever been observed or are even potentially observable are quite minute subpopulations of the species and are identified by their geographical location plus (unless that locus is extremely small) selected cultural, social, or psychological attributes such as nationality, socioeconomic status, and partisan identification. Many of the observations of individual behaviors are aggregated to from indices of social characteristics, which are then imputed back to individuals as modalities for their classification. Causation is mostly inferred statistically from correlations; statements about why people act as they are observed to have done are based either upon comparisons with rational models, or else upon what persons verbalize concerning their own internal thoughts and feelings (or upon the analyst's own thoughts and feelings, which he projects upon his subjects) as in most clinical psychological and cultural anthropological work.

This leaves human motivation, which purports to be a key variable for political behavioralism, in an anomalous position: surely motives and emotions relate to conditions that obtain *inside* an organism; yet we know such motives and emotions only on the basis of evidence external to the organism—because surely one's testimony to others about one's own motives and emotions is an act of external communication. Political scientists thus far have seemed somehow able to ignore the lesson learned by zoologists and physiological psychologists, who went through all this (with other animals, which could not talk but nevertheless seemed to be motivated) half a century ago; even European ethologists now concede the critical importance of checking their field (or zoo) observations, with neurological and endocrinological laboratory findings concerning the biophysics and biochemistry of animal motivational systems.[22] Our failure to follow suit has relegated the study of political motivation (and affectivity generally in political behavior) to the vitalistic level of the "spooks" and "ghosts" deplored by Arthur Bentley[23] in the year that James Bryce presided over the American Political Science Association.

As we have already seen, however, there are other aspects of political science in regard to which change already is under way to a significant degree, and bodes likely to grow exponentially. The direction of this kind of change is not primarily methodological, but is rather substantive and, above all, theoretical. In certain (and expensive!)[24] respects new skills are going to have to be acquired, and different types of equipment utilized, to be sure; but the most drastic changes will involve a shift in orientation toward (what will be for most political scientists, novel) cognate disciplines as wellsprings of insight, information, and inspiration for our own guidance in the better understanding of public policy issues and political behaviors.

It is perfectly understandable that, when political science began to move away from mechanical metaphors to biological metaphors of "systems" of political relationships,[25] groping for an overarching paradigm that might better guide inquiry into and interpretation of the manifestly complex and multifaceted empirical relationships of politics, lateral interdisciplinary ties were established with nineteenth century macrobiological theory at the same level at which Herbert Spencer (or, if one prefers a more up-to-date example, Talcott Parsons) sought to develop models of social systems.[26] It is understandable because it is so much more difficult to develop cross-ruffs linking the leading interfaces (viz., "cutting edges") of two disciplines, especially when the twain have not been closely interdependent in the past. But organismic levels systems theory is not where the action was in biology during the late forties,[27] at either Chicago or Michigan;[28] and certainly this has remained true of the non-Soviet scientific world during the past quarter of a century.[29] The action has taken place first in microbiology, biochemistry, biophysics, molecular biology, genetics (including genetic engineering), and CNS neurophysiology; and secondarily in comparative psychology and ethology (especially primatology), and ecology. These are the aspects of modern—that is, *twentieth* century—biology[30] that political science is going to have to cultivate.

3. *Postbehavioralism*

Coincident with the emergence of an expanding group of political scientists who take biology seriously at the level of professional, as well as—like all of their colleagues—at the level of personal, experience, there has occurred what was alleged to be a denouement in the continuing developement of political behavioralism. One feels a need for circumspection as well as a certain skepticism in alluding to such matters, when recalling (for example) that one of the leading spokesmen for political behavioralism was eulogizing the movement at the very time when it was approaching its peak in influence and popularity[31]—although perhaps Dahl's celebrated essay at the beginning of the sixties was intended to function as a Judas sheep, leaving out-group critics to follow the proverb that admonishes against flogging a dead horse. As the decade drew to a close, along came another celebrated essay by another progenitor, this second one swearing to what the first had claimed to be truth, and proclaiming the advent of a new era of *post*behavioralism.[32] But David Easton's speech was no arm-chair exercise in political philosophy; it was instead a political event in its own right, a document recording the view from the bridge at a time when "The New Revolution in Political Science" clearly constituted primarily the protests of a radical faction that called itself "The Caucus for a New Political Science;" and the Caucus was approaching an initial peak in its agitation within the profession during David Easton's term as President of the American Political Science Association.

Since then the initial and more generic Caucus has become fractionalized, of course, by competing (and more specialized) groups of blacks, females, chicanos, etc.; and many of the Caucus' leaders became coopted into the ruling circles of the Association, where they were kept busy (and quiet) by involvement in the gush of on-going busywork that the running of the organization entails. In 1969, however, the Caucus was a much more fluid and rapidly-growing propaganda movements, and it is inconceivable to me that one can appraise the presidential address of that year without considering that at least one of its no doubt many functions must have been the containment of what those on the Council at the time certainly looked upon as a threat to the organizational equilibrium of the Association,[33] whose political system was being subjected to the greatest stress it had ever experienced. The good news about postbehavioralism oozed like oil onto the troubled waters that had been roiled by the Caucus; it was comforting for dissenters to be told that their time already had come to pass. But the depressingly conservative turn of our national politics, our national economy, our professional employment market, our students, and indeed our lives generally during the past five years makes it apparent that what seemed in 1969 to be a revolution in the making[34] now can be recognized as a much more evolutionary mode of adaptation—as was dramatically demonstrated by the "Caucus" victory at the polls in the 1974 APSA election.

The intimations of the mortality of behavioralism rumored by Dahl and Easton proved (if for opposite reasons) to be somewhat exaggerated: both were rather premature. Behavioralism had not yet swept the field in 1961 any more than it can be said to have done so today; nor was it about to be displaced in 1969 by the frenetic activities of semi-skilled professional do-gooders. Behavioralism continues to exert a major, but by no means an autonomous, influence upon teaching and research in political science.

A more fundamental problem with the Easton formulation lies in the inappropriateness of the metaphor that he invokes: acceptance of the biopolitical approach by no means commits one to acquiesce in the relatively crude organismic paradigm that became embraced by much of the writing in empirical political science during the past generation;[35] and especially are we not committed to necromantic notions about epitaphs and systemic survival.

It is even dubious whether it is more than incidentally and occasionally helpful to analogize the governance of a discipline, through a professional organization with a membership roster in the thousands, as though the American Political Science Association (for example) were like some living organism, especially when no attempt is made to specify what kind of organism one postulates as a model.[36] (This, at least, is one proposition that a more intimate acquaintance with biology—even at the social sciential level of an undergraduate course in ethology or animal psychology—would tend to make intuitively obvious to a larger number of political scientists than currently seem to appreciate such prudence.) Indeed, it is possible that professional associations are better analogized to viruses[37]—or perhaps to werewolves—than to either animate or inanimate structures per se. After all, they don't exactly die; but then again neither do they really live.

4. *Prebehavioralism*

Whether as a virus, a werewolf, or a living organism in its own right, traditional political science certainly continues to be an integral part of our publication, our teaching, and our thinking as a profession; and it remains the only emphasis in the discipline that journalists understand, and hence to which they pay any attention. By traditional I mean those political scientists who look to law, history, and philosophy for their models and inspiration, as indicated graphically by Figure 1. The original figure from which Figure 1 is adapted relates specifically to a particular subfield

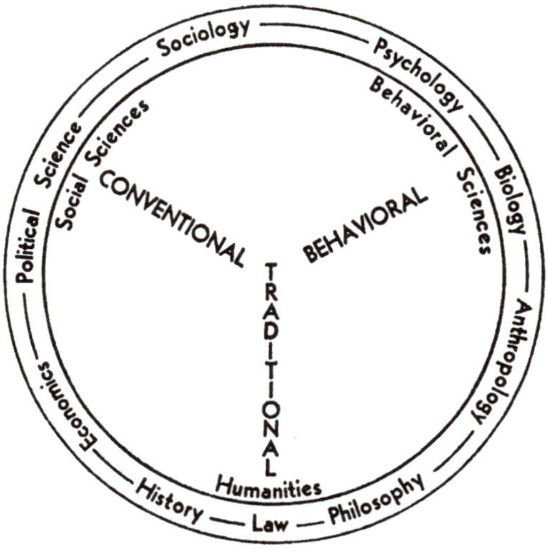

FIGURE 1. *Academic ideologies*

of political science; but its implications are perfectly general, as I thought at the time when I first proposed it almost a decade ago. It is notable that the professional discipline of biology marks the core of political behavioralism, and that biology is flanked by anthropology on the one side and psychology on the other, with sociology, political science, and economics positioned at successively greater distances from the biological core of behavioralism. In directing attention to my having thus precociously placed biology in the heart of political behavioralism, I do so in the spirit of a puckish remark of the late Mr. Justice Robert H. Jackson, that "sometimes one is tempted to quote his former self, not only to pay his respects to the author but to demonstrate the consistency of his views, if not their correctness."[38]

Among the issues of the *American Political Science Review* included in the sample discussed above is one that contains no less than three essays,[39] out of a total of a dozen articles, that are straightforward discussions of classical, historical, political philosophy, differing only in secondary bibliographical latency—not in method or style—from what was published in the same journal thirty (or sixty) years ago. Another of the sampled issues[40] includes an essay on constitutional law, critiquing the Supreme Court's opinions in the same way that remains commonplace in the hundred-odd American law school journals, and that was once (forty years ago) the standard of the *Review* itself—but that was back in the days when Corwin and Cushman and Swisher were leaders of the profession, and public law cut the thalweg of the discipline's main current (instead of backing and filling a few muddy bayous, as it now does.) The sampled issues of the *Review* also are sprinkled with a continuing display of institutional description, of "case studies" of unique historical sequences of decision-making processes, and of verbal metaphors that are proposed as "models" of various political processes.[41] But these occasional period pieces now provide the counterpoint, for which the point is provided by political behavioralism, which in the several variations that we already have observed continues to sound the principal theme and more loudly in the seventies than it had in the sixties.

The implications of biological theory, for the perpetuation of much of traditional political science, are much more revolutionary than those of political behavioralism as it has been understood heretofore. Our Scopes trial as a profession has yet to play out its drama, as the merest glance at where we begin our study of politics makes evident. Biological theory implies the rejection of the presumption that our political theory as a species began 2,500 years ago in Athens, or (alternatively) as described in "naturalistic" fables (whether optimistic like that of Rousseau or pessimistic like that of Hobbes), or according to the authoritative allocation of values in the even more popular fable of Genesis. The roots of political behavior go back not thousands but millions of years;[42] and political man did not spring (garbed in full civic regalia, and uttering a partisan war-cry) from the forehead of Socrates—as our teaching of the wellsprings of political philosophy might lead innocents to infer. The implications of contemporary research in physical anthropology, archeology, paleontology, and related sciences are going to jack political philosophy off its classical assumptions—once political scientists become better educated in, and start facing up to the facts of, biological life, including their own life history as a species.

It seems altogether likely that the apparent things[43] about political behavior, that could be learned through direct observation of one human by another, already have been learned during the course of the past half million years. It is also entirely possible that the Greeks of Periclean Athens (and for a century thereafter) had a particularly felicitous way of stating such observations, which can provide us with at least a benchmark against which to measure what we may have learned since then as a species. But we would not wish to rely upon either Hippocrates' or Aristotle's knowledge of human anatomy or medicine. It is at the level of folk wisdom (disregarding questions of general knowledge base, cultural dissonance, language, etc.) that Socrates excelled; and one can appreciate the attractive simplicity of confining political analysis to this level of understanding: it is indeed basic. But there are other things about human behavior that could not have become observed until the development of appropriate technologies: photography in general is one example, and high-speed photography of human facial changes is a more particular example.[44] Behavioral science has not hesitated to take advantage of what additionally could be learned about human behavior by adopting technologies that enhance the capacity to observe organisms in action. What remains to be learned will be primarily the previously *unobservable,* and this will consist mostly of events that take place within the human body, or of human organismic behavior in the context of environments that previously were inaccessible to observation (like the long political past that antedates the invention of writing, or that parallels the absence of documentary evidence about it).

II. Biobehavioralism

Peter Corning has pointed out that a fundamental challenge to political scientists was levied in "Christian Bay's assertion that a satisfactory political theory must be derived from an adequate understanding of the 'basic human needs,' as well as from man's overt and often variable preferences."[45] Bay was anticipated, in at least this respect, by James C. Davies in his *Human Nature in Politics*[46]—a book that borrows its title from Graham Wallas and its substantive approach from Abraham Maslow's theory of human needs (with credits going also to Freud and David Krech). Davies' book is a creative attempt to tease out of social psychology the "satisfactory political theory" that Bay has called for; but in precise accord with my own thesis in the present paper, events have demonstrated that social psychology—the guts of what has been perceived to be the political behavioral approach—is not enough.

Whatever their respective merits and styles as professional writing and contributions to political theory, it is now a matter of history that it was Bay's radical critique of political behaviorism rather than Davies' focus on basic human needs that sparked the protest movement within the profession that led to the organization of the Caucus and the professional confrontations of the late sixties. The approach fostered by the Caucus itself[47] featured well-meant shooting from the hip by ertswhile political engineers, both within and without the profession, which may not do too much harm but is also likely to do little good except by chance: our expectation of the probable pay-off for any therapy based on substantial ignorance of underlying causes (to say nothing of both direct and indirect effects) ought to be something close to zero improvement. The situation is not really that favorable, however; with only distressingly finite time and resources available for whatever constructive endeavors may yet be made to cope with some of the more pressing policy problems that face us,[48] the waste of professional time and resources (in retreats to phenomenology, parapsychology, yoga, and transcendental meditation, to say nothing of encounter groups and the other standard bill-of-fare featured in *Psychology Today*) has to be reckoned as downright harmful *socially,* whatever its private psychological benefits to the individuals who become so involved.[49]

An alternative approach to an adequate understanding of basic human needs is to study the way in which those needs are understood by the research scholars who know most about them, at a level of causation more fundamental than what can be inferred from observations of either what human organisms do, or what they can articulate about their internal states of being. This does not mean that we need to disregard either of the latter types of behavioral information; quite the contrary. But it does mean that other types of information, relating to independent observations of the operations of the neural, hormonal, and motor systems (among other internal ones) of humans must be taken into consideration if we are to become more realistic in our studies, of such phenomena as political action.[50] At the very least we need to understand—and to deal with in our own theories of political behavior—how the satisfaction of basic human needs affects the possibility and the modes of acting politically.

Before considering in the following section of this paper a paradigm that will attempt to explicate what these interrelationships are between our conventional analysis of political behavior, and biological analysis of human behavior, it is necessary to discuss the relevant sets of variables that will be used in constructing that paradigm. This section of the paper will examine first basic human needs from a biological point of view, followed by environmental sustenance, then some indicators of human development that figure in both social and biological theory, and it concludes with a discussion of biofeedback.

1. *Basic Human Needs*

From a biological point of view, the most basic need of any animal is to survive long enough to reproduce. It is the gene pool (to use the language of population genetics) and not the individual phenotype that natural selection operates, ceteris paribus, to preserve; but that can be done only through the successful adaptation of some minimally large population of phenotypes that do reproduce successfully and rear their young to the stage of self-sufficiency. To survive that long an individual human animal must continuously satisfy certain physical/chemical requirements which have long been crudely understood: air, body temperature maintenance, sleep, water, and food. Aristotle was quite aware of the need for air, but he could not have known that it is specifically oxygen that is needed, nor why and how it enters into metabolism. He was aware also of both sleep and dreams; but more recent observations of brain-wave fluctations are suggestive of support for a structural typology of sleep and a theory of its function for decisional purposes that became possible only with the development of the technology of the electroencephalogram.[51] Relatively sophisticated technical knowledge concerning most of these needs (adequate nutrition, potable water, and oxygen energetics) is indispensable for an understanding of what many biologists, and at least some political scientists, consider to be the most crucial problems of public policy confronting political science today; and these involve the ecologies of land, sea, and air.[52]

A fifth type of physical/chemical need is that for sensory stimulation. It is through the senses (tactile, visual, auditory, olfactory, gustatory) that an animal maintains contact with its environment, and such contact on a continuing basis is essential to survival. The evolution of internal homeostatic limits for animals presumes that appropriate stimulation will be forthcoming; hence the "vacuum activity" to which ethologists have directed attention, referring to extreme lowering of threshold for a behavior; and hence the probability that humans, like other animals, require a certain minimal continuing stimulation of at least several of their senses, quite apart from the question of the use of those receptors to provide environmental information needed by the animal for other purposes.

A sixth type of physical/chemical need is that of infants to be held, touched, and otherwise stimulated. Such care, which until very recent times was part of the human maternal role, appears to be necessary to the survival of infants up to about two years of age; and it appears also to be essential (although not directly to survival) to learning, growth, and security for older human children as well as for other primate juveniles.[53]

There is a problem about sex. As Davies points out, Maslow classified "sex" (evidently signifying sexual intercourse) as a physical need; but that seems to reflect his having to categorize it in relation to the even greater unsuitability of the other alternatives in the Maslowian typology, and it results in the blurring of one of the unique characteristics peculiar to the human species. Here is an instance where reference to zoology results in a much more humanistic judgment than one based upon the softer side of psychology. In animals other than primates it is not usually necessary (or possible) to distinguish between sexual interaction and sexual reproduction; but in primates these two functions can readily be separated, and especially is this true of humans (because of the loss of estrusity in human females), and the policy issue today is to what extent and for whom they should be completely separated.

It seems to be consensual, among social scientists and biologists, that sexual interaction is a human *social* need, which is important to good health (both mental and physical) but not otherwise or directly essential to survival of the interacting individuals. Much more controversial is the question of whether and how to limit sexual reproduction, with the primary objective of attaining zero or negative population growth on a world-wide basis, and eventually (i.e., in some future century) working toward a reduction in the absolute global level of the species population. Because neither voluntary nor democratic decision-making is considered feasible as a means to attaining global ZPG, the issue will be a difficult one for political scientists to help resolve—and they have as yet barely begun to take cognizance of the existence of the problem.

Other social needs, confirmed by both sociological and primatological research, include group association with and orientation to conspecifics. This means that humans need to touch, see, talk with, and otherwise to interact with other humans—for direct sensory gratification, for information about the environment in addition to what an individual can derive directly from his own sensory contacts, and for the security in relation to the environment that the individual finds in association with other humans. During infancy such needs are critical to the point of survival; subsequently they are necessary if humans are to become sufficiently skilled and involved in social transactions to be able to act politically.

Maslow's typology specified certain psychological needs, of self-esteem and self-actualization; but if an operationalized theory of psychological needs is to be developed, evidence to test it is at least as likely to come from social psychology as from physiological psychology or biology.[54] It is entirely possible that the satisfaction of such psychological needs, before a person is competent to act politically, is just as critical as is the prior satisfaction of the social needs described above. But only the physical/chemical needs are literally and directly prerequisite to the survival of individual adult humans.

2. *Environmental Sustenance*

There are two principal respects in which the environment makes possible both the satisfaction of basic human needs, and the practice of politics. The first and most fundamental is ecological, and involves energetics; the global distribution and other aspects of the species population; and the niches occupied by various subpopulations of humans in relation to other species with whom we share the biosphere. Less basic (in long-run terms of *species* survival) but at the same time more obvious in its impingement upon political behavior, and also more immediately critical for the survival of any particular individual, is the social environment, which educates him into a particular culture. To a greater or lesser extent the social environment facilitates the manipulation of an individual's psychophysical systems so as to attempt to influence that person's behavior (including political behavior).

Energetics is so painfully obvious as a critical dimension of public policy for all countries today that little argument seems necessary in support of the proposition that political science is likely to remain preoccupied, during any future that now seems imaginable, with problems of energy types, quality, production and distribution costs, dissipation (pollution), consumption priorities, and conservation. Indeed, I cannot imagine a viable politics (or political science) that will not *increasingly* be so preoccupied during any finite period of time that can possibly be a personal interest to any person assembled at the present conference, or who otherwise may read this remark.

Evidently the distribution of human populations in relation to the availability of resources for the satisfaction of basic needs, and in relation to different cultures, is likely to continue to be of concern to political scientists, in the future as now and as in the past. Other aspects of population dynamics, such as density differentials and differences in both the absolute sizes and growth rates of human populations, seem at least equally viable for political analysis. Except for growth rate, these have long been the stock in trade of traditional studies of international politics; but it is likely that a more technical level of knowledge, reflecting the escalating growth of demography as a specialized field, will characterize future political analysis of population policy.

The biological niche of Homo sapiens is a much less obvious matter of concern for most social scientists, and largely because they seem capable of indulging in the almost incredible smugness of feeling that they can take the niche of humanity for granted.[55] Such humanistic conceit is credible only on an hypothesis of sheer and utter ignorance of the manner in which the biosphere has operated historically, works at present, and seems most likely to evolve in the future.

The success of humans in the competitive exclusion of other living species (of both flora and fauna), particularly during the most recent ten to twelve thousand years since we began in a serious way to scarify the natural land surface with our agriculture, has by no means necessarily been adaptive for our species except from a point of view with as short a range as that. Our increasing capacity and tendency to eliminate, more often unwittingly or accidently than by design, other living species upon whom we previously had relied for sustenance—and now also each other, as usual but with the aid of the improved and more complex technologies of the latter half of the twentieth century—is an index of the extent to which our trophic niche, defined as the *functional* status of an organism in its community, is being redetermined as much by the indirect as by the direct effects of our predatory activities; and the fossil record is replete with evidence of extinct species whose predation was so successful that they themselves starved to death. The issue goes far beyond the restoration of token vegetation to, and the elimination of domestic pets from, urban areas; or even the apparent trends in the direction of human conspecific predation ranging from licensed hunting in season (under circumstances such that other hunters present more frequent targets of opportunity than the crops of ruminants or rodents available for harvesting) to the ubiquitous predations now characteristic of all large urban areas in the United States (where only humans can be and are hunted by each other, and at least in part because all other prey have been extermined).

Mankind cannot destroy the biological community of which he has been a part, without his degradation of that biological community returning as feedback to threaten the human political community; and I should like to give two examples which are textbook material for undergraduate zoology courses:[56]

a) "The most obvious example of an ecosystem that has been altered by man is [. . .] agricultural fields. Essentially, one species of organism is allowed to exist while other species are removed by tremendous expenditures of energy, herbicides, and pesticides, and the abiotic environment is controlled by extensive use of irrigation, fertilization, and tilling. [. . . T]he successful exploitation of natural ecosystems in order to increase the production of human food has led in most cases to a dramatic lowering of the fitness of the exploited ecosystems. Technology can allow these ecosystems to remain viable and productive, but only at considerable cost. There are clearly limits to which technology can overcome the tendency of ecosystems to revert to natural equilibrium, and the instability and uncertainty of the system's capacity to produce products useful to man increase greatly as these limits are approached."

b) "Current agricultural economics dictates that livestock should be fattened in feedlots rather than on the range. On the range, the urine and feces from the animals fertilize the land naturally to replace much of the nutrients removed by herbivory. In feedlots, however, the nutrients contained in the feed are removed from fields in their entirety, causing a reduction in soil fertility which must be made up with massive fertilization, and the excreta of the livestock are too concentrated to be utilized by plants in the feedlot area, and so are washed into nearby waterways, where they become pollution problems. This is a classic example in which the disruption of a normal biogeochemical cycle has led to the deficiency of materials in one ecosystem and a surplus in another, reducing the fitness of both."

In the past political science has assumed the complete beneficence of both the fattening of livestock for market, and the raising of grain with which to do it. Perhaps that complacency has been unwarranted, and has skewed our assumptions about the political economics of food policy, at both the national and international levels of public policy making.

The effects of social environment tend to be much better recognized as pertinent to political science, as a sampling of the recent literature in political behavior (above) and in biopolitics (below) demonstrates. Political socialization, and particularly as it relates to the modes and quality of political learning among schoolchildren, has emerged as a distinctive subfield of the discipline of both political science and social psychology—in political science since the beginning of the sixties, and somewhat earlier than that in social psychology. This work is well known, and what is most necessary to note here is that political scientists to date have paid precious little attention to the developmental psychobiology of political learning. I presume that as our interests in the subject deepen and broaden, both we and our sociologist cohorts are going to have to take into consideration the biological aspects of the learning process just as psychologists already are doing. Moreover, we shall sooner or later have to broaden our focus in another respect: learning is accelerated during, but not monopolized by, childhood; and in any case questions of the reinforcement and loss of learning are also part of the relevant subject.[57] So both political learning among adults, and the eventual loss in both the quality and quantity of political knowledge and ability to accept new political learning among the elderly[58] (or "desocialization," as it is increasingly being described) are a part of the study of normal political education.

There is a marked shift in both emphasis and interest, when we turn to the experimental manipulation of psychophysical systems, as a means of influencing political learning. One very common method of so doing is through the customary activities of the medical profession: the dispensation of either drugs or surgery is designed to change the ways in which certain of an individual's bodily systems are operating; and the execution of that experiment often entails uncontrolled (or uncontrollable) side-effects for the person's political behavior (as by affecting his attitudes, mobility, appearance, or personality). We have recently been passing through a period during which many persons have experimented with their own minds and bodies, by ingestion of drugs subject to little control of either quality or quantity; typically the effects of such experiments have been to lessen political participation: "Tune in, turn on, and drop out !" Other experiments have been conducted on a social scale by public health programs, and in this form they have been a part of political science as long as there has been such a subject; the leading Supreme Court decisions in regard to compulsory vaccination or sterilization go back half a century or more.[59] The floridation of water supplies—though still controversial in some outback regions, such Hawaii—is now generally countenanced like chlorination, although it must be admitted that the latter has been in innocence of serendipitious discoveries such as the recent detection of carcinogens as a by-product of the chemical interaction between chlorination and pollutants in the metropolitan drinking water supplies of New Orleans and Indianapolis. It is hardly comforting to think so, but this may prove within another decade or so to have become a major factor in the reduction of population densities in urban centers— and especially those with modern water purification systems—throughout the world. Other novel subjects, such as sickle-cell anemia, present perplexing political issues;[60] and the range of such questions is going to expand, with demands for positive preventative medicine emanating from crowded populations who are becoming aware that they are experiencing a plethora of biogenic pathologies, as fallout from the urban physical/chemical environment.[61]

Quite another form of experimental manipulation is operant conditioning. It happens that the chief proponent, an unsuccessful and presumably frustrated writer of fiction during his younger years, thinks of himself as a political philosopher. So it is not surprising that B. F. Skinner's ideas have finally begun to attract some serious critique by political scientists, as we shall note presently. What is relevant here is that both his technical research and his popular writings have direct implications for political science, and their thrust is in the direction of proliferating relatively simple (though generally, laboratory) techniques for controlling human behavior. The practice of Skinnerism on human subjects to date seems to have been largely confined to his own prototypical daughter and a few hundred other infants raised in Skinner boxes, a few dozen patients in veterans hospitals, and sundry other captive and relatively helpless institutionalized populations. But the potentiality for expansion is tremendous, particularly if political decisions during the next generation are supportive of future life styles (viz., cities afloat, or completely underground, or in space stations) that tend to further reduce some of the analogical differences that now seem still to distinguish humans from pigeons and rats.

3. *Biosocial Indicators*

There are difficult *biological* problems in regard to many key "social" variables (which are imputed categories for classifying individuals). Among the indicators of human development that are central to political behavior are race, age, sex, and intelligence; but political scientists study them strictly on a sociological level because that is the (limited) degree of sophistication to which our data (and presumably, therefore, our understanding) extend. Race no longer is a viable concept for use in biology, except in discussion of human evolution and in regard to the distribution ranges of certain genetically determined aspects of blood chemistry that are important to public health and nutrition; but it remains of critical importance to the social and cultural identifications of most persons throughout the world, and political scientists work with a research literature that reflects overwhelmingly the presumption of a generation ago: that half a dozen nominally distinguished categories provide an adequate basis for policy and analysis alike in regard to human subpopulation genetic variations.[62] The possible impact on political science of the alternative premise supported by modern biology, that the relevant gene pool with which public policy should be concerned is that of the species, is staggering to contemplate.

Similarly with sex: the ongoing drive for redefinition of sex roles goes far beyond an increase in political participation by females (or in their practice of political science), which is a level at which sex still is dealt with as a natural dichotomy and its chief problem is defined as equalizing (in so far as possible) cultural values to be associated with the two halves of the split variable. Modern biology is concerned with a proliferation of policy spin-offs from human reproduction, ranging from qualitative changes in populations, to biological engineering; and developmental psychobiology studies specific sexual differences (including those relating to the central nervous system) that may affect both political attitudes and political behavior; but none of these matters (except for population growth) is involved in the usual treatment of sex in contemporary political science literature.

The work in political socialization is from a methodological point of view straightforward survey research, but it does treat age as a developmental construct (and not simply as an index to constitutional eligibility for voting, public office holding, and trial in the regular criminal courts). There are, however, other (and possibly more important) implications of aging, for the quality of political attitudes and participation, as a direct consequence of biophysical and biochemical changes—all deleterious from the point of view of optimality of function—that are products of aging and that bear directly upon perception, memory, judgment, and other behavioral subcomponents of political decision-making. In any society (like the United States and other industrialized countries) whose age/sex population "pyramid" has become virtually a parallelogram (i.e., with as many persons over fifty as under twenty, and with more older females than males) and with prospects for even greater top-heaviness in the closing decades of this century, it might seem important to study political desocialization (viz., loss of learning, decrease in participation) both quantitatively and qualitatively, and in regard to both affective and effective qualities. The political problem is there, whether we choose to study it or not.

Political scientists seem to have managed pretty well to have avoided becoming involved very much in the Jensen-Shockley heresy, which concerns a direct attack at the genetic level of argumentation upon the possibility of establishing racial equality. Of course the question of intelligence does enter into the socialization studies of children; and there is one recent book—not by a political scientist and not yet (at least) reviewed in the *A.P.S.R.* so as to have showed up in the sample that I shall discuss below—which does deal with the question of the extent to which genetic variation in IQ (if not in intelligence) affects social and political status.[63] We should be reminded, I suppose, that Harold Lasswell always has stressed the intelligence function as a key value in his grammar of professional obligation in policy research:[64] but political scientists seem to have regarded a concern for (to say nothing of the practice of) intelligence as undemocratic. In American academia the measurement of intelligence (like the measurement of performance) is not in vogue. Nevertheless, substantial differences in human learning ability and creativity obtain and remain relevant to political attitudes, participation, and decision-making, although we have no choice but to continue to treat intelligence as a constant (rather than a variable) unless and until we attempt to acquire a better understanding of how it does affect political behavior.

Health, likewise, remains unexamined in any systematic way, notwithstanding its evident importance to the performance of certain political roles (of which those of President of the United States, and Supreme Court Justice, are only among the most conspicuous). But health may also be directly relevant to mass political attitudes as well as to participation, as several recent papers suggest. One of these is a research design on the effects of John Kennedy's chronic illnesses, upon his foreign policy decisions as President.[65] A second is a study of a large sample of high school students, relating their health (and body image) to their political knowledge, attitudes, and participation; while a third study analyzes the relation between the physical fitness of college students and their attitudes towards international politics.[66] Other research indicates the importance of health as a condition prerequisite to effective mass political participation,[67] which thereby serves to limit the possibility of either political democracy or political revolution throughout much of the non-industrialized world today—just as failing health is functioning also as a reinforcing constraint upon the participatory behavior of those persons in industrialized societies who are enjoying the "golden years" that increasingly are becoming characterized as a time of "poverty, fear, and malnutrition."[68]

4. *Biofeedback*

Biofeedback refers, in this discussion, to observations of an individual's on-going bodily processes, processes which may be operating either at the level of biophysical-chemical systems, or at the level of biophysical-chemical systems interacting with cognitive systems.[69] Such observations may, in the latter case, be made by the individual of himself; or the observations can be made, in either case, by other persons' monitoring of the individual's physiological systems. It is also possible for other persons to make observations of his behavior, from which they, in turn, make cognitive inferences concerning what is (or has been) on-going in the physiological system and/or cognitive systems of the subject individual. Virtually all of political science knowledge about political motivation is based upon only two of these types of evidence: i.e., either *a)* that of a subject individual's own verbal articulations about how he feels, or why he wants to do something; or else *b)* upon other persons' observations of and inferences from his behavior (including his speech behavior, upon which type *a* data depend). A major part of the work in political behavior—indeed, one is tempted to say "most"—has been respondent testimony about self-motivation, or inferences about motivation based upon observed actions. Evidently, independent observations of the variations in operation of the psychophysical systems of subjects of political analysis, and particularly when such data relate directly to theories of motivation grounded in biological (rather than social) events, ought to be extremely important in helping to appraise the degree of confidence warranted by much of our present knowledge base about political behavior.

The validation of political motivational theory at the level of biological theory has one great advantage, deriving from the circumstance that the biological theory is based almost entirely on comparative research, whether ethological or laboratory in method; whereas socio-psychological (and derived political) motivational research is both confined to a single species and then confounded by a babel of cross-cultural dissonance resulting in a focus upon a very limited set of variables out of which to attempt to construct a viable theory of political motivation. However well (or poorly) such a theory might correspond to our present understanding of political culture, at least we should expect that it will be consistent with what otherwise is known to be true of the motivation of animals generally, mammals in particular, and especially of primates. (To the extent that unresolved inconsistencies might become apparent, the task then would become to guide inquiry at the biological level, to ascertain the particular events in human evolution that account for the deviation(s) in human morphology that make human motivational processes, in this respect, different [from all other primates, or mammals, or animals, or whatever]; conversely, it would become incumbent to focus inquiry at the social science level upon aspects of human motivation that seem to be peculiarly human.)

An example of what presently appears to be a morphological idiosyncracy of our species is found in the theory of aggression.[70] In most other animals there is non-cognitive, psychophysical-level (probably neural and biochemical, in almost all instances) inhibition of the *killing* of adult conspecifics. Humans (even unlike other primates) have lost, and apparently completely, any inhibition response to signals of appeasement and submission. Some evidence, from cultural anthropological work on surviving remnants of neolithic cultures in New Guinea, suggests that a *possible* explanation of how the inhibitory loss may have been species-adaptive is that the loss was a useful device in maintaining population control among small bands of humans, where competitive exclusion (for niche space at the *habitat* level) kept the overall level of local populations of the species within existing levels of variation of supporting resources. In any case existing restraints upon the killing by humans of each other are completely cultural rather than biological; and the cultural inhibitors do not operate very reliably. Conceivably this unreliability reflects contradiction among the norms more than (or rather than) ineffective learning: in virtually all societies of which we have historical evidence, the "Thou shalt not kill"s are balanced by an equal and opposing array of "Kill!"s, depending upon the social circumstances.[71] At all levels of governance, or the lack thereof, violence is of sufficient political concern today that the question may soon be raised of the possibility of a technical solution (i.e., at the level of biological engineering: estrogens as well as carcinogens in public water supplies ?) to what is clearly a social, and political, problem.

A complicating biological factor in our efforts to cope with social, including political, violence may well prove to be that we are using a single word to describe the output of what may be both causally and operationally a variety of *different* biogenic systems of aggression. It has been proposed, for instance, that animal aggression generally *a)* includes several types which, although individually distinct, can co-exist in complex relationships of mutual inhibition and reinforcement; *b)* involves typally distinct stimuli (or "releasers," in the more quaint terminology derived from continental ethology); and *c)* operates through typally *independent* neural substrates and hormonal systems. At this level the theory may find considerable support notwithstanding many species-specific differences in the details of repertoires of both stimuli stereotypy and behavioral action sequences. The presumption certainly must be that humans are (like other animal species) only a special case, and that disinhibition of intraspecific killing among humans is an empirical detail which in no way contradicts the general theory.

We are not likely to make much progress with either world peace or criminal justice reforms (to take two subfields of the discipline that remain fashionable) if we limit ourselves in our understanding of human aggression to either what political historians tell us about the careers of a relative handful of leading personages, or what social psychologists tell us about the frustrations of our pent-up storages of hydraulically engineered reserves of aggressive energy.[72] Moyer's physiological model of aggression[73] proffers what is probably a better fit to even our present empirical data than the rationalistic models of game (econometric) theory on the one hand or the irrationalistic models of Neo-Freudian social theory on the other, the two stools between which falls our contemporary understanding of the theory and practice of both war and crime. There is a not inconsiderable amount of support in zoological research, for example, for the proposition that sexual and aggressive behaviors are both physiologically and genetically interdependent in primates,[74] which raises the question of possible interactive effects between political policies designed to promote world peace[75] and those intended to reduce population levels.

Similarly, we can consider that the study of biofeedback can function as a means for the articulation of biological with political theories of behavior in many other ways, not only in regard to motivation theory but in diverse other respects as well. A continuing and important effort in this direction, that began in work by John Wahlke and Milton Lodge at the State University of Iowa during the sixties,[76] and is now being carried out by Lodge and his associates in the Laboratory for Behavioral Research at the State University of New York at Stony Brook,[77] is an investigation into the correlation, between conventional verbal or ordinal responses to sociopsychological stimuli (slogans, pictures with prima facie political content), and psychophysical measures of arousal (galvanic skin potential, pulse, heart rate, and blood pressure). The measured relationships between the cognized and the autonomic responses are so close that more recent work has focused on the methodological problem of the development of scales, designed to measure the cognitive response, that are anchored in autonomic systemic bench marks; and because the latter *are* measurable on interval (and often, ratio) scales which parallel those long in common use in behavioral survey research, those scales of political behavior now can be given an interval-level interpretation. This is a methodological improvement that adds tremendously to the power (in the sense of statistical efficiency) of work done along these lines. It seems fair to add that the results of the initial efforts that have been published thus far barely scratch the surface of what may be possible to do, both from the point of view of the technology of the observation of psychophysical responses, and the articulation of those responses with the social science theory of political behavior.

Two developments in recent biopolitical research should be mentioned: studies of body image, and of posture. The work on body image hypothesizes that there is an important relationship between the qualitative appraisal that an individual makes of his own self as a discrete, phenotypic human organism; and his attitudes toward and (therefore) participation in the political process.[78] Preliminary findings are suggestive of a correlation sufficient to be mutually supportive with other research (on health and nutritional minima as prerequisites to effective political behavior), and indicate that personal body images tend to be projected onto the body politic.[79] Evidently, body image requires self perception, and is therefore to be associated (like self perceptions of arousal, emotion, and motivation) with that category in our typology of biofeedback. The other recent development lies (like laboratory monitoring of psychophysical systems) in the other category, of analyst observation of biofeedback, and it concerns bodily posture as an indicator of psychophysical states and changes.[80] The possibility of using body language in a serious way for the observation and analysis of political behavior ought to be of considerable interest to political scientists because of the widespread potential that it offers for field (as distinguished from laboratory) investigations.

III. A Paradigm of Biopolitical Behavior

Let us turn now to a consideration of how it is possible and useful to interrelate the conventional paradigm of political behavior with a radically different but nevertheless complementary paradigm of biological behavior.

The political behavioral paradigm is shown in Figure 2. Political

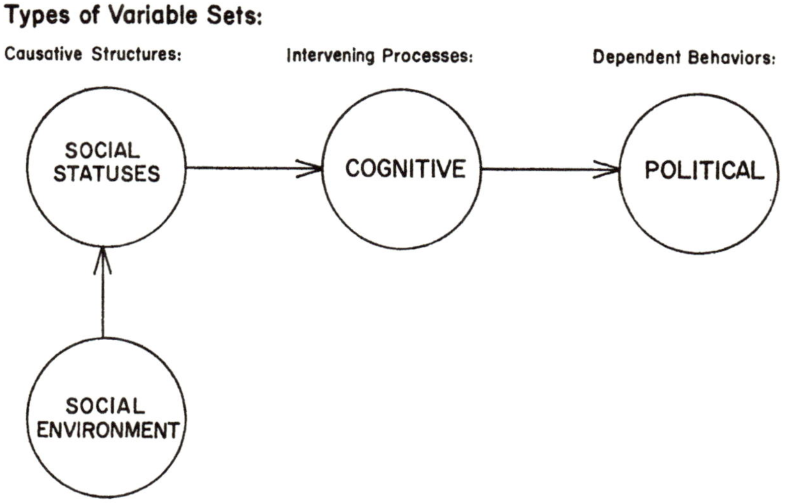

FIGURE 2. *The social science paradigm of political behavior*

actors are classified, for purposes of data observation and aggregation, according to a set of indicators of their attributes that are deemed relevant for analysis (party, age, sex, socioeconomic status, etc.); these inputs of the system of political action have themselves been determined by the social environment. Political behaviors are classified in terms of a set of equally conventional action modes (policy choice, vote, role performance, speech, etc.); these are the outputs of the system. Intervening between social attributes and political action are the cognitions of the individual actors (beliefs, attitudes, decisions, preferences, etc.); by making conscious choices among competing alternatives political actors determine what their behavior will be.[81]

As I have pointed out elsewhere, such a paradigm postulates a (social) *psychological* model of decision-making; and this lies modally between the more strictly *logical* model of traditional political science, and the *non-logical* models that are suggested by clinical psychology (on the one hand) and comparative psychology (and biology) on the other hand.[82] The social indicators are (admittedly crude) indices to learning and experience, and the indicators function as surrogates for the total life history and socialization of the individual. According to this psychological model (and the paradigm of political behavioralism), differences in political behavior are explained by differences in conscious choices as to how to behave, which in turn are explained by differences in the life experiences of the actors.

The biological paradigm of behavior is sketched in Figure 3. It directs attention to the non-logical influences—at least, from the point of view of the logic of political behavior—upon all behavior, and to the hierarchy of survival requirements, the satisfaction of which is preconditional to indulgence in political behavior. According to the biological paradigm, human needs have to be satisfied through sensory and appetitive interaction with an environment that is both partly natural and partly social; and different kinds of needs find sustenance in differing parts of the environment. Such biological characteristics as age and sex are reciprocally engaged in interactions with both the natural and the social environment: how either sector of that environment will affect individuals differs according to their stage and type of development, and how they will respond to that environment is partially determined by such aspects of their development. Similarly, an animal's psychophysical systems are directly affected by age, sex, health, and the other biological characteristics. Appetitive behavior involves searching of either the natural or the social environment (or both), and hence the feedback link indicated in Figure 3; while the satisfaction of physical-

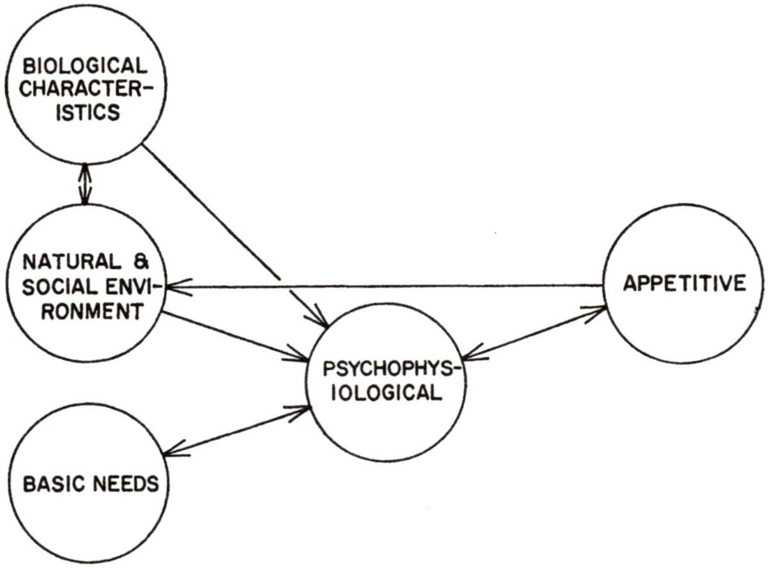

FIGURE 3. *The biological paradigm of animal behavior*

chemical needs necessarily entails feedback through psychophysical systems to the organic causes of those needs. So the biological paradigm states that needs autonomically activate psychophysical systems which introduce appetitive behavior, the consequence of which is to cause the animal to probe its environment, which in turn provides further stimulation for the animal.

Figure 2 puts humans more in the stance of gods than of animals, and hence is incomplete and inadequate precisely to the extent that humans *are* animals. Figure 3 describes human behavior as animal behavior, and it is incomplete and inadequate precisely to the extent that humans *are* different from all other animals. So Figure 4 attempts to put them together, so that humans can better be studied as and for what they are—not gods, but also the only animals that were characterized—and quite properly—by Aristotle, as "the *political* animal."

Figure 4 depicts the psychological system of political behavioralism as (as in one sense it is) superior to, but continuously interacting with, the more fundamental system of biological behavior. The statement of a human need occurs initially not at any level of conscious thought, but rather involves autonomic invocation of psychophysical systems, which in turn activate appropriate appetitive behaviors; those appetitive behaviors

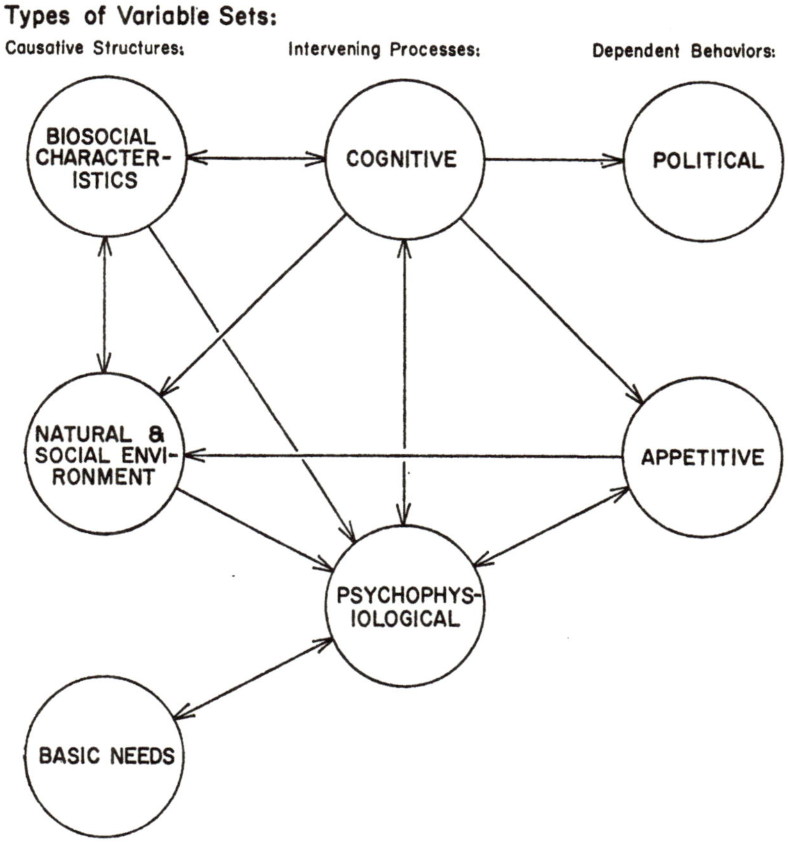

FIGURE 4. *A life science paradigm of political behavior*

necessarily involve interactions between the person and the environment, and at some point of intensity either the behaviors or the needs motivating them become sufficiently amplified by the psychophysical systems to become perceived as *bio*feedback to cognitive systems of the human, who may then choose to impose various (perhaps cultural) restraints or reinforcers (as the case may be) upon the appetitive behaviors. In relation to basic needs, such restraints never can be more than a temporary reordering of preferences because the feasible time scale for postponement of basic physical-chemical satisfactions is strictly determined by the operating limits of interacting psychophysical systems, which if exceeded result in the disablement or death of the organism and hence the elimination of both the necessity and the possibility for any kind of choice-making. In addition to that limited and partial kind of cognitive intervention in appetitive behavior, cognition affects one's biosocial characteristics, because (for example) sex identification and attitudes toward age and aging are important feedback to the further development of these facets of a human organism. Moreover, cognitions about the environment can lead to choosing behaviors that will affect the environment as feedback to it.

The set of characteristics that are designated as "social" in Figure 2, and as "biological" in Figure 3, are redesignated as *biosocial* in Figure 4 because those with which we are concerned here (age, sex, race, intelligence, and health) belong clearly in both realms of discourse: all have a clear and direct biological significance which is determinative of the social statuses that are derived from them. The environmental variables provide an indirect link between basic needs and biosocial characteristics; and consciously perceived *bio*feedback serves as one important link between psychophysical and cognitive systems, although direct two-way interaction between these two sets of systems (without biofeedback through self-conscious awareness) is much more important to behavior generally, and possibly also to political behavior. Biofeedback refers here strictly to perceived physiological systemic effects upon cognitive systems; the reverse link, postulated by humanistic psychology (with cognition affecting or "controlling" physiological systems) lacks the degree of consensual support among physiological psychologists, zoologists, and other biologists that does undergird the other relationships denoted in Figure 4. No direct link is shown between political and appetitive behavior, which constitute discrete and alternative modes of action. (Of course it is possible for a dinner at the White House or a Washington cocktail party to function empirically and simultaneously as *both* appetitive and political behavior; but to suggest this is a semantic quibble, and raises no question of analytical importance.)

What clearly needs to be done, in moving beyond political behavior to biopolitical behavior, is to design research that will explore the vertical as well as the horizontal relationships depicted in Figure 4. That will necessitate consideration of the interfaces between the biological and the social sciences, and will point political science in the direction of a study of political behavior which, for the first time, can deal with political life at the level it is lived and not as an exercise in puns, rhetoric, or intellectual dilettantism.

IV. Research in Biopolitics and Biopolitical Behavior

In this final section of the paper, we shall first consider a sample of the literature on, or relating directly to, biopolitics, as this is seen through what has been published, reviewed, or advertized contemporaneously in the *American Political Science Review*. The paper concludes with some examples of how a biological approach might be developed in relation to the on-going political behavioral enterprise.

1. *Status Quo*

A search of eleven recent issues of the *Review*[83] turned up relatively few biopolitical items—some half a dozen articles, most of which are only partially or marginally relevant—and all of which are included in the frequencies attributed to political behavioral research, published in the *Review*, as described in the introductory section of this paper. It is notable that these articles all fall exclusively within the initial rubric of "Mass political behaviors:" two are studies of attitudes (political thinking among adolescents as a function of complex influences, and a study of the relationship between childhood and adult attitudes toward politics);[84] two are inquiries into the causation of electoral voting (mobilization of black voters in the South as affected by apathy, fear [measured by occupational vulnerability (sic ?]) and discrimination; and effects of generational differences [which have at least something to do with age differentials] on class-party voting);[85] and two are studies of political participation in the form of urban protest and violence (the potential for political violence, including effects of race and age; and the rate and direction of change, as influences on the potential for political violence).[86] Whatever their merits in relation to the intentions of their respective authors, these half dozen articles do not add up to provide even a fragmentary sample of the metes and bounds of biopolitical research during the early seventies; nor were they—of course—selected and published with that objective in mind. They are best viewed, no doubt, as a projection of the cognitive mapping of the managing editor.

The book review and advertising pages of the same journal provide a more broadly ranging indication of recent research which, although not necessarily authored by political scientists, is deemed to be of interest to them; and in order to facilitate the explication of the paradigm just presented in Figure 4, I have undertaken to classify these books, which discuss political behavior on the basis of variables that articulate directly with biological research, in terms of the typology of sets of variables introduced in the second section of this paper: biopolitical indicators, environmental sustenance, biofeedback, and political behaviors.

Biosocial Indicators

Of the five indicators of biosocial development included in our typology (age, sex, race, intelligence, and health), sex and race have attracted the greatest attention of political scientists in recent years, at least as such matters are reflected in the relative numbers of books which have been reviewed or advertised recently in the *Review*. The political interest in sex has focused upon the equalization of opportunity for political participation, which has been understood to imply both the elimination of special distraints upon women and the undertaking of more positive action to encourage greater female participation in politics. In the present sampling there are two books by women and one by a man;[87] most of the sexual politics writing has been done by women—in perfect accord with the "liberal-interest-group" theory of politics that is so obnoxious to so many self-proclaimed postbehavioralists.

The books on race are divided, with the earlier ones emphasizing racism (i.e., white discrimination against blacks, serving both directly and indirectly to limit black participation in the political process); while more recent work focuses upon the contemporary political era and analyzes the processes and effects of increasing political participation by blacks.[88] Much less attention has been given by political scientists to the effects of aging upon political attitudes and participation, or the policy problems that have been attendant upon the changing demographic structure of industrial/urban societies; only one item was noted.[89] Conceivably the availability of federal research funds (through the Commissioner on Aging of the Department of Health, Education, and Welfare) at a time when both private and public foundation funds in support of conventional sociopsychological research on political behavior are much less readily available than during the sixties, may have some impact upon the definition of political science research priorities, not only in regard to aging but in the more general field of population policy research which is currently funded (as during the past several years) by both Rockefeller and Ford, as well as the program explicitly on "Political Aspects of Populations, Family Planning, and Reproduction Research."[90] And one is aware, of course, of more extensive interest in the political aspects of aging than happened to turn up in this particular sampling process;[91] several political scientists at the University of Southern California, for example, have adjunct appointments for teaching and research in the program of the Gerontology Center there; similar arrangements are in effect at the University of North Carolina, Chapel Hill; and no doubt there are other joint programs that do not happen to have been much publicized as yet, or at least not well enough to have come to my attention.

No book focusing upon either intelligence and politics, or health and politics, turned up in this sample.

Environmental Sustenance

The bulk of the references noted fall in the category of environmental sustenance, the majority of these related to the public policy aspects of ecological problems, and most of the latter consisted of general discussion, including several which adopt a global perspective.[92] Much of the early literature is reviewed in a bibliographical essay (which was published in the *Review*, but in its book review section).[93] More recent writing includes a book jointly authored by a political scientist and a biologist,[94] another which focuses upon the political process of environmental policy-making,[95] and an undergraduate text intended for courses in political science or economics.[96] There are half a dozen books on population policy, including one by biologists[97] and several by a group of political scientists who have specialized in population research,[98] among others.[99] The works on energetics include two that discuss pollution,[100] another that focuses on a decisional category—that of "non-decision-making," which is reminiscent of the "negative order doctrine" of administrative law a generation ago—applied to municipal policy-making in regard to air pollution;[101] and there is only a single study of energy sources.[102] No study of niche politics appeared in the sample.

There was something of an escalation of the previously noted interest of political scientists in the political socialization of children, with four new books.[103] In regard to the other subcategory of the social environment, the experimental manipulation of psychophysical systems, nothing showed up in this sample, but I believe it should nonetheless be noted that there was a revival of interest in Skinnerian behaviorism, evidently in response to the publication in 1971 of his *Beyond Freedom and Dignity*.[104] A critique was based on a symposium that convened at the Center for the Study of Democratic Institutions;[105] and there were also three papers presented at a panel session of the Midwest Political Science Association in 1973,[106] plus the introductory "framework" section of a paper presented at the Ninth World Congress of International Political Science Association in Montreal later in the year.[107]

Biofeedback

No relevant items were noted.

Political Behavior

No biopolitical books dealing with political attitudes appeared in the sample, and even the ones that dealt with political participation and that were concerned with political violence were written from a social science rather than a biological point of view. These four books represented, however, an intensification of interest in the subject, and also something of an extensification, ranging as they do from assassinations to ideologies;[108] they are noted because of the linkage that may develop between such work and the biological-comparative psychological work on aggression. The sample includes also a book review essay which discusses several recent works dealing with chemical and biological warfare; in that essay and the books that it reviews the linkage already has been made because it is inherent in the subject matter.[109]

2. *Prospectus*

In conclusion, I shall indicate a few of the ways in which political research can use biological theory in order the better to understand, and therefore conceivably to help resolve, important and contemporary political problems. The examples that I shall suggest are in no sense intended to be exhaustive, but I do consider them to be representative of the current thinking of biologists, and also political scientists with a biological orientation, while at the same time avoiding redundancy with what already has been said in this paper.

The study of violence—both public and private, actual and potential—always has been of central concern to political scientists. Violence is of central concern to zoologists too, because of the large proportion of animal species (including Homo sapiens) whose survival in this sense depends upon their successful practice of violence. A critical question is whether violence among ourselves is essential also to the survival of our species, and if so, to what extent it can be reduced; and if not, how it can be eliminated. The value preferences implicit in the question stem, of course, from humanitarian rather than biological premises; and it is entirely possible that we might choose, if it were conclusively proved that human aggression is not only innate but critically adaptive for species survival, to act politically to eliminate or reduce violence amongst ourselves even with the knowledge that by so doing we had to accept as a consequence the extinction of our own species. We can claim, that is to say, the humanist right to die like men rather than to live as some other (and sufficiently less desirable) form of life. But any such decision—no matter which way it goes or by whom or what process made—ought to be made on the basis of the best biological knowledge that we can acquire, and not on the basis of our ignorance concerning ascertainable empirical evidence. (I consciously have avoided saying the "facts" [whatever the sacrifice in euphony] which in this context are best thought of as probability values of one type or another.)

The road to a useful (for engineering purposes) biological theory of human aggression is undoubtedly long and tortuous; and Moyer's physiological theory is far from having been tested and accepted by biology to the extent that we might be warranted in viewing it as something more than an interesting hypothesis; but it will serve my present purpose as an example of how we might proceed. Of Moyer's eight categories, two (sexual aggression, and maternal aggression) seem prima facie to be less obviously related to *political* violence than the remaining six: territorial, predatory, instrumental, inter-male, fear-induced, and irritable. An initial step in using Moyer's theory might be to examine the possible relationship of each type of aggression, to such different types of violence as war, revolution, rioting, and inter-ethnic. We might wish to examine also certain other types of violence, such as motor vehicular; although the form of violence in that instance is private, its consequences are very much involved in public policy. It would be desirable to control, for each set of aggression/violence typal relationships, for such *biological* variables as age, sex, health, and intelligence—by which I mean that we would want to use physiological measures (rather than relatively crude social indices) of these variables. Another approach would be to examine the effect of population density upon the set of typal relationships; the pioneering pilot studies of the effect of crowding upon war both reported negative findings,[110] but these were based upon designs to test quite different theories, besides which a washing out of statistical variance for methodological reasons could account for the conclusions reached. In any case one would want to investigate typal relationships involving other types of violence such as urban civil and inter-ethnic,[111] and indeed also motor vehicular and intra-familial.

Another form of political behavior that has long been of interest to political scientists is leadership, but our analyses have always treated the biological organisms of political leaders as though they were empirical blanks or statistical constants. The exception has been case studies of individual leaders in which matters of health, sexual relationships, and the like often are discussed; but such findings are left at the level of idiosyncratic historical events and not as data that could test a theory of political leadership. Recent work in primatology and human evolution has produced a widely accepted hypothesis about the structure of human social and political relationships in the long pre-agricultural era that stretches back half a million years (at least) from about 12,000 B.P.; and this "hunting band" thesis has been explicitly proposed as a model for contemporary political behavior, by Lionel Tiger and Robin Fox. What they call the "basic biogrammar" of humans is a set of propositions about dominance relationships, as a function of certain biological statuses (especially of sex and age), in a small social group.[112] Some further work needs to be done to put certain of their key concepts (e.g., "bonding") into operational (and therefore, of course, testable) form; but there already has been a substantial critique of their book and their ideas,[113] and whether or not their model (in the pristine form in which they have stated it) would find empirical support, its testing would in any event provide us with some better measure than we now have of the authors' assumption that their key variables are biogenic. According to Tiger and Fox no change at all should have occurred, because the time span of twelve thousand years is far too short for significant genetic change in the species' gene pool; and they deem it impossible (on grounds of Lamarckism)[114] for cultural changes to have affected what are hypothesized to be genetically determined behaviors. A more general approach to the use of ethological theory for purposes of political analysis has been proposed by Roger Masters, who distinguished between analogical and homological comparison in regard to three levels of spatial analysis (individual, group, and species) and also three levels of temporal analysis (structure, process, and evolution).[115] Masters' distinctions are helpful, not only in placing the hunting band thesis within the context of a much broader conceptualization of ethological research—the hunting band thesis implies homological analysis of group processes (social behavior)—but more importantly in the guidance it can provide political scientists in where and how to look in ethology (and comparative psychology) for work that will be most relevant to the empirical problems in which they are interested.

A quite different approach is provided by ecological theory; and not just in regard to substantive problems of the *physical* environment, but also in regard to social environments and political behavior. An especially promising possibility is proffered by niche theory, as applied to the analysis of political competition. The niche theory, it will be recalled, applies at several analytical levels: habitat (or geographical), trophic (or functional status: role in the food chain), and multidimensional (or hypervolume, which concerns measurability).[116] Evidently niche theory can be used to analyze various levels of competition, including that between individuals, political factions, and political parties; and in terms of units of government, both horizontally and vertically, with respect to subdivisions of local government, cities, counties, states, and national governments. Specific principles, such as that of competitive exclusion, appear to have a potentially broad application to many different types of political competition, although not by any means to all of those specified above. Furthermore, the group selection theory, which remains controversial among evolutionary biologists, will make it possible to examine many of the above types of political competition from a quite different perspective than political scientists are accustomed to utilize; for example, we might note Corning's discussion of politics conceptualized as a collective survival enterprise.[117] Still another ecological approach is that suggested by Manfred Wenner,[118] who adapted from Odum[119] a classification system for interaction analysis between two populations (with categories of neutralism, competition, parasitism, predation, amensalism, commensalism, protocooperation, and symbiosis [or mutualism]). As in the cases of niche theory and group selection theory, it seems evident that population interaction analysis may be useful in examining a broad range of political relationships at all levels of governance.

A basis concept for behavioral genetics is that of the gene pool—the hypothetical aggregation, at some finite instant in time, of all of the genes embodied in the extant phenotypic organisms of a particular species. It is not, of course, conceivable that an empirical gene pool ever can be established or observed; but the concept is nonetheless very useful in facilitating genetic analyses. As Corning has pointed out, there has been a rising interest in the use of the metaphor "symbol pool," for studies of culture;[120] and we can consider its possible application in studies of political culture. From the point of view of the development of the species, the gene pool is the survival unit,[121] the diversity and redundancy of which largely will determine the rate of change in species adaptability vis-à-vis a rapidly changing environment. Corning has proposed a theory of "human selection" to explain the evolution of culture (as an analogue of natural selection in the evolution of the human gene pool);[122] and hence the possibility arises for a completely novel reexamination and appraisal of political culture from the perspective of fundamental biological theory.

As a final pair of examples, let us consider one aspect of political process, and one of political roles. Public events on a global scale during the past year have tended to highlight what should perhaps always have been obvious: that the costing of public policies in terms of a fiscal metric (e.g., The Dollar) is a very misleading and incomplete way to attempt to appraise the price that will be paid for political activity (or inactivity)—in terms of human survival, which we can translate alternatively into a schedule of human lives, or a revised probability value for the survival of the species. Odum has remarked the obvious fallacy of continuing to reckon the resources necessary to the satisfaction of basic human (physical/chemical) survival needs—such as water and air, to say nothing of natural vegetation and deposits of fossil hydrocarbons—as free goods; he has suggested also an explicit operational measure of "power density," defined as a thousand calories per square meter per day [kcal/(m^2) (day)], to be used as a measure of the limit of requirements for a planned unit drawing upon a life-support system in which it is imbedded.[123] It may well be worth attempting to discover what kind of political decisions would be implied by such a criterion, as applied to empirical problems in fields such as public planning and public budgeting.

Neither the female liberationists nor their male chauvinist colleagues have undertaken to examine sex roles in relation to political behavior, from the perspective of a biological analysis of human sexual differences. There is sufficient evidence now available concerning various genetic and physiological (including hormonal and CNS neural) differences among humans,[124] in relationship to their development at various stages of the life cycle, that it is both feasible and desirable to investigate the possibilities for differing ranges of political role performances in the light of biological sexual differences. Masters has warned against the fallacy of postulating a simple transferal of biological differentials at the level of individuals, to presumptions about role performance.[125] He gives the example of Golda Meir and Indira Nehru Gandhi, as contradictions of the generalization that males tend to be more aggressive than females in most respects; and of course we can resolve the contradiction with a cultural explanation: the sexes differ in their means (in the sense of averages) in regard to aggression, although there is overlap between their respective distributional ranges, and only highly aggressive females can possibly achieve positions of such preeminence within dominance hierarchies of the sort posited, say, by Tiger and Fox. That may be correct; but still left open is the question whether, given a political system in which more equal participation by females is both possible and actual, what then might it be possible to say concerning "role speciation"—that is, the assignment of political roles by sex according to the statistically predictable optimality of performance? The available empirical samples of female chiefs of state may be much too small to permit direct testing with adequate data of the proposition that a substantial increase of female participation in positions of national political leadership at the highest levels would enhance chances for international peace. But that does not preclude political scientists from working on the theory, with laboratory, simulation, and other techniques of experimental research in which they already are quite skilled; and it is not inconceivable that the payoff may be worth the effort.

Notes

1. See, for example, H. J. Eysenck, *The Biological Basis of Personality,* Springfield, Ill., Charles C. Thomas, 1967; and D. P. Kimble, *Psychology as a Biological Science,* Pacific Palisades, Calif. Goodyear, 1973.
2. H. Callan, *Ethology and Society: An Anthropolitical View,* New York, Oxford University Press, 1970, attempts to synthesize ethology and social anthropology; E. D. Chapple, *Culture and Biological Man: Explorations in Behavioral Anthropology,* Holt, Rinehart, and Winston, 1970, is an ambitious endeavor to integrate cultural anthropology and behavioral biology.
3. For further discussion of some of the reasons why sociology has been slow in developing a subfield of social biology (and the journal of that name is, of course, a recently rechristened *Eugenics Quarterly,* to which sociologists occasionally contribute but in which they play a minor role), see R. Means, "Sociology, Biology, and the Analysis of Social Problems," 15, 1967, pp. 200-212 (in which his analysis of over an hundred sociology texts, turning up negative evidence of interest in biology, is reported at p. 202). However, now there is at least one introductory text in sociology that explicitly embraces an approach adapted from human evolutionary theory in biology: G. Lenski, *Human Societies: A Macrolevel Introduction to Sociology,* New York, McGraw-Hill, 1970; and Lenski has pointed out that he believes that "the rapprochement between the biological and social sciences is long overdue. Human societies *are* part of the biotic world, and by denying or minimizing this fact we impoverish both theory and research." See his Letter to the Editor, *Contemporary Sociology* 1, 1972, p. 306. See also D. W. Ball, "The Biological Bases of Human Society," ch. 5 in: J. D. Douglas (ed.), *Introduction to Sociology,* New York, The Free Press, 1973, pp. 118-138. These two examples should perhaps (to scramble the similes slightly) be viewed as further evidence of the proclivity of political science for sucking the hind teat scientifically—at least, in the pecking order of the social sciences—because there is no such introductory text, or even chapter in any introductory text, in political science to the best of my knowledge.
4. K. E. Boulding, "Economics as a Not Very Biological Science," ch. 19 in J. A. Behnke (ed.), *Challenging Biological Problems,* New York, Oxford University Press, 1972, pp. 357-375; but cf. G. Tulloch, "Biological Externalities," *Journal of Theoretical Biology* 33, 1971, pp. 565-576.

5. Of course we political scientists do have T. L. Thorson's *Biopolitics,* New York, Holt, Rinehart and Winston, 1970, which is many things—second-hand philosophy of science, exegesis upon non-cognitive notions about knowing, and other political transcendentalism dressed up in an evolutionary false face (borrowed primarily from Teilhard de Chardin)—but it is not about biopolitical behavior in any sense discussed in this paper, save perhaps in his critique of Easton's political systemizing, which he castigates as an organismic theory that goes to great pains to include no organisms in its political system. But Thorson is all hung up with his spooky dichotomization of biology and science, between which he thinks a choice is necessary (p. 96); and *he* concludes—his book concludes with another long quotation from Teilhard—with the confession that by biopolitics he means "politics understood by man *as evolution becomes conscious of itself*" (p. 208; emphasis added). I contend that the italicized words are forthrightly metaphysical. Of the same ilk is W. J. Thorbecke, *A New Dimension in Political Thinking,* New York, Oceana, 1965, which is even more inspired by Teilhard de Chardin. Unfortunately, political science has books that pretend to be biological in approach but that in fact are not so at all; and this is almost certainly worse than to have no "biopolitical" books.
6. Caldwell's essay appeared shortly after the award of the Nobel Prize to James D. Watson as co-discoverer of the structure of DNA and the genetic code, consequent to a process of human as well as scientific interaction, of which Watson himself has provided a revealing sociopsychological account in his book *The Double Helix,* New York, Atheneum, 1968.
7. L. K. Caldwell, "Biopolitics: Science, Ethics, and Public Policy," *The Yale Review* 54, 1964, pp. 1-16. For a precocious precedent effort by an English novelist, see M. Roberts, *Bio-Politics ; An Essay in the Physiology and Pathology and Politics of the Social and Somatic Organism,* London, J. M. Dent, 1938. Although Robert's book is now out of print, Caldwell has written a summation and critique of it (1970, 25 p., mimeo).
8. Cf. Caldwell's "Problems of Applied Ecology: Perceptions, Institutions, Methods, and Operational Tools," *BioScience* 16, pp. 524-527.
9. R. Pranger, "Ethology and Politics: The Work of Konrad Lorenz," paper presented at the Annual Meeting of the Southern Political Science Association, New Orleans, La., 1967; A. Somit, "Toward a More Biologically-Oriented Political Science: Ethology and Psychopharmacology," *Midwest Journal of Political Science* 12, 1968, pp. 550-567; and C. Adrian, "Implications for Political Science and Public Policy of Recent Ethological Research," paper presented at the Second International Sinological Conference, Taipei, Taiwan, The China Academy, 1969.
10. M. Haas, "Toward the Study of Biopolitics: A Cross-Sectional Analysis of Mortality Rates," *Behavioral Science* 14, 1969, pp. 257-280. See also two papers by public health specialists: E. S. Rogers / M. Yamamoto / H. B. Messinger, "Ecological Associations of Mortality in Japan and the United States: A Factor Analytic Study," paper presented at the Eleventh Pacific Science Congress, Tokyo, 1966; and E. S. Rogers / H. B. Messinger, "Human Ecology: Toward a Holistic Method," *Milbank Memorial Fund Quarterly* 45 (1), January 1, 1967, pp. 25-42.
11. R. Stauffer, "The Biopolitics of Underdevelopment," *Comparative Political Studies* 2, 1969, pp. 361-387.
12. J. C. Davies, "The Psychobiology of Political Behavior: Some Provocative Developments," paper presented at the Annual Meeting of the Western Political Science Association, Honolulu, Hawaii, 1969.
13. Somit, "Toward a More Biologically-Oriented Political Science...," *op. cit.*
14. The relevant literature through 1971 has been discussed in A. Somit, "Review Article: Biopolitics," *British Journal of Political Science* 2, 1972, pp. 209-238.
15. A. Somit / J. Tanenhaus, *American Political Science: A Portrait of a Discipline,* New York, Atherton, 1964, p. 58.
16. H. Eulau / J. G. March (eds.), *Political Science,* Englewood Cliffs, NJ, Prentice-Hall, 1969, ch. 1. An alternative (and if anything, even more official) perspective, but in this case of the development of political behavioralism during the fifties and sixties, is provided by the Committee on Political Behavior, 1949-196[4, and] Committee on Governmental and Legal Processes, 1964-1972, *A Report on the Activities of the Committees,* New York, Social Science Research Council, 1973. Also relevant are two commentaries on these reports: A. Somit, "Reports of Two Key SSRC Committees: Back to the Drafting Board," and G. Schubert, "Sauce for the Gander, or, Putting Political Behavior to Work for Political Science: A Proposal for Implementing the Somit Critique of the Reports of Two Committees of the Social Science Research Council," both in *PS* 8 (1), Winter 1975, pp. 25-27. One step ahead of

them came an appropriately optimistic commentary on the reports by the co-author of (the latter) one of them: A. Ranney, "The Committee on Political Behavior, 1949-64, and the Committee on Governmental and Legal Processes, 1964-72," *Items* 28, 1974, pp. 37-41, in which he observes that the Social Science Research "Council's impact has nowhere been clearer or more powerful than in the study of political behavior and institutions, especially but not solely as practiced within [the] discipline of political science. The early committee played a seminal role in the discipline's 'behavioral revolution' in research philosophy and methods, and the later committee did much to turn the attention of political scientists to problems of immediate social concern" (p. 37), and that "prior to 1945 the behavioral approach had very little impact on what political scientists did despite the advocacy and prestige of the 'Chicago school'; the post-1945 'behavioral revolution' has succeeded at least to the extent that behavioralism is now one of the discipline's most important outlooks, though certainly not its only outlook; it has not driven out all other approaches, nor is it likely to do so; it is equally unlikely that some counterrevolution will roll political science methodology back to where it was before 1945; and the activities of the Committee on Political Behavior constituted one of the prime forces leading to the present state of affairs" (p. 39). A less sanguine view of these same events is presented in C. A. McCoy / J. Playford (eds.), *Apolitical Parties: A Critique of Behavioralism,* New York, Crowell, 1967.

For a more dispassionate interpretation of the transition from social science to behavioral science, with emphasis upon interdisciplinary integration during the middle decades of this century, and including an analysis of an earlier Social Science Research Council committee report, see an essay co-authored by a political scientist and two psychologists, M. Landau / H. Proshansky / W. H. Ittelson, "The Interdisciplinary Approach and the Concept of Behavioral Science," ch. 1 in: N. F. Washburne (ed.), *Decisions, Values and Groups,* New York, Macmillan, 1962, pp. 7-24.

17. Robert Dahl's donning of the toga of Marc Antony in 1961—for he came, Dahl said, to bury political behavioralism, not to praise it—a subject to which we shall return presently, is explicitly contradictory of the panel's finding that there was in 1969 a *subfield* of political behavior. Dahl had said that behavioralism already at the beginning of the sixties was diffused throughout *all* fields of the discipline: see his "The Behavioral Approach," *American Political Science Review* 55, 1961, pp. 763-772, at p. 770.
18. Quite independently of the present task, I have contributed my mite toward the cause of catalyzing some change in the direction of greater modernity, in regard to at least this particular anachronism: see my Letter to the Editor, *PS* 7, Fall 1974, pp. 449-450.
19. One might also be skeptical about the wisdom of having Yalepersons back-to-back in the role of managing editor.
20. Certain questions of judgment necessarily arise as to the procedure employed: e.g., the *Review* has featured controversial confrontations of opinion in recent years, often in the pattern of thesis/antithesis/counterthesis, and then perhaps off into the "Communications" section of subsequent issues. I have counted such a sequence of writings as a single article.
21. Cf. F. E. Yates / D. J. Marsh / A. S. Iberall, "Integration of the Whole Organism: A Foundation for a Theoretical Biology," ch. 6 in: J. A. Behnke (ed.), *Challenging Biological Problems,* New York, Oxford University Press, 1972, pp. 110-132.
22. For a recent survey of the state of this research, which suggests how far psychology has moved—at least in this Jekyllian side of its schizoid disciplinary personality—from the vitalism upon which it once relied, see S. P. Grossman, *Essentials of Physiological Psychology,* New York, Wiley, 1973.
23. The cop-out prescribed by Bentley, to construct a science of politics that ignores affectivity, no longer preoccupies as many political scientists as it did a generation ago during the early days of political behavioralism. For an authentic articulation of the logic of Bentley's position, we need only recall Jack Peltason's advice to "turn our attention to the [bureaucracy] as a facet in the group struggle and relate the activities of [officials] to that of other groups," so that "we can begin to develop a political science [. . .] without trying to 'outhistory' the historian, 'out-law' the lawyer, or 'out-psychology' the psychologist." See his "A Political Science of Public Law," *Southwestern Social Science Quarterly* 34, 1953, pp. 51-56, at p. 56.
24. An experimental laboratory is a different kind of proposition, from the points of view of either public or private foundations, university administrators, and regents or legislators, than the traditional governmental research bureau—even when the latter has been redefined in social science (political behavioral) terms. And "harder" scientists are not going to welcome with open arms political scientist *entrée* to such academic baubles and spangles;

time-sharing is going to have to be earned initially by a few dedicated political scientists who will already somehow have acquired the requisite training.

25. J. A. Robinson, "Newtonianism and the Constitution," *Midwest Journal of Political Science* 1, 1957, pp. 252-266; and M. Landau, "On the Use of Metaphor in Political Analysis," *Social Research* 28, 1961, pp. 331-353.
26. D. Easton, *A Systems Analysis of Political Life,* New York, Wiley, 1965; K. Deutsch, *The Nerves of Government,* New York, The Free Press, 1964; W. Mitchell, *The American Polity: A Social and Cultural Interpretation,* New York, The Free Press, 1962; G. Almond, "Comparative Political Systems," *Journal of Politics* 18, 1956, pp. 391-409; and cf. Roberts, *Biopolitics. . ., op. cit.* See also M. Landau, "On the Use of Functional Analysis in American Political Science," *Social Research* 35, 1968, pp. 48-75.
27. I have discounted, but I have not overlooked von Bertalanffy and the general system theorists, for a discussion of whom one may consult J. Stephens, "Some Questions About a More Biologically Oriented Political Science," *Midwest Journal of Political Science* 14, 1970, pp. 687-707, his title to the seeming contrary notwithstanding. The general thrust of Stephens' argument is diametrically opposed to my own in the present paper.
28. See D. Easton, *A Framework for Political Analysis,* Englewood Cliffs, NJ, Prentice-Hall, 1965, pp. xii-xiii.
29. Except in one subfield of biology: ecology. See, e.g., E. P. Odum, *Fundamentals of Ecology,* 3rd ed., Philadelphia, Pa., W. B. Saunders, 1971; Howard T. Odum, *Environment, Power, and Society,* New York, Wiley, 1971; and W. B. Clapham, Jr., *Natural Ecosystems,* New York, Macmillan, 1973.
30. F. Crick, *Of Molecules and Men,* Seattle, Wash., University of Washington Press, 1966; J. Monod, *Chance and Necessity: An Essay on the Natural Philosophy of Modern Biology,* New York, Knopf, 1971; and P. Handler (ed.), *Biology and the Future of Man,* New York, Oxford University Press, 1970.
31. Cf. Dahl, "The Behavioral Approach", *op. cit.*
32. D. Easton, "The New Revolution in Political Science," *American Political Science Review* 63, 1969, pp. 1051-1061.
33. Cf. D. Easton, "Limits of the Equilibrium Model in Social Research," in: H. Eulau / S. J. Eldersveld / M. Janowitz, *Political Behavior: A Reader in Theory and Research,* Glencoe, Ill., The Free Press, 1956, pp. 397-404.
34. See my "The Third Cla't Theme: Wild in the Corridors," *PS* 2, 1969, pp. 591-597.
35. Including some of mine: *Judicial Policy-Making: The Political Role of the Courts,* Glenview, Ill., Scott, Foresman, 1965, especially ch. 1.
36. The Portugese Man-of-War comes readily to mind; consider: "The Portugese Man-of-War is a jellyfish that floats on the surface [. . . it] is not really a single animal, but a group of animals attached to a hollow float [. . .] that is filled with gas [. . .] Hanging from the float are long stringlike filaments called *tentacles.* These tentacles acts as arms and are used to grasp food. They contain a poison that [is] dangerous to man [. . .] touching them will [cause] painful welts, or even shock and prostration that could be fatal. All the animals that are a part of one float make up what is called a *colony.* Each animal in the colony has a different job to do. Some of them reproduce their kind. Others find food, while still others protect the colony against enemies. The structure of each animal determines the job it must do." (*The World Book Encyclopedia,* XV, 1965 ed., at p. 620; emphasis in the original.) Perhaps we have not been nearly imaginative enough in working out the implications of some of our most popular metaphors. Cf. Roberts, *Bio-Politics. . ., op. cit.,* ch. 2, especially pp. 17-18.
37. Although perhaps not, because that leaves us to reconcile the discomforting observation that "for the foreseeable future the only function of viruses is to destroy higher forms of life." (Sir F. MacFarlane Burnet, a Nobel prizewinning biologist, as quoted by Julius Stone, "Knowledge, Survival, and the Duties of Science," *American University Law Review* 23, 1973, pp. 231-261, at p. 234.)
38. See his *The Supreme Court in the American System of Government,* Cambridge, Mass., Harvard University Press, 1955, p. 82. Figure 1 is adapted from my "Academic Ideology and the Study of Adjudication," *American Political Science Review* 61, 1967, pp. 106-129, at p. 108. The original figure appears also in a companion article, "Ideologies and Attitudes, Academic and Judicial," *Journal of Politics* 29, 1967, pp. 3-40, at p. 7.
39. Vol. 68, March, 1974, pp. 11, 78, and 153. Issues including two such articles are by no means uncommon, e.g., Vol. 67, March, 1973, pp. 128, 142.
40. Vol. 66, December, 1972, p. 1226.
41. E.g., Vol. 68, June, 1974, pp. 473, 561, and 701; and Vol. 68, September, 1974, p. 1171.

42. M. A. Edey (ed.), *The Missing Link,* New York, Time-Life Books, 1972; M. H. Fried, *The Evolution of Political Society,* New York, Random House, 1967; D. Pilbeam, *The Ascent of Man,* New York, Macmillan, 1972; and L. Tiger / R. Fox, *The Imperial Animal,* New York, Holt, Rinehart and Winston, 1971.
43. I speak advisedly here, choosing the terminology of those who prefer a literary phenomenological approach, like the disciples of the late Leo Strauss. For examples of such discourse in regard to political behavioralism, see W. Berns, "Law and Behavioral Science," *Law and Contemporary Problems* 28, 1963, pp. 185-212; and H. Storing (ed.), *Essays on the Scientific Study of Politics,* New York, Holt, Rinehart and Winston, 1962.
44. S. S. Tomkins, "The Primary Site of the Affects: The Face," ch. 7 in his *Affect Imagery Consciousness,* Vol. 1. *The Positive Affects,* New York, Springer, 1962, pp. 204-242, at p. 206; and cf. H. Leventhal / E. Sharp, "Facial Expressions as Indicators of Distress," and R. V. Exline with L. C. Winters, "Affective Relations and Mutual Glances in Dyads," Part 6 in: S. S. Tomkins and C. E. Izard (eds.), *Affect, Cognition, and Personality: Empirical Studies,* New York. Springer, 1965, pp. 296-350. Cf. H. Haas, *The Human Animal: The Mystery of Man's Behavior,* New York, Dell ed., 1972.
45. P. A. Corning, "The Biological Bases of Behavior and Some Implications for Political Science," *World Politics* 23, 1971, pp. 321-370, at p. 339, quoting C. Bay, "Politics and Pseudopolitics: A Critical Evaluation of Some Behavioral Literature," *American Political Science Review* 59, 1965, pp. 39-51, at p. 40.
46. New York, Wiley, 1963.
47. See V. Van Dyke, review of: M. Surkin / A. Wolfe (eds.), *An End to Political Science: The Caucus Papers,* New York, Basic Books, 1970; in *American Political Science Review* 65, 1971, pp. 793-794.
48. Cf. G. Hardin, *Exploring New Ethics for Survival: The Voyage of the Spaceship Beagle,* New York, Viking Press, 1972; and for a contrary view, see E. Vlachos, "Doomsday and Ecocatastrophe: Dystopia Today," paper presented at the Eighth World Congress of Sociology, Toronto, 1974, and political scientist B. L. Crowe's "The Tragedy of the Commons Revisited," *Science* 166 (3909), November 28, 1969, pp. 1103-1107.
49. Among the really quite talented professional psychologists whose careers provide the kind of data that one would want to examine in such a cost-benefit analysis are Dr. John C. Lilly, who managed to convince himself that dolphins were talking to him before he moved on to Esalen; and Dr. Timothy Leary, concerning whom one is entitled to say that his public life since he left Harvard almost blows one's mind.
50. It should be remarked that James C. Davies' more recent work has been strongly in this direction, and outstandingly so among political scientists, as exemplified by his paper for the present conference, "Ions of Emotion and Political Behavior: A Prototheory," see above pp. 97-125.
51. See W. E. McAlpine, "Information Reduction Processes and Politics," paper presented to the Eighth World Congress of the International Political Science Association, Munich, 1970 (hereinafter cited as "IPSA, 1970").
52. Also a somewhat larger number of economists: see, e.g., L. G. Hines, *Environmental Issues: Population, Pollution, and Economics,* New York, Norton, 1973, especially ch. 9-12; and for an interdisciplinary symposium (including economist contributors, among others) see P. B. Downing (ed.), *Air Pollution and the Social Sciences: Formulating and Implementing Control Programs,* New York, Praeger, 1971, of which ch. 4 is by R. O. Loveridge, on "Political Science and Air Pollution: A Review and Assessment of the Literature," pp. 45-85.
53. L. I. Gardner, "Deprivation Dwarfism," *Scientific American* 227 (1), July 1972, pp. 76-82; and P. H. Mussen / J. J. Conger / J. Kagan, *Child Development and Personality,* 3rd ed., New York, Harper and Row, 1969, pp. 228-236.
54. For an attempt to do this, see Davies, *Human Nature in Politics, op. cit.,* pp. 45-60; and cf. J. Nash, *Developmental Psychology: A Psychobiological Approach,* Englewood Cliffs, NJ, Prentice-Hall, 1970.
55. See E. P. Odum, *Fundamentals of Ecology, op. cit.,* pp. 214, 234-236, 510-516.
56. Clapham, *Natural Ecosystems, op. cit.,* pp. 234, 236.
57. Cf. D. Jaros, "Biochemical Desocialization: Depressants and Political Behavior," *Midwest Journal of Political Science* 16, 1972, pp. 1-28.
58. See N. E. Cutler / V. L. Bengtson, "Age and Political Alienation: Maturation, Generation and Period Effects," *The Annals of the American Academy of Political and Social Science* (415), September 1974, pp. 160-175. See also N. E. Cutler, "Aging and Generations in Politics: The Conflict of Explanations and Inference," in: A. R. Wilcox (ed.), *Public Opinion and Political*

Attitudes, New York, Wiley, 1974, at pp. 440-462; and D. C. Schwartz, *Political Alienation and Political Behavior*, Chicago, Ill., Aldine, 1973.

59. See the "public law" decisions of the United States Supreme Court in: *Buck* v. *Bell*, 274 US 200, 1927; and *Skinner* v. *Oklahoma*, 316 US 535, 1942 (sterilization); and their precedent, *Jacobson* v. *Massachusetts*, 197 US 11, 1905 (vaccination).
60. M. S. Frankel, "Political Responses to Controversial Issues in the Development of Biomedical Technologies," paper presented at the Seventieth Annual Meeting of the American Political Science Association, Chicago, 1974.
61. See A. Chase, *The Biological Imperatives: Health, Politics, and Human Survival*, New York, Holt, Rinehart, and Winston, 1971.
62. The same political scientists who proclaim that "race" is an unscientific concept with no operational meaning—although "ethnic" is quite satisfactory—are also persons who insist that race be accepted implicitly as a basis for representation in the councils, and explicitly in the organization of and allocation of budgets to the committees, of the American Political Science Association.
63. R. F. Herrnstein, *IQ in the Meritocracy*, Boston, Mass., Little, Brown, 1973. See also H. J. Eysenck, *The IQ Argument: Race, Intelligence, and Education*, LaSalle, Ill., Library Press, 1971.
64. Among many possible examples, see his *The Future of Political Science*, New York, Atherton, 1963.
65. P. Bernstein / D. C. Schwartz, "A Note on the Impact of Health on Presidential Decision Making," paper presented to the Ninth World Congress of the International Political Science Association, Montreal, 1973 (hereinafter cited as "IPSA, 1973").
66. D. C. Schwartz,/J. Garrison/J. Alouf, "Health and Body Image Correlates of Political Attitudes and Behaviors," "Health Processes and Body Images as Predictors of Political Attitudes and Behaviors: A Study in Political Socialization," IPSA, 1973; and T. C. Wiegele/ S. Plowman/R. Carey, "International Crisis, Cardiorespiratory Health, and Political Attitudes: A Literature Review and a Pilot Study," IPSA, 1973.
67. Stauffer, "The Biopolitics of Underdevelopment," *op. cit.*
68. Cf. J. F. Gubrium, *The Myth of the Golden Years: A Socio-Environmental Theory of Aging*, Springfield, Ill., Charles C. Thomas, 1973.
69. For an introduction to the subject, see Grossman, *Essentials of Physiological Psychology, op. cit.*; and see several works by S. Fisher: with co-author S. Cleveland, *Body Image and Personality*, Princeton, NJ, Van Nostrand, 1958; his *Body Experience in Fantasy and Behavior*, New York, Appleton-Century-Crofts, 1970; and a more recent work of his written for a mass audience, *Body Consciousness: You Are What You Feel*, Englewood Cliffs, NJ, Prentice-Hall, 1973. I am aware of the reversed and more esoteric meaning that humanistic psychologists associate with the term "biofeedback;" and my preference here for using the word in its more zoological signification should be understood as explicitly in pursuance of K. Burke's advice on the "Stealing Back and Forth of Symbols," in his *Attitudes toward History*, New York, New Republic Press, 1937, Vol. 2, pp. 229-230.
70. For a review of the biological literature, see P. A. Corning / C. H. Corning, "Toward a General Theory of Violent Aggression," *Social Science Information* 11 (3/4), June-August 1972, pp. 7-35; and P. A. Corning, "Human Violence: Some Causes and Implications," in: C. R. Beitz / T. Herman, *Peace and War*, San Francisco, Calif., W. H. Freeman, 1973.
71. For an approach, at the level of cultural change, to the social problem of intraspecific violence, see the "Proposal for Civil Disarmament" in my "The Rhetoric of Constitutional Change," *Journal of Public Law* 16, 1967, pp. 16-50, at. pp. 44-50.
72. For a reproduction of Lorenz' somewhat Rube Goldberg-ish drawing, see A. Manning, *An Introduction to Animal Behavior*, Reading, Mass., Addison-Wesley, 1967, p. 65.
73. K. E. Moyer, *The Physiology of Hostility*, Chicago, Ill., Markham, 1971, especially ch. 3 through 5. See also the discussion in P. A. Corning, "An Evolutionary Paradigm for the Study of Human Aggression," in: M. Nettleship / *et al.* (eds.), *War: Its Causes and Correlates*, in press.
74. Manning, *An Introduction to Animal Behavior, op. cit.*, remarks that "there is plenty of evidence that the ease with which external stimuli can release an aggressive response is increased by *testosterone*, the male hormone, which also increases sexual responses." (Emphasis in the original.) Sexual aggression is one of Moyer's eight categories of animal aggression, although it is notable that it was not initially so and it remains the least well-defined (by him, at least) of his types. With explicit reference to humans, John Dearden has pointed out that "males dominate and are aggressive in all areas of behavior, an exception being that females

are more verbally aggressive," a finding that has been confirmed for several other primate species: see his "Sex Linked Differences of Political Behavior: An Investigation of their Possibly Innate Origins," *Social Science Information* 13 (2), April 1974, pp. 19-46, at p. 28.

75. See K. Larsen, "Aggression and Social Cost," *Peace Research Reviews* 5 (1), January 1973, pp. 1-104, at pp. 77-104.
76. J. Wahlke / M. G. Lodge, "Psychophysiological Measures of Political Attitudes and Behavior," *Midwest Journal of Political Science* 6, 1972, pp. 505-537.
77. M. Lodge / B. Tursky / J. Tanenhaus / D. Cross, "The Cross-Modal, Multiple Indicator Analysis of Political Behavior: The Case of Psychophysiology," Stony Brook, NY, State University of New York at Stony Brook, Laboratory for Behavioral Research, Report 1, no date [1973]; M. Lodge / B. Tursky / J. Tanenhaus / D. Cross, "The Development and Validation of Political Attitude Scales: A Psychophysical Approach," Report 2, no date [1973]; and M. Lodge / D. Cross / B. Tursky / J. Tanenhaus, "The Psychophysical Scaling and Validation of a Political Support Scale," Report 3, 1974.
78. Schwartz, "Health Processes and Body Images...," *op. cit.;* and P. Shubs, "Political Correlates of Self-Body Image," IPSA, 1973.
79. One is reminded of Harold Lasswell's famous definition of political behavior: that it is what happens when private motives are displaced onto public objects and then become rationalized in the public interest. See his *Psychopathology and Politics,* New York, Viking Press, 1960 ed., pp. 261-262.
80. D. C. Schwartz / N. Zill, "Psychophysiological Arousal as a Predictor of Political Participation," 1972, mimeo.
81. See my "Behavioral Jurisprudence," *Law and Society Review* 2, 1968, pp. 407-428, Table 1 at p. 415.
82. *Ibid.,* Table III at p. 417. At least to this extent a biological approach functions as an alternative to a Freudian one.
83. Neither were any biopolitical articles included in the twelfth (December 1974) issue which appeared after this paper was presented, nor among the forty-six "Articles Accepted for Future Publication"—virtually equivalent to what could be published during the entire next year (1975)—listed at pp. 1479-1481 of that issue.
84. Vol. 67, March, 1973, p. 161; and June, 1973, p. 415.
85. Vol. 67, December, 1973, p. 1288; Vol. 68, March, 1974, p. 93.
86. Vol. 66, September, 1972, p. 928; Vol. 67, June, 1973, p. 514.
87. C. Andreas, *Sex and Caste in America,* Englewood Cliffs, NJ, Prentice-Hall, 1971; J. Kirkpatrick, *Political Woman,* New York, Basic Books, 1974; and W. H. Chafe, *The American Woman: Her Changing Social, Economic and Political Role, 1920-1970,* New York, Oxford University Press, 1972.
88. L. J. Friedman, *The White Savage: Racial Fantasies in the Postbellum South,* Englewood Cliffs, NJ, Prentice-Hall, 1970; and J. Kovel, *White Racism: A Psychohistory,* New York, Vintage, 1971; E. S. Greenberg / N. Milner / D. J. Olson, *Black Politics, The Inevitability of Conflict: Readings,* New York, Holt, Rinehart and Winston, 1971; and H. Walton, Jr., *Black Politics: A Theoretical and Structural Analysis,* Philadelphia, Pa., J. B. Lippincott, 1972.
89. Z. Blau, *Old Age in a Changing Society,* New York, New Viewpoints, 1973.
90. Sponsored by the Population and Reproduction Grants Branch, Center for Population Research, National Institute of Child Health and Human Development, Bethesda, Md. 20014.
91. For example, F. R. Eisele (ed.), "Political Consequences of Aging," *Annals of the American Academy of Political and Social Science* (415), September 1974, pp. 1-212.
92. L. Caldwell, *Environment: A Challenge for Modern Society,* Garden City, NY., Natural History Press, 1970; and *In Defense of Earth: International Protection of the Biosphere,* Bloomington, Ind., Indiana University Press, 1972; H. Sprout / M. Sprout, *Toward a Politics of the Planet Earth,* New York, Van Nostrand Reinhold, 1971; and R. A. Falk, *This Endangered Planet,* New York, Vintage, 1971.
93. C. O. Jones, "From Gold to Garbage: A Bibliographical Essay on Politics and the Environment," *American Political Science Review* 66, 1972, pp. 588-595.
94. D. Pirages / P. R. Ehrlich, *Ark II: Social Response to Environmental Imperatives,* San Francisco, Calif., W. H. Freeman, 1973.
95. W. A. Rosenbaum, *The Politics of Environmental Concern,* New York, Praeger, 1973.
96. G. J. C. Smith / J. J. Steck / G. Surette, *Our Environmental Crisis: Its Biological, Economic and Political Dimensions,* New York, Macmillan, 1974.
97. P. H. Ehrlich / A. H. Ehrlich / J. P. Holdren, *Human Ecology: Problems and Solutions,* San Francisco, Calif., W. H. Freeman, 1973.

98. R. L. Clinton, *Population and Politics: New Directions in Political Science Research,* Lexington, Mass., D. C. Health, 1973; R. L. Clinton / R. K. Godwin (eds.), *Research in the Politics of Population,* Lexington, Mass., D. C. Heath, 1972; and R. Clinton / W. Flash / R. K. Godwin (eds.), *Political Science in Population Studies,* Lexington, Mass., D. C. Heath, 1972.
99. N. Chamberlain, *Beyond Malthus: Population and Power,* Cambridge, Mass., Harvard University Press, 1972; and L. R. Brown, *In the Human Interest: A Strategy to Stabilize World Population,* New York, Norton, 1974.
100. C. Davies, III, *Politics of Pollution,* New York, Pegasus, 1970; and D. F. Paulsen / R. B. Dehnardt, *Pollution and Public Policy: A Book of Readings,* New York, Dodd, Mead, 1974.
101. M. A. Crenson, *The Un-Politics of Air Pollution: A Study of Non-Decision-making in the Cities,* Baltimore, Md., Johns Hopkins University Press, 1971.
102. D. H. Davis, *Energy Politics,* New York, St. Martin's Press, 1974.
103. C. F. Andrain, *Children and Civic Awareness,* Columbus, Ohio, Charles E. Merrill, 1971; R. M. Merelman, *Political Socialization and Educational Climates: A Study of Two School Districts,* New York, Holt, Rinehart and Winston, 1971; H. Tolley, Jr., *Children and War: Political Socialization to International Conflict,* New York, Teachers College Press, 1972; and D. Jaros, *Socialization to Politics,* New York, Praeger, 1973. The earlier literature is listed in the bibliography to K. P. Langton, *Political Socialization,* New York, Oxford University Press, 1969, pp. 185-210.
104. By B. F. Skinner, New York, Knopf.
105. H. Wheeler (ed.), *Beyond the Punitive Society; Operant Conditioning: Social and Political Aspects,* San Francisco, Calif., W. H. Freeman, 1973.
106. M. W. Watts, "B. F. Skinner and the Technological Control of Social Behavior" *American Political Science Review* 69, 1975, pp. 214-227, 230-237; D. D. Dabelko, "B. F. Skinner and Political Authority;" and S. W. Panyan, "B. F. Skinner's Design for Control: Totalitarian, Humanistic or Democratic ?", Papers presented to the Midwest Conference of Political Scientists, Chicago, Ill., 1973.
107. M. Lodge / B. Tursky / J. Tanenhaus, "A Bio-Behavioral Framework for the Analysis of Political Behavior," at pp. 1-15.
108. W. J. Crotty, *Assassinations and the Political Order,* New York, University Press, 1972; Fred R. von der Mehden, *Comparative Political Violence,* Englewood Cliffs, NJ, Prentice-Hall, 1972; H. Hirsch / D. C. Perry, *Violence as Politics: A Series of Original Essays,* New York, Harper and Row, 1973; K. W. Grundy / M. A. Weinstein, *The Ideologies of Violence,* Columbus, Ohio, Charles E. Merrill, 1974.
109. G. H. Quester, "Chemical and Biological Warfare," *American Political Science Review* 68, 1974, pp. 1285-1291.
110. S. Bremer / J. D. Singer / U. Luterbacher, "The Population Density and War Proneness of Nations, 1816-1965," *Comparative Political Studies* 6, 1973, pp. 329-348; and S. Welch / A. Booth, "Crowding as a Factor in Political Aggression: Theoretical Aspects and an Analysis of Some Cross National Data," *Social Science Information* 13 (4/5), August-October 1974, pp. 151-162.
111. Typically these two types are inextricably fused (and therefore, according to the theory presented here, confused) in the analysis; see, for example, J. R. Feagin / H. Hahn, *Ghetto Revolts: The Politics of Violence in American Cities,* New York, Macmillan, 1973.
112. L. Tiger, "Dominance in Human Societies," *Annual Review of Ecology and Systematics* 1, 1970, pp. 287-306; and also G. Maclay/H. Knipe, *The Dominant Man: The Pecking Order in Human Society,* New York, Dell, 1972, a perfectly fowl book that appears in the same series, Delta Books on Anthropology and Ethology, as the paperback edition of *The Imperial Animal, op. cit.*
113. L. Tiger / R. Fox, "Animal, Venerable, Imperial: An Adventure Story," *The Columbia Forum* 3[?], Fall 1973, pp. 26-31; and cf. M. Fried, "Mankind Excluding Woman," *Science* (N.S.) 165, August 29, 1969, p. 884, and my "Biopolitical Behavior: The Nature of the Political Animal," *Polity* 6, 1973, pp. 240-276.
114. For a political analysis by a biologist, of Soviet Lysenkoism, see I. M. Lerner, *Heredity, Evolution and Society,* San Francisco, Calif., W. H. Freeman, 1968, ch. 21: "Genetics and Politics," at pp. 277-286.
115. R. D. Masters, "Functional Approaches to Analogical Comparisons between Species," *Social Science Information* 12 (4), April 1973, pp. 7-28.
116. For an example of ecological analysis at all three levels, see R. H. MacArthur, *Geographical Ecology: Patterns in the Distributions of Species,* New York, Harper and Row, 1972.
117. See the discussion in P. A. Corning, "Evolutionary Indicators: Applying the Theory of

Evolution to Political Science," University of Colorado, Institute of Behavioral Genetics, 1971, p. 38.

118. M. Wenner, "Symbiosis and Politics: Notes toward the Use of Biological Models in Political Science," IPSA, 1970 (revised), especially p. 6.

119. Odum, *Fundamentals of Ecology, op. cit.*, p. 211.

120. See his "Human Nature Redivivus," in: J. R. Pennock / J. W. Chapman (eds.), *Human Nature NOMOS* XVII, New York, Lieber-Atherton, 1975 [6, in press]. My own reference is to a manuscript copy of this paper, section 6.

121. Masters, "Functional Approaches...," *op. cit.*, p. 115.

122. See his "Politics and the Evolutionary Process," pp. 253-289 in: T. Dobzhansky/*et al.* (eds.), *Evolutionary Biology*, Vol. 7, New York, Appleton-Century-Crofts, 1975 [4]. See also Roger Masters' discussion of the striking parallelism in the structure of human langage, and that of DNA, in his "Genes, Language, and Evolution," *Semiotica* 2, 1970, pp. 295-320.

123. Odum, *Fundamentals of Ecology, op. cit.*, pp. 45-46, 300-301. For an earlier step in this direction, which remains tied to fiscal measurement but is right on target in spirit and approach, see K. W. Kapp, *The Social Costs of Private Enterprise*, Cambridge, Mass., Harvard University Press, 1950 (also now available in a paperback edition, Schocken, 1970).

124. See R. G. D'Andrade, "Sex Differences and Cultural Institutions," in: E. E. Maccoby (ed.), *The Development of Sex Differences*, Stanford, Calif., Stanford University Press, 1966.

125. R. D. Masters, "On Comparing Humans—and Human Politics—with Animal Behavior," IPSA, 1973, p. 10. For a sampling of the empirical research, see R. B. Zajonc (ed.), *Animal Social Psychology: A Reader of Experimental Studies*, New York, Wiley, 1969; and also his *Animal Social Behavior*, New York, General Learning Press, 1972; and W. H. Thorpe, *Animal Nature and Human Nature*, Garden City, NY, Doubleday, 1974; and see especially the magnificent new opus by Edward O. Wilson, *Social Biology*, Cambridge, Mass., Harvard University's Belknap Press, 1975.

ROGER D. MASTERS

The Impact of Ethology on Political Science

AMONG THE AREAS of contemporary biology relevant to the social sciences, ethology (the comparative study of animal behavior) has lately received considerable attention. In addition to suggesting that biology, rather than physics, is the epistemological norm for the social sciences, ethology promises to make substantive theoretical and empirical contributions to political science. Konrad Lorenz's theories, as modified by subsequent research, suggest an O-S-R model of behavior that is both more general and more fruitful than the conventional S-R approach. From an ethological perspective, politics can be defined as "behavior which simultaneously partakes of the attributes of dominance and submission (which the human primate shares with many other mammals) and those of legal or customary regulation of social life (characteristic of human groups endowed with language)." Such a definition suggests new empirical propositions, based on concepts and phenomena studied in animal populations, which can be applied to politics without denying that humans differ from other species. Especially notable are the relation of "attention structure" to political power, the role of non-verbal communication (kinesics) in leadership behavior, and the spread of "moods" through human societies. Preliminary empirical data from a study of the 1972 American Presidential campaign illustrate the fruitfulness of several of the seventeen researchable hypotheses proposed.

Customarily, modern political scientists have not viewed biology as especially germane to their discipline. Recently, however, it has been argued that biological variables and theories are highly relevant to political science.[1] The term "biopolitics" has gained increasing currency, and a number of panel discussions and conferences have been devoted to the subject.

For most political scientists, however, it remains to be seen what—if any—utility is to be found in "biopolitics." While the scholars whose work could be described by this term cover a range of methods, theoretical approaches, and research interests, many relate ethology (the study of animal behavior) to human politics. If "biopolitics" is to emerge as a viable scholarly approach, it is therefore especially appropriate to assess the contribution of ethological research to the study of politics.

Three areas in which ethology seems important to the work of political scientists should be distinguished: epistemological or methodological norms, theory, and empirical propositions. In all three, I would like to argue that contemporary ethology in general—and the approach of Konrad Lorenz in particular—is making an important contribution to contemporary political science.

I. Epistemological and Methodological Implications: Physics vs. Biology as the Paradigm for a Science of Politics

From the first serious attempts to create a rigorous *science* of politics in the modern sense (as distinct from an "art" or "philosophy" of politics), it seems fair to say that the prevalent model of a true science has been either mathematics or physics. Among political theorists, from the mechanism of Hobbes and eighteenth century philosophers like Helvetius to the nineteenth century positivism of Comte, physics increasingly became the standard of what would be scientific in a science of politics. Indeed, one eighteenth century group which pretended to have formulated a science of politics—the physiocrats—symbolizes this tradition in its very name.

Despite the vagaries of Social Darwinism—in any event a perspective most often dismissed as an ideology—twentieth century political science has been dominated by the assumption that a fully developed political science would resemble physics. One need only recall the influential writings of Arthur F. Bentley, whose image of a "parallelogram of forces" served not merely as a metaphor for his own epistemology, but as the methodological basis for over a generation of scholarship (Bentley, 1908; Truman, 1951; and the extensive literature on the "group theory of politics"). Even the names of such phenomena as political *movements, pressure* groups, *power,* political *stability,* and party *realignment* reflect the tacit assumption that, of the natural sciences, physics is the scientific ideal toward which empirical political science should aspire.

In many ways, this preconception was ironic. Long after the discoveries of quantum mechanics led physicists like Bohr and Heisenberg to challenge a positivistic epistemology (Bohr, 1958; Heisenberg, 1958), the influence of the Vienna Circle combined with the behaviorist tradition of American psychology to perpetuate—at least in the United States—the norm of physics as the deterministic science par excellence. Be that as it may, even in recent years political scientists have assumed—by implication if not explicitly—that their methodological ideal is classical physics (e.g., Goldberg, 1963).

Although the study of ethology and its application to human behavior has not been the only factor in challenging this attitude, the popularized works of Robert Ardrey, Konrad Lorenz, and Desmond Morris have reflected a movement from physics to biology as the scientific model to which political science should aspire.[2] To be sure, what has often been called the "biological revolution"—and notably the extraordinary advances in molecular biology and biochemistry—have contributed to a renewed interest in the biological sciences. Moreover, it was George Gaylord Simpson (1969, p. vii), an exponent of the neo-Darwinian "synthetic" theory of evolution rather than an ethologist or biochemist, who asserted that "biology [. . .] and no longer mathematics, is now the queen of the sciences."

There are, it might be added, numerous considerations which converge to support this epistemological and methodological shift. Like biology—and unlike classical physics—the social sciences study *populations* of organisms that change over time. Like biology—and unlike classical physics—time is an essentially irreversible variable of decisive importance in most of the phenomena analyzed by political scientists. Like biology—and unlike classical physics—the perfectly controlled experiment is difficult if not impossible in political science.[3] Like biology—and unlike classical physics—some form of teleological or functional reasoning seems inherent in political life.[4] Finally, like biology—and unlike classical physics—political science studies complex systems (human societies) which are self-replicating organizations of information.[5] If nothing else, the convergence between biology and what has come to be called "structuralism" in anthropology and linguistics (e.g., Stent, 1972) suggests the importance of the parallels between biological and social science.

It can be argued, however, that the widespread interest in and respect for biology would not in itself have led political scientists to take a biologist like Simpson seriously when he asserted: "I am content to define the social sciences as those branches of biology dealing with organisms that have language" (1969, pp. 29-30). Rather, the emergence of ethology as a sub-field of biology devoted to the comparative study of animal behavior, and especially its popularization by authors who included human behavior in their comparisons, has apparently encouraged many social scientists to consider more seriously the kinship of their disciplines to biology.

For example, discussion of the propriety of "analogies" between humans and other species has stimulated a better understanding of the concept of analogy in comparative morphology and ethology—and therewith a more balanced assessment of the debates between Lorenzian ethologists and American behaviorist psychologists.[6] Comparison of human behavior—and particularly such political phenomena as dominance hierarchies, aggression, socialization, group fission, and territoriality—with animal behavior has therefore begun to contribute to a reconsideration of physics as the epistemological norm of political science.[7]

Whatever the theories or empirical propositions that political scientists may borrow or derive from ethology, this shift in perspective may have exceptionally profound effects on the discipline. At the risk of using a word rendered trite by overuse, at this level it is entirely possible that political science is in the process of what Kuhn called a change of paradigm.

II. Theoretical Implications: Behaviorism vs. Lorenzian Ethology

1. *An Historical Note*

The comparative study of animal behavior has taken two forms in this century, each of which has had a marked influence on political science. The first is generally known as the "behaviorist" movement, and had its origins in American experimental psychology. The second—inspired by the theoretical works of Konrad Lorenz and the naturalistic observation of animals—is usually called ethology.

In practice, these two schools have converged to a surprising degree in the last decade (e.g., Dethier / Stellar, 1970, pp. 98-99). For example, Harry Harlow—who began his career as a behaviorist psychologist working on classical conditioning (Harlow, 1939)—has more recently focused on the behavioral comparisons between primates and humans (Harlow, 1971). Many ethologists, particularly in the younger generation, find the once bitter debates between American behaviorists and Lorenz or his European colleagues (e.g., the discussions in Autuori / *et al.*, 1956) to be sterile and outmoded.

It is nonetheless characteristic of the relations between academic fields that a debate now largely settled on home territory continues to reverberate in neighboring disciplines. As a movement in American political science, behaviorism fully came into its own in the 1950's (Dahl, 1961). By that time, Lorenz had been attacking the methods and theories of behaviorist psychology for two decades (see his papers of the 1930's, which have only recently been translated; Lorenz, 1970, 1971). Yet such is the nature of interdisciplinary frontiers that Lorenz's ethological theory had little if any role when American political scientists debated the virtues of behaviorism after the Second World War.[8]

A difference in the organization of academic disciplines in Europe and the United States contributed to the tendency of political scientists to ignore Lorenz's work even when they followed experimental comparisons between humans and animals by such American behaviorists as B. F. Skinner. Whereas political science had long been an autonomous discipline in the United States, the study of politics in Europe only began to free itself from faculties of law or sociology after World War II (at least in many continental universities); Europeans were not generally attracted to the use of scientific methodology in the study of politics until they came in contact with American political scientists in the 1950's and 1960's. Hence methods developed in the "harder" social sciences tended to spread in American political science before they were adopted in Europe.

At the same time, comparative animal behavior was also in a different academic situation on the two sides of the Atlantic. In the United States, this field had long been a sub-field of psychology called "Comparative Psychology" (e.g., Ratner / Denny, 1964). And since Americans generally classify psychology as a social science, the experimental work of American behaviorists—particularly in the area of conditioning—was well known to many political scientists. In Europe, by contrast, the study of comparative animal behavior tended to develop as a sub-field of zoology. Even where American biologists studied animal behavior, moreover, they tended to emphasize physiological mechanisms and processes, unlike European specialists who were often trained in comparative morphology and therefore turned their attention to the relevance of behavioral traits in phylogenetic classification.

These historical details are perhaps not irrelevant to the peculiar way that ethology came to the attention of political scientists. Non-technical "best-sellers" by an American dramatist (Robert Ardrey) and an English ethologist (Desmond Morris) combined with the translation of Lorenz's popularization of his own theories to attract attention to behavioral comparisons between humans and other animals. Hence American social scientists did not become aware of ethology by means of academic training and scholarship, based on familiarity with the theories and experimental observations of European ethologists; instead, literary and quasi-journalistic reports directed to a mass public helped diffuse a perspective which had previously been largely unknown in the United States.

This historical circumstance explains several peculiarities of the discussion of ethology by American social scientists over the last ten years. Although Lorenz was widely viewed as the theoretical founder of European ethology, even before he received the Nobel Prize in 1973 along with Tinbergen and von Frisch, his position was first known and widely attacked on the basis of his popularized exposition in *On Aggression* (e.g., Montagu, 1968). With a few exceptions (e.g., Lorenz, 1956, 1965), Lorenz's scientific papers were not even published in English or French until *after* his thought had been widely discussed and frequently criticized.

Even the few technical writings by Lorenz which were available in English at the time of the translation of *On Aggression* were generally ignored by the social scientists who attacked him. To my knowledge, the only hostile review of *On Aggression* that also focused on *Evolution and the Modification of Behavior* (1965) was by T. C. Schneirla (Montagu, 1968, pp. 59-64), an eminent comparative psychologist who had long known and debated Lorenz (e.g., Autuori / *et al.,* 1956, pp. 387-439 et passim). Hence most American political scientists first learned of the ethological comparison of human behavior to that of other animals with little or no consideration of Lorenz's scientific theory.

This historical note explains why discussion of the implications of comparative animal behavior for political science took the form of a debate between American behavioral psychology and Lorenzian ethology. To be sure, this debate—which in any case never prevented the gathering of observational descriptions of animals in field settings—has now been largely superseded among specialists. But the hypotheses and empirical data of contemporary ethologists cannot be fully understood without reference to their theoretical basis. And since almost all specialists in comparative animal behavior accept at least some aspects of Lorenzian theory, it is particularly important to show how the ethological theory of Lorenz differs from the behaviorist approach so well known by American political scientists.

2. *Lorenz's Theory of Animal Behavior*

For American behavioral psychology, all behavior—whether of humans or lower animals—has long been analyzed in terms of a "stimulus-response" (S-R) model.[9] Variations in the behavior of organisms are thus viewed as different *responses* to changed environmental conditions, perceived as stimuli or causes. Although organisms are presumed to have drives or motivations, the latter are treated as diffuse sources of activity (e.g., hunger, sex) which do not explain differences in the behavior of individuals or species. Indeed, according to one extreme view, it was denied that any animal behavior could be described as innate, on the grounds that conditioning or variable "development" is the cause of *all* animal behavior (e.g., Kuo, 1967).

Lorenz's ethology represents a fundamental attack on the S-R model.[10] Lorenz set out to study animal behaviors which were species-specific characteristics, varying in terms of phylogenetic inheritance rather than individual conditioning. As a result, Lorenz focused on those behaviors he called "Fixed Action Patterns," which are preprogrammed genetically and merely triggered by an environmental configuration (the "releaser" and "innate releasing mechanism"). Rather than diffuse or generalized drives which could not account for the specific behaviors observed, Lorenz postulated quite narrowly circumscribed motivations or appetites corresponding to discrete stages in a complex sequence leading to a "consummatory behavior" (e.g., eating, copulation). Although Lorenz has never denied the existence or importance of conditioning processes, he sought to specify their exact role in behavior by showing that many traits simply cannot be explained with the S-R model.

One could describe Lorenz's theory as an O-S-R model of behavior, since he begins from inherited properties of the organism rather than from environmental stimuli. Indeed, one of Lorenz's major contributions has been to focus attention on the problem of why a stimulus is a stimulus—i.e., why an organism responds to a particular environmental configuration at a given time.[11] Whereas behaviorists like Skinner tend to treat the organism as a "black box" between the stimulus and response (S-O-R), Lorenz ***begins*** from the characteristics of the organism as a prior determinant of the stimulus capable of influencing its behavior (O-S-R).

Lorenz thus divides what had formerly been called instincts or "drives" into at least three distinct components: motivation or appetite; innate releasing mechanism (IRM) and releaser; and observable behavior or Fixed Action Pattern. But although these three components of innate behavior correspond roughly to the O-S-R sequence, Lorenz (1971, p. 206) has explicitly pointed out that it would be a mistake to treat his theory in such simplistic terms. On the contrary, his main contribution to the study of behavior has probably been his emphasis on its complexity—and more particularly his demonstration that many behaviors represent lengthy sequences of interaction between an organism's appetites, environmental stimuli, and behavioral responses.

These behavioral chains are often initiated, according to Lorenz, by the organism itself (rather than by some environmental "cause"). Hence Lorenz's theory assumed the importance of endogenous stimuli, continuously generated by the central nervous system. Such endogenous stimulation triggers appetitive behavior, which Lorenz views as a tendency of the organism to seek the environmental setting corresponding to a specific activity or motor coordination.[12] Performance of this motor coordination is, however, normally inhibited by the central nervous system until the organism perceives a "key stimulus" or innate releasing schema.

Since motivations or appetites are related to specific behaviors, they can form hierarchical sequences. In a simplified case, endogenous stimuli produce a primary appetite which, on recognition of a first releaser (IRM), leads to a motor coordination which serves as a second endogenous stimulation, arousing a "more specific appetite" and search for a second releaser triggering a further motor coordination (Lorenz, 1971, p. 207). This chain usually narrows the range of environmental stimuli to which the organism responds, until the final action or consummatory behavior completes the sequence.[13]

Lorenz often described the primacy of endogenous stimulation and the resulting appetite to perform a given consummatory behavior with what he called a "hydraulic" model (Lorenz, 1956). He suggested that an animal is constantly accumulating an "action specific energy" or "action specific potential," much as a basin would be progressively filled by a leaking faucet; in this model, the appropriate key stimulus or releaser functions as a valve, allowing the accumulated energy to flow into the performance of an appropriate motor coordination.

As evidence, Lorenz cited a phenomenon he called "vacuum behavior:" on occasion, an animal deprived of the appropriate environmental releasers will perform the corresponding motor coordination with no apparent external stimulus whatsoever (Lorenz, 1971, p. 204). Under these circumstances, it would appear that an S-R model is unable to account for the response, since no environmental stimulus is present. Although the hydraulic model Lorenz originally used to illustrate this process has subsequently been heavily criticized and generally abandoned, the underlying phenomenon of endogenous stimulation within the central nervous system has been experimentally demonstrated (Hass, 1970); hence many American experimental psychologists have now accepted Lorenz's fundamental point that behavior originates with endogenous stimuli (e.g., John, 1972).

As has been indicated, Lorenz has never denied the existence of conditioning or its great importance in mammalian behavior (Lorenz, 1965). On the contrary, his conception of complex behavior chains led to an emphasis on the "intercalation of learning and instinct" whenever components of a behavioral sequence are individually learned rather than genetically programmed (Lorenz, 1970, esp. pp. 261-292). Such diverse phenomena as imprinting (early fixation of behavior), classical conditioning, and conditioned modification of internal states (Garcia / *et al.*, 1974) can thus be integrated with Lorenz's theory without denying—as some American behaviorists originally did—that many animals inherit not merely drives, but the specification of those environmental stimuli which trigger a "species-specific" behavior.

Unfortunately, *On Aggression*—Lorenz's best known work—is somewhat misleading as a presentation of this theoretical framework. In popularizing his scientific findings, Lorenz sometimes blurs the distinction between physiological causes and biological or selective functions: innate appetites, releasing mechanisms, or motor coordinations tend to be confused with the adaptive consequences which made them selectively advantageous (Masters, 1973a). As a result, when Lorenz analyzed the adaptive functions of aggression, some American readers thought that he spoke of aggression as a unitary drive (in the behaviorist sense of that term). In comparative animal behavior, such a confusion is extremely dangerous. Like simplistic teleology—e.g., the argument that the "purpose" of the nose is to hold up eyeglasses—a confusion of the causation of social behavior with its adaptive function leads to the notion that a functional attribute of behavior is directly inherited as an "instinct."

Since this formulation is abstract, a concrete example will help. In common usage, one often speaks of an "instinct of self-preservation." Such a concept is rendered meaningless by Lorenz's scientific writings: survival is an outcome of adaptive behavior, and as such serves as a functional concept in biology. But genetically transmitted behavioral components, whether innate appetites, releasing mechanisms, motor coordinations, or simpler movements like the unconditioned reflexes of Pavlovian conditioning, are always less general than "survival." An animal may inherit the orienting reflex (called a "taxis") of moving away from extreme heat, or a positive phototaxis (a tendency to move toward light), because these responses increased the probability of survival in that species' evolution. But it makes no sense to say that a human who moves his hand away from a flame does so "because of his instinct for survival." Nor can one say that the moth who flies into the candle flame lacks such a vague "instinct."

In the same way, endogenous stimuli, releasing mechanisms, or motor coordinations which contribute to successful reproduction may be innately programmed without the existence of a single "sexual instinct" (Morris, 1956). The annual return of salmon to their spawing grounds is due to a complex of highly specific processes, not to a single "territorial" drive (Fontaine, 1956). What can be inherited is a component of behavior, ultimately traceable to the biochemical properties of the proteins produced in particular cells according to the genetic information in the chromosomal DNA. The functional adaptation, such as survival or reproduction, is a *consequence* of this genetic information, but is not itself innate or instinctive as Lorenz redefined these terms: "A definite and self-contained function of an organism, such as feeding, reproduction, or self-preservation, is never the result of a single cause or a single drive." (Lorenz, 1966, p. 86.)[14]

When approached in this light, Lorenz's ethological theory, especially as modified by more recent experimental findings, is broad enough to explain both the data of experimental psychology, and field observations of animals in their natural habitats. Moreover, because Lorenz showed the need to distinguish bodily states, innately programmed perceptual mechanisms, and motor coordinations—all of which had been simplistically lumped into the vague concept of "drives"—his work is consistent with more detailed experimental findings in neurology and the physiology of behavior (for a review, see Handler, 1970, ch. IX-X).

Compared to traditional behavioral psychology, Lorenz' ethological theory has the immense advantage that it avoids a sharp dichotomy between nature and nurture, and can explain the complex interrelationships between genetic, physiological, and social factors in behavior. Despite the frequent criticism that ethological research presumes genetic causation of behavior, this approach has even suggested an environmentally modifiable etiology for human pathologies such as autism, which have otherwise been unsuccessfully diagnosed as congenital (e.g., Tinbergen, 1974). Far from implying a determinist or reductionist model of behavior, especially among species with a highly complex central nervous system (such as primates and humans), Lorenz's theory is well suited to statistical or stochastic analysis, and follows the general epistemological conceptions of contemporary biology.[15] Hence it is not inappropriate to emphasize the potential contribution of Lorenz's theories to political science, even though "biopolitical" research utilizes many findings in experimental psychology or field observation that need not be directly derived from his work.

3. *Ethology and Social Behavior*

One aspect of Lorenz's ethological theory which makes it particularly useful for political science is his approach to social behavior. Since Lorenz focused on animals in natural or quasi-natural settings, he was less likely than experimental psychologists to analyze the behavior of isolated individuals. Using the O-S-R model sketched above, Lorenz has devoted much of his career to studies of the way one animal's behavior serves as the "releaser" or stimulus for another's response.

Social interaction is thus readily included in Lorenzian theory without thereby *reducing* it to a mere product of individually conditioned responses in a group of animals. On the contrary, Lorenz was able to show how evolution could produce chains of social behavior in which the motor coordination of one individual—e.g., attack movements used in aggressive encounters—became the releaser or key stimulus for apparently unrelated behavioral responses by others—e.g., mating (Lorenz, 1970, esp. pp. 188-210; 1971, esp. pp. 140-153; Morris, 1956). He used the term "ritualization" to describe a behavior which evolves a new social function as a releaser, and thus plays a different role in social interaction than in individual activity.

The popularization of Lorenz's work has emphasized ritualized signals of dominance and submission, which tend to regulate much social interaction among the vertebrates (Lorenz, 1966). But ritualized behaviors often play crucial roles in other areas, such as courtship and mating, nest building, rearing of the young, food gathering, and group movement. Hence Lorenz not only provides a theoretical framework within which innate and individually acquired components of behavior can be carefully analyzed, but relates this perspective to *social* behavior.

Most field studies of mammalian behavior, even when conducted by researchers who did not share Lorenz's theoretical approach, can thus be understood in his terms. This is even true of the newly developing approached called "sociobiology," or "socioecology," which in the long run may have a greater influence on the social sciences than Lorenz's original work. Recent research in this field has analyzed more directly the social patterns in various species as they relate to ecological niches and evolutionary change. Instead of beginning from individual behavior, whether conceptualized on the S-O-R model of behaviorists or the O-S-R model of Lorenz, ethologists are treating the "social systems" of different animals as the primary object of study.

One reviewer has summarized developments in this area as follows:

> By combining studies of animal behavior with those of ecology, sociobiologists are beginning to understand why different social systems have evolved. [. . .] These investigators observed that the social systems of different bird species in the same environment are often similar and that the social systems of related species in unlike environments are often different. These observations stimulated students of animal behavior to search for general rules governing the ways that environments affect social systems. (Kolata, 1975, p. 156.)

In this perspective, one can analyze social structures which range from "closed" groups or "colonies" to populations of individuals living in "isolation" (except for a period of infant dependence on the mother, mating, or brief aggressive encounters). For example, Barash (1974) has shown that among the marmots—a family of species of small rodents including the woodchuck—variations in social patterns are highly correlated with the environmental setting, not only in comparisons between *different* species, but also between different populations of the *same* species. Similar findings have been shown for primate social organization (Rowell, 1969; Kummer, 1971).

Since this level of research emphasizes animal social structure as it relates to ecological setting, sociobiology might seem essentially different from the study of innate components in behavior using Lorenz's approach. Such a distinction probably reflects the variables being emphasized rather than a fundamental theoretical change: "Since reconstruction of the evolution of social systems from field studies and analyses of parental care and other forms of interactions based on genetic theory appear to be founded on the same basic premise, many investigators predict that the two lines of research will converge to provide a unified theory of sociobiology." (Kolata, 1975, p. 157, cf. Wilson, 1975.)

As the field of ethology develops, therefore, it becomes increasingly likely that human social life will be compared to that of other species. Many social scientists have emphasized the importance of work on primate behavior, since it is presumed that humans share behavioral traits with living primates (cf. Somit, 1968; Davies, 1970). But some have also argued that early hominid behavior is illuminated by that of hunting carnivores (e.g., Van Lawick-Goodall, 1970), and increasing use is being made of functional comparisons between humans and other mammals, not to mention social insects.

Ethologists have argued, for example, that such culturally variable norms as incest taboos fulfill biological functions satisfied by different causal processes elsewhere in the animal world (Bischoff, 1972; Wickler, 1972). The regulation of human behavior by political institutions has its functional analogue (if not phylogenetic origin) in dominance behavior among other species, especially primates (Fox, 1967; Tiger / Fox, 1971). Similarly, culturally variable spacing mechanisms in human societies have functional similarities elsewhere in the animal kingdom (Hall, 1969; Lorenz / Leyhausen, 1973; Vine, 1973). Much research in political science can be illuminated by ethological approaches, though care is needed to avoid confusion between biological analogies and vague metaphors (for a review, see Masters, 1973b).

From a theoretical point of view, however, the crucial point is that social behavior is a natural characteristic of human beings. Whatever the interest of the specific empirical propositions that can be derived from ethology and verified among human populations (e.g., Section IV below, p. 209) political theory is influenced in the first instance by the simple realization that society is a widespread phenomenon in the animal world—and hence that explanations of human society as something totally conventional or man-made are inconsistent with biology (Hummel, 1971; Masters, 1975). Insofar as human politics *is* different from the behavior of dominance and subordination in other mammals—and it surely is—such differences can only be discovered by a study of comparative animal behavior.

III. An Ethological Definition of Politics

Before turning to several examples of political behavior which can be illuminated by ethology, however, one major contribution of this approach should be noted. From the perspective of comparative animal behavior, human politics can be redefined in such a way that the relationships between political science and other academic disciplines in both the natural and social sciences will be clarified. Moreover, the proposed definition will suggest not only why ethological research has a particular relevance to the study of politics, but why it is subject to abusive generalizations which cause confusion and error.

Many other social species, including virtually all primates, exhibit some form of social dominance behavior. The existence of leadership, submission, and competitive rivalry for status—often described by the technical term "agonistic behavior" (Altmann, 1967)—obviously suggests parallels with politics. Especially among primates, where observers have noted the transmission of high social status from parents to offspring (Koford, 1963, p. 147) and the formation of coalitions among dominant males (DeVore, 1965), social organization seems to reflect processes similar to human politics.

While the perspective of ethology encourages the examination of possible functional analogies between humans and other species, the danger of the reductionist fallacy has often been noted (Montagu, 1968). As Simpson's definition of the social sciences (cited above, p. 199) indicates, speech or symbolic language can be shown to distinguish human behavior from that of other animals (Masters, 1970). Can one therefore define human politics so that comparisons with other species are possible without denying the biological uniqueness of our own?

I have suggested elsewhere that a biopolitical definition—unlike prevailing definitions of politics by political scientists—will do precisely this. According to this proposal, political life is the arena of "agonistic" behavior directed to the establishment, maintenance, or change of social rules.

> One can therefore define politics more precisely as behavior which simultaneously partakes of the attributes of dominance and submission

(which the human primate shares with many other mammals) and those of legal or customary regulation of social life (characteristic of human groups endowed with language). Politics is not merely what ethologists have called "agonistic" behavior (Altmann, 1967): competitive rivalry for dominance exists in sports, on school playgrounds, and in business without thereby deserving the name "politics." Nor is all behavior governed by legal norms in itself political: as cultural anthropology teaches us, legal or customary rules govern childhood, marriage, and the entire range of social life.

> Political behavior, properly so called, would seem to be those actions in which the rivalry for and perpetuation of social dominance impinges upon the legal or customary rules governing a group. As such, political science has a peculiar status, for it lies at the intersection of ethology and anthropology—or, more broadly, at the point where the social and natural sciences meet. Indeed, this definition of politics may help explain why political theorists, at least before the middle of the ninetheenth century, were almost always concerned with the definition of human nature and the relationship of nature and society (Masters, 1975, pp. 34-35).

In other words, political behavior is by definition a set of human actions which cannot be completely explained *either* by comparisons with other animals *or* by analysis of the conventional norms elaborated by a human culture.

This definition indicates, at the outset, why popularizations of recent ethological research have been of particular interest to political scientists. Territorial behavior among animals, and especially the aggressive defense of home space, seems so obviously relevant to warfare between human societies, not to mention private property, that the possibility of new insights was hard to ignore. But territorial behavior, which is highly variable even among the primates, also varies from one human society to another. Because our territorial boundaries are defined in terms of linguistic or cultural symbols, human warfare includes rivalry over cultural norms as well as physical space, and hence cannot be reduced to a simple manifestation of animal territoriality (which is any case a functional category and not an instinct).

As this example indicates, the political element in modern warfare is precisely the *combination* of the aggressive or territorial behaviors analogous with some animal behavior *and* such cultural phenomena as nationalism, advanced military technology, and competing economic or strategic interests. Note that, in themselves, nationalism, military technology, or economic interests are not, in a precise sense, *political*—but each may become so when invested with the properties of what ethologists call agonistic behavior.

A final remark concerning the proposed definition. Unlike many alternative ways of defining politics, the definition of politics as agonistic behavior directed toward and embedded in cultural and linguistic structures can specify *both* phenomena within an organized state *and* those in a social system without political institutions. Hence this definition can apply to the politics of the modern nation state, to international politics, or to the politics of pre-literate or "primitive" societies without formal governments (cf. Masters, 1964). Since political systems can be identified even where it is impossible to speak of a single "society," alternative definitions of politics in terms of the "state" (Almond, 1956) or a "society" (Easton, 1953; Parsons, 1959) would appear to be less parsimonious and fruitful than the formulation proposed here.

IV. The Empirical Relevance of Ethology to Political Science

Methodological or epistemological considerations, broad theoretical implications, and definitional elegance are of far less interest to most political scientists than substantive empirical propositions. In a very real sense, the foregoing examples of the relevance of ethology to political science will strike many as a mere promise that "biopolitics" can contribute to the discipline. Without specific examples of researchable areas not otherwise studied, the claim that ethology can or should be of concern to political scientists is likely to produce little more than a yawn.

It is ironic that much of the research by "behavioral" political scientists has focused on *indices* of behavior rather than behavior itself. For example, when variables like income, sex, social status (SES), education, or age—as well as attitudes and opinions—are correlated with voting choice, the behavioral act of voting itself is not directly studied; similarly, when political scientists analyze the influence of personality, economic interests or social structure on leadership, they seek independent variables which could be used to explain variations in political behavior. An ethological approach, as the above definition suggests, does not challenge such research insofar as it reveals the effect of cultural and social factors in politics. But it suggests that the conventional approaches of political scientists can be enriched by also considering, in a more direct and observational way, political acts themselves.

Examples of research based on ethological theory which could be conducted by political scientists will be useful. In an earlier paper (Masters, 1972), I spoke of two aspects of agonistic or dominance behavior which have attracted the attention of ethologists: gestural symbols of dominant or submissive status, and "mood convection" (i.e., the spread of motivations or attitudes through an animal group). Both have political analogues that have not received much if any attention among political scientists. In addition, recent work on the role of "attention structure" in the dominance behavior of primates suggests a way of analyzing the political process which might explain the sometimes rapid and puzzling shifts in popularity and legitimacy. Description of these areas for possible research will therefore illustrate how ethology could influence empirical political science.

1. *Attention Structure, Dominance, and the Loss of Power*

The pattern of dominance in an animal group is not solely a question of which individuals are "victorious" in agonistic or threat encounters. On the contrary, the establishment, maintenance, and loss of dominant status seem highly related to the capacity of an individual to serve as the focus of attention by other members of the group. In an important article, Chance (1967) described this phenomenon as "attention structure," and showed that it explains animal social organization more fully than models of a dominance hierarchy based solely on threat or conflict.

While this concept will be readily understood by those who have followed some of the research on group dynamics in social psychology, an ethological frame of reference emphasizes the importance of "attention structure" in human politics. Shifts in political power may be associated, to a degree that has been underestimated, to changes in the pattern of attention on the part of citizens. Hence this phenomenon may be a clue to the way leaders gain—and lose—political power.

It is well known, for example, that "name recognition" is a necessary prerequisite in American political campaigns. On a theoretical level, this common observation seems to reflect the inability of a politician to establish dominance if he is not the center of public attention. While political scientists have of course noted this factor, shifting patterns of attention in the media are easily verified empirically. For example, it has been argued that McGovern converted his ability to attract more attention into increased popular support during the spring of 1972.[16] Such a shift in media coverage is obvious in Table 1, which presents preliminary research on the visual

TABLE 1. *Relative photographic "coverage" of Nixon and McGovern in* Time *and* Newsweek *during early stages, 1972 campaign*

Month	*Nixon Photos*	*McGovern Photos*	*Total Nixon and McGovern*	*Nixon Photos as % of Monthly Total*
January	23	3	26	88.5
February	9	1	10	90.0
March	22	2	24	91.7
April	5	5	10	50.0
May	27	11	38	71.0
June*	2	2	4	50.0
Totals	*88* (78.6%)	*24* (21.4%)	*112* (100%)	

* First week only.

Source: all issues of *Time* and *Newsweek* magazines between January 3, 1972 and June 12, 1972. (For further discussion, see below, p. 218 and Table 2.)

images of Nixon and McGovern during the 1972 Presidential campaign; while this data is partial and merely suggestive, it is presented as an example of an easily reproducible method for studying changes in attention.

While electoral campaigns in Western democracies serve the function of providing an arena in which rivals compete for public attention, this process could be compared in political systems where the focus of attention within a political elite or closed party is decisive in contests for power (e.g., the USSR).

It should be relatively easy to develop research designs which measure the "attention structure" of different political arenas. Kremlinologists have long done something of this sort, attempting to judge the relative status of individual members of the Soviet hierarchy by their location in public meetings of various sorts. More precise measures, of course, would be based on the source of communications to which individuals in a ruling or elite citizen body paid attention.

For example, in a system like the United States, in which the mass media provide the relevant arena for many purposes, one could study statistically the source of communications and their relative weight in the press, TV news, and the like. As a research hypothesis, it could be argued that *the more an individual is the source of communications which provide the focus of public attention, the more likely he is to establish or retain dominance.*

Nixon's loss of power between 1972 and 1974 could be used to test this hypothesis. If attention structure plays a role in human politics analogous to that among primates, one would expect that the televised hearings of the Ervin Committee in the Summer of 1973 played an unexpectedly large role in undermining Nixon's dominance. Similarly, in the final stages of Nixon's loss of power, the television coverage of the House Judiciary Committee's hearings in the summer of 1974 would have had an effect that was disproportionate to the information conveyed by the sessions themselves.

One could test this hypothesis by creating measures of the extent to which Nixon was effectively serving as the focus of positive attention, controlling the flow of political communications in the media. Events like the TV coverage of Congressional hearings, or the sequence of Watergate revelations by other members of Nixon's entourage, would thus reflect a shift in attention structure away from Nixon himself and toward other actors in the political arena. These changes in attention structure could then be compared to shifts in Nixon's public opinion ratings, or to expressed attitudes toward Nixon by other politicians (particularly in his own party).

Attention structure is thus an excellent example of a concept developed by ethologists which can readily be studied in human politics. In so doing, moreover, one is led to propose hypotheses which focus directly on the dynamics of the political process and are readily verifiable; while consistent with public opinion survey research, quantification of the sources and saliency of political information measure more directly the impact of the media—and provide a possible means of explaining shifts in attitudes. To be sure, it could be said that "attention structure" is merely a fancy name for common sense, since according to conventional political wisdom "it doesn't matter what the press says about you as long as they spell your name correctly." But, as I shall indicate, the same theoretical frame of reference which accounts for the importance of this conventional judgment also provides insights into many phenomena that are not studied by political scientists.

2. *Gestures, Dominance, and Authority*

Ethologists have shown that social dominance in many species is rarely challenged: frequently, members of a social group adjust their movements to those of dominant individuals, so that one can literally speak of *leader*ship in bands of gorillas (Schaller, 1965), baboons (Hall / DeVore, 1965), patas monkeys (K. R. L. Hall, 1967), and many other non-primate and primate species (Lorenz / Leyhausen, 1973). Even when other members of the group move about or interact independently, they often pay continued attention to the reactions of a dominant male or males, and adjust their behavior accordingly (e.g., Van Lawick-Goodall, 1971).

The behavioral acts which symbolize dominant status for other members of an animal group—i.e., in Lorenz's terms, the dominance rituals—are therefore a significant factor in much animal social behavior. To be sure, among some primates there are physiological correlates of dominance status—notably male sex, age, and physical attributes such as the "silver back" in gorillas. But gestural cues also play a vital role in confirming or gaining dominant status, particularly in species where social groups include a number of adult males.

Application of the insights of ethology to human non-verbal communication has already begun to emerge in other areas of the social sciences (for a bibliography, see Davis, 1972). Sometimes called "proxemics" (Hall, 1959, 1969) or "kinesics," such research has usually focused either on the public actions of ordinary members of society (e.g., Goffman, 1971) or on gestural and postural cues in the private lives of individuals (e.g., Fast, 1970). But there is no reason to assume that non-verbal signals are absent in politics—and, as the remarks below will indicate, much impressionistic evidence points toward its importance.

Following the proposed definition of human politics, it is interesting to wonder whether gestures of dominance, appeasement, or submission are relevant components in establishing or confirming the authority of human leaders. Here, several researchable issues can be distinguished: First, on a purely descriptive level, what are the gestures, postures, and facial expressions which signal dominance or subordination among humans? Second, insofar as leaders vary in their personal styles of behavior, are there *differences* in political effectiveness that can be correlated with non-verbal gestural cues? And third, it is possible that these ritualized behaviors can help explain cognitive attitudes by revealing the relationship between "gut reactions" and verbalized responses (such as the attitudes studied in opinion research)? A word on each of these possibilities will provide a good example of how political scientists could profit from ethological research.

Descriptive Observation of Political Dominance

It is well known that the social rituals surrounding human rulers vary widely from one culture to another. In many civilizations, the ruler can be approached in public only in ritualized settings—e.g., with the ruler seated on a dais, subordinates approach, often with some gesture of submissiveness (standing with heads bowed, kneeling, or prostrate).[17] In other cultures, including contemporary Western ones, political leaders very often face large public audiences standing behind a raised podium, while subordinates are seated (though often the audience stands momentarily on the leader's entry and applauds).

Whatever the convention, however, settings in which political leaders formally relate to a large public audience seem to be ritualized in such a way that the dominant individual is sharply distinguished from those subordinate to him. Ethological research suggests an explanation in the functional role of attention structure as a correlate of dominance (Chance, 1967): if the leader normally confirms and exercises his status by serving as the focus of attention, a ritualized behavior pattern that sharply and symbolically distinguishes the leader from others permits immediate identification and reinforcement of status.

Highly formalized settings, such as the Presidential Press Conference, are however relatively rare events. In attempting to describe the role of gestures in human politics, other phenomena are probably more important. In particular, individuals who are competing for dominance obviously exhibit non-verbal cues in their political behavior; for example, the gestures of candidates for public office—whether when kissing babies, shaking the hands of factory workers, or standing in a motorcade—can be observed and compared not only in any one country, but cross-nationally.

At this purely descriptive level, humans display one gesture that merits a remark. In sports—a characteristic arena of non-political agonistic behavior—victory often releases a gesture which is so spontaneous that it may even be innate: both arms raised high, often a springing in the air or at least standing to the fullest height, half-smile showing the upper lip, wrinkled eyebrows and a "surprise" stare. This response has analogies with some components of the aggressive "display" gestures of gorillas (Schaller, 1965, p. 150, 218) or chimpanzees (Van Lawick-Goodall, 1971, pp. 122-129); like some other facial gestures, therefore, it may be a response with a phylogenetic origin common to humans and some primates (Van Hoof, 1973).[18] In any event, what I would call the human *triumph gesture* (cf. Lorenz, 1966, pp. 175-215) is surely widespread in politics and can be observed in political candidates in many societies (e.g., Figures 1-3).

Other non-verbal gestures associated with the public acts of political candidates and leaders deserve careful observation and analysis. While the "triumph gesture" seems at first to be an aggressive or agonistic signal, analogous displays among primates often have a component that reassures subordinates and facilitates social grouping (Van Hoof, 1969). In any case, what ethologists call "appeasement" rituals are of great importance in human politics: like grooming, hand-touching, and lip-smacking in chimpanzees or gorillas, the politicians, hand-shaking and baby-kissing should probably be characterized as signs of reassurance, confirming the non-antagonistic intention of the leader's dominance.[19]

In addition to political rituals in public, the study of gestures in private settings would be valuable. In bureaucratic situations, for example, the interaction of leaders and their staff assistants is often subtly influenced by the spatial arrangement of the participants and their postural behavior. Hitler's device of requiring visitors to cross a large office to approach his desk is well known, and is used by many high officials; at a staff meeting, the dominant individual often arrives last—and usually receives a ritual greeting from his subordinates (who typically rise in the US). Seating arrangements of staff meetings are especially significant because of the role played by eye contact in regulating attention structure and dominance (Chance, 1967; Callan / Chance / Pitcairn, 1973).

Such non-verbal gestures are also highly important in political encounters between rivals, especially in international relations. The delicacy of seating arrangements and even table-shapes in diplomatic negotiations or social functions is proverbial. Less well known is the role of gestural cues which may help establish dominance in summit meetings and negotiating sessions. But since non-verbal cues often greatly influence private interactions, there is no reason to exclude them from political science—especially since direct encounters between Heads of State have taken on increased significance in recent years.

Beyond description of the non-verbal behavior of political men, research in this area could include laboratory experiments on the way gestures transmit images of dominance or subordination. It should be possible to identify a number of political styles, including the out-going search for gestural contact with subordinates, the minimal execution of dominance or appeasement gestures, and an inhibited avoidance of exhibiting such non-verbal cues. Moreover, this entire range could well be studied not only in public encounters, but in private and social spaces as well; following Hall (1969), we would probably find rather different gestures and psychological effects in each context. As a result, it might be possible to develop a typology of political "styles" with reference to individually varied mixtures of the repertoire of gestures across the range of political settings.

In this way one could distinguish between politicians who are at ease in social or private settings but inhibited or incompletely dominant in public interactions, those who are inhibited in private or social settings but exhibit dominance in public spaces, those who easily display dominance in all three conditions, etc.[20] For example, on an entirely superficial and impressionistic basis, it would appear that George McGovern is more fully at ease—and more likely to display dominance behaviors—in social spaces than in public, whereas the reverse may well be true of Richard Nixon.

Differential Effectiveness of Political Rivals

Although we lack even a systematic description of the role of rituals and non-verbal cues in political behavior, mere observation alone is hardly the end of science; most political scientists would like to *explain* behavior. One possible contribution of ethology lies precisely in the differences of non-verbal behavior among political figures. While it is impossible to present more than some impressionistic hypotheses, several examples of the relevance of gestures in establishing, maintaining or losing authority can be suggested.

Individuals vary widely in their style of behavior in public. Some are ill at ease, and enter a crowded room with slightly stooped shoulders and a nervous look of apprehension; others show a more fully erect stature and confident gaze when joining an already formed group or moving to address a crowd. Such interpersonal differences are also exhibited by politicians, and may contribute to an explanation of their careers.

For example, in the 1972 American Presidential election, George

McGovern was defeated more thoroughly than might have been expected merely from the analysis of traditional political factors such as partisan identification, socio-economic status, etc. (Polsby, 1974, pp. 17-51). I would suggest, as an hypothesis, that at least one contributing factor (though of course not the sole cause) was McGovern's inability to establish himself as a sufficiently dominant male who could be identified as a believable authority figure by many voters. While evidence of this perceived "weakness" of McGovern can be found in such data as opinion polls or Letters to the Editor (e.g., *New York Times,* October 1, 1972, p. 14 E; October 12, 1972, p. 44), how is it to be explained?

The conventional interpretation emphasizes McGovern's indecision in the Eagleton affair and the subsequent naming of a new Vice-Presidential candidate, and other similar events. But why wasn't McGovern's defeat of the regular Democratic organization treated as a sign of personal strength, and his behavior toward Eagleton a measure of sensitivity to another human being? Ethology teaches us the importance of attention structure, and forces us to question the conventional interpretation: during the campaign, why did journalists and the public devote so much *attention* to the Eagleton affair, and so little to Watergate? Could this difference be related to images of each candidate's dominance, images that created differing predispositions toward Nixon and McGovern?

To get evidence of the differential perceptions of the strength or dominance of rival candidates, traditional methods of attitude analysis are highly useful. For example, even superficial reading of Letters to the Editor during the 1972 campaign reveals not only that Nixon supporters perceived him to be "strong" and McGovern "weak", but that McGovern supporters sometimes implicitly admitted that their candidate was "weaker" than Nixon.[21] Content analysis, perhaps combined with semantic differential techniques, could well be used to confirm and quantify this phenomenon, indicating more precisely the extent to which different images of power, strength or dominance were attributed to the two candidates.

If my hypothesis is correct, this kind of research would reveal *different perceptions of the personal dominance of leaders among many American voters.* I would also hypothesize that *such differences would be particularly marked among those voters whose level of cognitive information was relatively low, or who were cross-pressured.* If both hypotheses were confirmed, this "image" variable of perceived dominance might go far to explain election *outcomes;* as is well known from the literature on voting (e.g., Campbell / *et al.,* 1960), partisan identification or other socio-economic variables may explain the largest proportion of voter choices, but they do not fully account for the decisions of those with no partisan identification who actually tend to decide most elections.

But even if verified, these hypotheses do not explain how voters establish different perceptions of the dominance of rival candidates. It is here that non-verbal cues may play a role—though of course not the *only* role—in political campaigns. Again on a superficial basis, examination of pictures of the two candidates suggests that Nixon used certain dominant gestures far more frequently than did McGovern. To cite but one example, which suggests the different impressions each candidate made upon the electorate, consider the following detail: in the *Washington Post,* detailed campaign coverage was generally on a page with a "logo" showing pictures of the two candidates; Nixon was depicted in the full triumph gesture, whereas McGovern was shown with a smile (Figure 17). The only time I saw McGovern campaigning, he greeted the crowd by smiling and waving one hand in a way characteristic of social settings; McGovern's campaign photographer saw him use the "V" gesture only twice—and once it was to indicate the number *two!*

That Nixon may have been somewhat more adept at the triumph gesture than McGovern does not, of course, do more than suggest a personality variable in politics. But this variable is obviously important, especially since Nixon himself was not always effective in conveying dominance throughout his political career. In the 1960 campaign, it is often said that Kennedy's performance in the televised debates did much to establish him as a credible rival to Nixon. But detailed comparisons of the two, particularly during their campaign swings, might contribute to an explanation of why Nixon so consistently lost support during that campaign or in the course of his successful race in 1968 (when his large lead in the polls almost disappeared by election day). To explain this, I would hypothesize that while Nixon exhibited dominance gestures in public settings more frequently than McGovern, Nixon himself has generally been ill at ease in social settings, and his public displays are more consciously contrived—and hence less effective—than the dominance behavior of Kennedy (or even Humphrey).

Since systematic evidence is lacking, a bit of anecdotal data will have to suffice. In discussing the above hypothesis with a noted conservative editor in 1973, I was amused at his reasons for a deep dislike of Nixon. This editor was invited to a dinner at the White House early in Nixon's first term, and described the President's entry as follows:

> We guests were gathered in the East Room, where we were allowed one cocktail (no more) while awaiting the President's arrival. The trumpeters from the Marine Band, dressed in silly costumes, blew "Hail to the Chief," the doors opened, and Nixon entered with the visiting Head of State and his wife. Nixon's shoulders were partly stooped, and he half-waved at the assembled dignitaries, nodding his head from side to side with a sickly smile. I had the feeling that he was ill-at-ease and weak, and I've never gotten over a dislike of Nixon since.[22]

What is surprising in this account is that the gestures described seem so similar to some of those used by McGovern during the 1972 campaign—and very much unlike some of the more studied "triumph gestures" exhibited by Nixon himself when campaigning.

The effects of non-verbal communication are, of course, difficult to assess: few voters are invited to the White House, and most have no more than a passing occasion to see candidates in person. Media coverage thus "mediates" the transmission of gestural cues, not only because it is through the press or TV the citizens have their largest exposure to images of candidates and leaders, but because journalists and editors select the pictures they will present. Conceptually, therefore, the model of how gestural cues might influence voting choice is rather complex, and must take into account both media selectivity and reference group perception as well as the variables traditionally studied by political science (Figure A).

Using this model, it would be interesting to compare the cognitive perceptions of the "strength" or dominance of political rivals with pictorial evidence of their gestures. Needless to say, this is not the only variable in politics: after all, presumably McGovern used similar gestures in his successful Senatorial campaigns and in the 1972 debacle. However, as McGovern himself said (during an off-the-record visit to Yale University as Chubb Fellow), his electoral success in South Dakota was very much linked to the smallness of the state, and his ability to walk down the streets of small towns greeting many notables by name and shaking their hands. *Presidential elections may focus rather more on personality attributes of dominance than those of Senators or Representatives, for whom reassuring gestures like the handshake may be more appropriate (especially in small constituencies).*

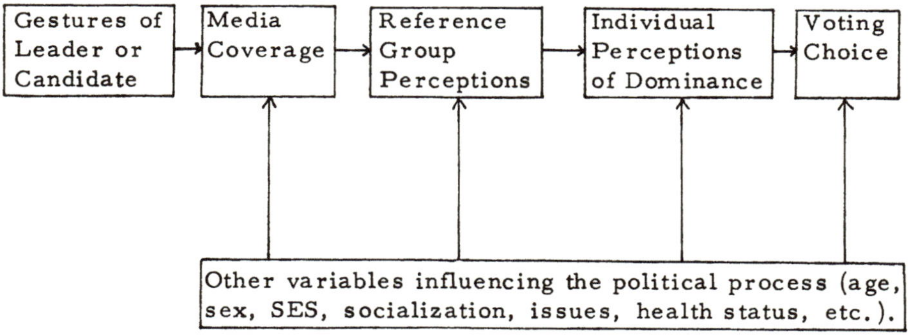

FIGURE A. *The role of non-verbal communication in politics*

The model in Figure A assumes that the perceived dominance of a candidate is a residual factor in the formation of voter attitudes: presumably here as elsewhere, the phenomena of "cognitive dissonance" might explain why fervent supporters of a candidate do not notice non-verbal cues or screen them out, whereas they influence voters who do not begin with a clearly defined cognitive set such as party identification. In other words, the hypothetical role of political dominance behavior would be a reinforcement of cognitive attitudes among supporters or confirmed opponents, whereas such gestures might actually influence decisions among marginal voters.

To verify this hypothesis, the perceived "weakness" or "strength" of individual candidates could first be correlated with partisan identification and expected vote early in a campaign. To do this, standard opinion polling could be used, although content analysis of newspaper editorials, Letters to the Editor, etc., might also be illuminating. The non-verbal communications of candidates could be measured by sampling photographs in newspapers and magazines as well as TV news sequences. Such data would be doubly significant, since the unconscious selection of images by the media not only influences the electorate, but reflects the responses of editors to the candidates.

Preliminary analysis of photographs during the early stages of the 1972 American Presidential campaign indicates the feasibility and interest of such research. All photographs of Nixon and McGovern appearing in *Time* and *Newsweek* magazines between January 3, 1972 and June 12, 1972, have been coded for differences in 15 postural cues (e.g., eyebrows and forehead: raised, normal, or lowered). While four researchers reached a high degree of agreement in the coding, the small size of the sample of photographs (N=112) did not warrant statistical measurement of coding reliability.

TABLE 2. *Selected dominance gestures during the early stage of the 1972 presidential campaign*

	McGovern		*Nixon*	
	N	*% McGovern Photos*	*N*	*% Nixon Photos*
Arm raised	4	16.66	3	3.41
Both sets of teeth visible	2	8.33	3	3.41
Top teeth visible	17	70.83	33	37.50
Total	24	100	88	100

Source: all photographs of McGovern and Nixon in *Time* and *Newsweek* magazines, January 3, 1972 to June 12, 1972.

Table 2 presents some of the data from this sample. Although pictures of Nixon were more than three times as frequent than those of McGovern (cf. the discussion of "attention structure" above), more interesting is the proportion of pictures of each politician in which components of dominance gestures appear. Photographs of McGovern which exhibit at least one raised arm or a dominant smile were far more frequent than similar gestures in photographs of Nixon. In the quest for a Presidential nomination, such gestures probably seem more relevant to the media (and may be more frequently exhibited) than in the continued exercise of power. While further research in this area must obviously be done, the preliminary data thus indicate that the historical and political context may be at least as important, if not more so, than personal idiosyncracies in non-verbal political communication.

Further evidence of the importance of changes in the non-verbal images of leaders is provided by press coverage of Nixon during his last two years in office. Cover photographs from *Newsweek* during this period (Figures 4-15) show Nixon exhibiting an increasing frequency of submissive or worried gestures or postures. In addition, this series reveals how pictures of those challenging Nixon's status not only appear with greater frequency, but show facial expressions of threat which contrast sharply to the images of Nixon. Finally, editorial hostility to Nixon is graphically symbolized by the make-up of these covers, which use such devices as a frame surrounding Nixon's head (Figures 8 and 15), printing superimposed on his face (Figures 8, 10, 13, 15), silhouette and back views (Figures 10 and 11), or an overexposed photograph (Figure 13)—all of which convey non-verbal messages that could clearly influence reader perceptions of dominance.

Assuming more observation and classification of political gestures, there is no reason in principle why images of candidates could not be coded much like the content analysis of verbal messages. In this way, one could see whether candidates with high percentages of markedly dominant gestures also project cognitive impressions of strength or dominance, and whether changes in non-verbal cues are correlated with the evolving political context. Ultimately, such research might thus not only correlate the subconscious responses of voters with their conscious attitudes, but also verify whether the influence of such non-verbal gestures on voter preference is highest among those with the least information or partisan commitment to a candidate.

Physical Proximity to Leaders: "Gut Reactions" and Verbal Opinions

As the above example should make clear, non-verbal data has an advantage in political research precisely because it is non-verbal—and hence can be used to verify measures of opinion derived from traditional methods. In this, the proposed focus on political gestures is parallel to recent research using psycho-physiological measures to validate opinion scales (Lodge / Tursky / Tanenhaus, 1973). In addition, such research would link political science more directly to both ethological studies of dominance and to research in cognitive dissonance in social psychology.

This example therefore illustrates the definition of politics offered above, showing in a more concrete way how the agonistic behaviors of dominance, submission, and reassurance studied in ethology interact with the cognitive or culturally learned symbols which are the traditional focus of the social sciences. But it also turns our attention to the "gut reactions" in political life which are most widely *experienced*—even by the political scientist whose scholarship takes little notice of them.

A further phenomenon in this area, while not a matter of non-verbal gestures, is worth mentioning because it gives rise to a number of related empirical hypotheses. It concerns a behavioral characteristic of great potential significance, though to my knowledge it has never been discussed in contemporary political science. This phenomenon is the feeling of physical *pleasure* which is felt by a subordinate when in close proximity (Hall's *personal* distance) to a dominant political leader.

This sensation can be described as a "tingling" feeling of extreme pleasantness, not unlike a similar feeling of pleasure occurring in the context of highly successful interaction with equals or subordinates. Since the phenomenon has not been analyzed as far as I know, it is only possible to say that when pressed to recall experience of close personal proximity, a number of acquaintances have felt such a sensation. This hardly constitutes scientific proof of the existence of a phenomenon, much less an explanation of its importance. But the postulation of such a "gut reaction" to the close approach to dominant individuals is hardly unreasonable as an inference from observed behavior, quite apart from purely subjective experiences.

Among primates, it has often been observed that subordinates will sometimes seek to approach dominant males and remain close to them. However that may be, in contemporary politics the phenomena of crowd behavior in the presence of political leaders deserves far more attention than it has ever received (e.g., Figure 3). Just why it is that so many adults will stand for hours just to gain a glimpse of a President or a Prime Minister? Why do supposedly reasonable adults push and shove in dense crowds, striving eagerly to touch—and above all to shake the hand of—dominant individuals? And why do leaders, often ignoring the concerns of their security chiefs, engage eagerly in what the French aptly describe as a *bain de foule?*

From the perspective of ethology, one could suggest that *leaders who enter crowds of supporters, shaking hands and greeting them, thereby reinforce their dominance by triggering submissive responses among citizens;* as with other primates, the gestures of the politician or statesmen in these circumstances are often appeasement rituals which reassure the subordinate and facilitate the formation of social bonds. Moreover, if from the perspective of the members of the crowd, this experience releases a pleasurable feeling of close proximity to the leader, a complementary "gut reaction" of pleasure occurs in the applauded leader. Relatively unstructured contact with crowds of potential or actual supporters, which often seems to consume inordinate amounts of the politician's time, may therefore be an important manifestation of non-verbal behavior in politics.

These impressionistic remarks could provide the basis for a more systematic analysis of the role of non-verbal communicative gestures in politics. It could be argued that free contact of leaders with crowds reflects that aspect of political interaction described as "gut reactions:" the politician who moves through crowds of potential or actual supporters can judge the extent of their enthusiasm and support—and, of course, attempt to generate new adherents. Conversely, isolation from such contact with citizens presumably creates a different emotional state for the leader as well as a different basis of information concerning reactions to his political status.

From this perspective, one could almost say that the politician is primarily a judge of the "gut reactions" of a social group. Traditionally, the art of politics has been defined in cognitive terms (i.e., the interests or policies or party represented by an individual). Often, leadership is discussed as the ability to judge the wisdom or feasibility of alternative solutions to public problems. While these aspects of leadership are obviously relevant, the capacity to sense the degree to which followers will support a dominant individual—a capacity which ultimately involves accurate estimations of "gut reactions"—is of tremendous importance, and may explain why so many intellectuals turn out to be weak leaders (even when they can identify the nature and solutions to political problems).

This in turn suggests a further substantive hypothesis concerning the phenomenon of the politician's contact with crowds. If the politician or leader, to be effective, must judge the current extent of submission or deference to his leadership, and if contact with ordinary citizens provides a means of estimating this submissiveness by "reading" non-verbal cues, then a leader who isolates himself from the society must find alternative means of measuring popular support (e.g., Lyndon Johnson's continued reference to public opinion polls). Moreover, if repeated exposure to the public in more or less ritualized settings is a means of maintaining dominance status, isolated leaders must find alternative ways of continually showing their superiority.

As a general hypothesis, I would suggest that *the more isolated the leader, the more he must rely on bureaucratic or legal devices for assuring his dominance.* In a sense, this follows from the definition of politics proposed above: insofar as the non-verbal gestures and signs of dominance are not utilized by a human leader, increased reliance on purely cultural aspects of the power relationship would seem to be necessary. But the hypothesis is not primarily a deductive one. Rather, it suggests an explanation for different styles of politics within a given society—and perhaps even for differences in political regimes.

For example, in the US one could profitably study the different frequency of public appearances of various Presidents, and attempt to correlate it with the tendency to resolve conflict either by bureaucratic "management" or by the impersonalized legal process. The contrast between Presidents like FDR, Kennedy, or Lyndon Johnson—for whom political appearances as well as frequent meetings with other members of the elite were in various degrees both enjoyed and used to maintain support—and a President, like Nixon, might explain in a more scientific way the conventional judgments of the dangers of "isolation," so often noted in the latter's career.

Moreover, this hypothesis suggests that *the more radical the changes sought by a leader, the more likely that he will both attempt to engineer massive public manifestations of support (such as the Nuremberg "festivals" of the Nazis) and maximize the bureaucratic structures of power.* Conversely, when a society is undergoing minimal social change, so that the strain on obedience is minimized, a style of leadership which relies primarily on strictly legal processes, with neither personal appearances before a large public nor bureaucratic manipulation and control, becomes more successful. Hence it might be possible to correlate different styles of political life with their consequences not only for the public behavior of leaders, but the growth of bureaucracies and the management of support (including the use of totalitarian techniques).

Needless to say, however, the leader cannot *always* be in close physical proximity with all of his followers. Indeed, the temporary nature of most contacts between followers and politicians suggests a sharper definition of the interaction between non-verbal signals of dominance and cognitive political attitudes. As a hypothesis, I would argue that *the closer an individual is to a dominant male or leader, the less extreme the contradiction between their cognitive opinions.* This hypothesis concerns the transitory effect of dominance on subordinates during the moments of their physical proximity to a leader.

Impressionistically, many people will admit that they feel somehow inhibited in expressing their grievances or opinions when close to the "boss."

While conventional attitude surveys cannot measure this phenomenon readily (since opinion polls are not taken while the respondents are close to a political leader; and then compared to others taken in homes), experience in political or diplomatic settings often reflects the hypothesis suggested.[23] In negotiations, as a matter of fact, this situational variable often produces a major difficulty: having made concessions at the bargaining table (often under the pressure of a mediator or leader who is to some extent dominant), the representative of one or more of the parties must return to his own constituency and attempt to "sell" the policy or agreement to which he agreed.

As a general rule, *the greater the concession made by the representative of a group in bargaining or negotiating, the more he must either increase his own dominance within his group or lose support.* Such a pattern reflects the contradiction between the dominance of an individual within a social group (be it a labor union, a political party, or a social movement), and the dominance interactions of that same individual at a higher level. The often made criticism that a leader has "sold out" the interests of his union or party in return for status at the national level—like the equally frequent charge that a statesman has surrendered his "principles" in international negotiations—might thus reflect the tendency to minimize conflicting cognitive opinions when interacting face-to-face with other dominant individuals.

3. *"Mood Convection," Political Change, and Revolution*

In comparative studies of animal behavior, much attention has been given to cues that serve as social "releasers" or "key stimuli" that trigger group responses. The process by which such releasers convey a behavioral disposition or mood within a group has been called "mood convection" by Lorenz (1961). For example, when a flock of birds takes flight, the coordination of their behavior follows the spread of a "mood" or "tendency" signalled by recognizable "intention movements." Since the cues acting as releasers can be culturally transmitted or individually learned, even in other species, there is no reason to assume that the phenomenon is absent or unimportant in human politics (compare Figures 2 and 16).

An example will indicate the importance of "mood convection" among humans. Many observers were surprised by the extent of the popular reaction to Richard Nixon's decision to dismiss Archibald Cox as Special Prosecutor in the Watergate affair. Nixon and his advisors were said to have been surprised by what the press called the "firestorm" of protest. More recently, a similar—and equally unexpected—explosion of discontent was triggered by President Ford's pardon of his predecessor.

Mere extrapolations of prior attitude surveys do not seem to explain such events; on the contrary, it is customary to explain shifts in opinion as a response to such salient decisions as the dismissal of Cox or the pardon of Nixon. But to understand the massive and sudden explosion of discontent—or the equally significant "rallying" behind a leader and his policies (e.g., the shift of American attitudes after the bombing of Pearl Harbor), careful study of the process of "mood convection" would seem appropriate.

While the human analogues of this aspect of animal social behavior are probably highly relevant to a number of aspects of political life, sudden communication of a mood which incites previously passive citizens to action is particularly relevant to explanations of political change. "Mood convection" would seem to be especially significant during revolutions. Not only would this phenomenon help us to understand what Brinton (1957) describes as the stage of "fever" in a revolution; it may also help explain the function of public "festivals" as a means of expressing and then channelling the explosions of energy associated with sudden political upheavals.

At the very least, therefore, studies of "mood convection" in human politics might be related to better description of the rapid spread of protest activity (e.g., the wave of strikes and student unrest in France in May, 1968—or the more recent series of prison riots in France during the summer of 1974). And ultimately, the functional significance of this process might help us to understand why some political movements flare up and disappear as passing moments of unrest (e.g., student protest in most of the industrialized nations of the West during the late 1960's), whereas other "firestorms" release the massive shifts in political power we call revolutions.

Research strategies for studying "mood convection" in politics would, of course, vary according to the particular example being studied. Sudden shifts in the mood of a crowd sharing a physical space (be it a national party convention in the United States, or students listening to speeches in a lecture hall at the University of Nanterre) are presumably triggered by different cues than a mass public reacting to the evening's TV news. Anthropological techniques of observation and interview would probably be more fruitful, at least as a first step, than data analysis of past events.

One can, nonetheless, imagine experimental designs which could explore the mechanisms involved. For example, since television is a prime medium for conveying political information, one could examine the cues that release varied responses in the viewer. To control for variations in individual political actors and the cognitive content of their messages, an actor could be videotaped giving the same speech but with different gestural cues, different settings, or different emotional intonations. Samples of viewers, comparable in terms of conventional demographic and attitude characteristics, could then view each of the experimental videotapes under conditions approximating the home TV setting.

Conventional questionnaires on viewer response, or psychophysiological data on responses (Lodge / Tursky, 1975), could be used to assess the role of different cues in producing a disposition to engage in direct political action. With a large enough sample of experimental videotapes (presumably based in part on exploration of the range of gestural cues), at least some of the components of human "mood convection" could be illuminated. Similarly, however, one could vary the cognitive message given by an actor using identical gestural or non-verbal cues—and thereby use this experimental design to measure the role of verbal symbols in "mood convection."

These examples indicate that adoption of an ethological frame of reference does not imply a rejection of existing research techniques or empirical findings in political science. They do, however, suggest some rather different areas in which research can be done, providing a more direct approach to the actual behavior of politicians and citizens alike. Within the context of empirical political science, therefore, there is good reason to believe that contemporary ethology provides an illuminating range of interesting and feasible hypotheses for empirical research.

V. Conclusions

I have argued that the influence of ethology on contemporary political science can be assessed in three broad areas: methodology or epistemology, theory (including the definition of politics), and empirical research. At the first level, ethology supports the increasing tendency to see biology—and not classical physics—as the model for a science of politics.

On the second, Lorenz's ethological theory—which can be schematically described as an O-S-R model of behavior—appears to be more general and more fruitful than the stimulus-response (S-R) framework of American behaviorist psychology. Although Lorenz's approach will increasingly be modified as sociobiologists analyze the relation between animal social systems and ecological settings, human society can no longer be divorced from nature and treated as a purely man-made creation. Moreover, comparisons between humans and other species lead to a redefinition of politics as behavior in which dominance-subordination (agonistic) interactions, characteristic of many animals, are combined with cultural or legal regulation of human social life.

On the third level, ethology suggests new avenues of empirical research. It points to political behaviors that have hitherto been little studied if at all: the gestures and non-verbal cues of leaders and followers, their responses in crowd settings, the sensations accompanying close physical proximity to leaders, the rapid spread throughout a society of sharp changes in mood triggering unexpected political activism, and the role of "attention structure" in the political arena.

An ethological approach to these phenomena suggests the following empirical hypotheses, for which feasible research strategies have been indicated.

1. As in other species, to establish and maintain dominance a human leader must serve as the focus of attention (in the arena on which political power depends).
   - *a)* Candidates who cannot become the focus of public attention only win when mere partisan identification is sufficient to insure election.
   - *b)* Establishing "name recognition" is normally more difficult for new entries in the political arena than for incumbants, and is more important in primary campaigns than in formal elections.
   - *c)* The percentage of the electorate with a strong partisan identification is positively correlated with the percentage unable to identify by name its legislative representative.
   - *d)* The percentage of the electorate expressing satisfaction with a

Figure 1
Newsweek
July 24, 1972

Figure 2
Time
Nov. 20, 1972

Figure 3
Newsweek
June 24, 1974

Figure 4
Newsweek
Nov. 13, 1972

Figure 5
Newsweek
July 23, 1973

Figure 6
Newsweek
Sept. 10, 1973

Figure 7
Newsweek
Oct. 29, 1973

Figure 8
Newsweek
Nov. 15, 1973

Figure 9
Newsweek
Nov. 26, 1973

Figure 10
Newsweek
May 20, 1974

Figure 11
Newsweek
June 17, 1974

Figure 12
Newsweek
July 15, 1974

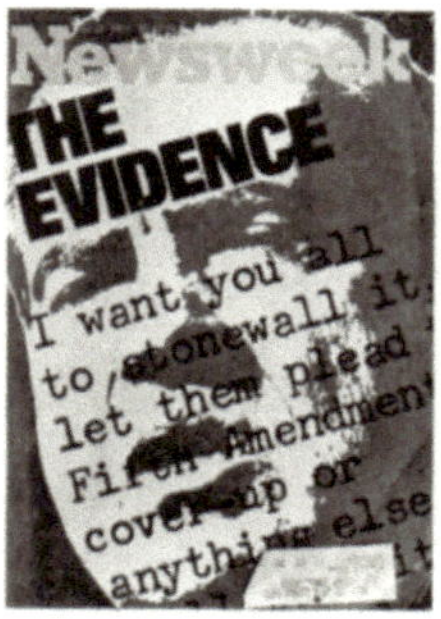

Figure 13
Newsweek
July 22, 1974

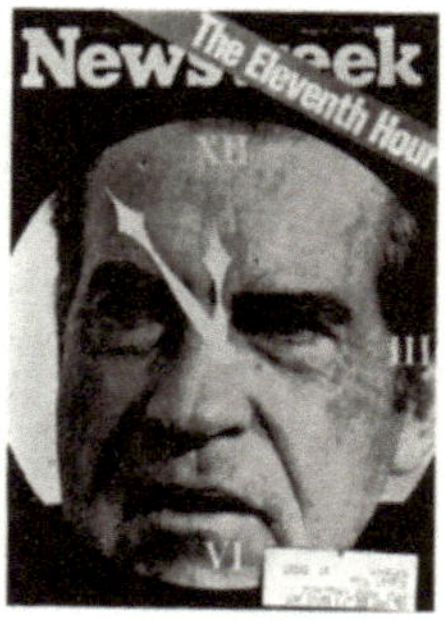

Figure 15
Newsweek
Aug. 12, 1974

Figure 14
Newsweek
July 29, 1974

Figure 16
"V" gesture at
Peace Rally

Figure 17
Washington Post "Logo"
Campaign 1972

leader's conduct in office is correlated with the frequency that he is the source of communications in the political arena.

e) A leader who definitively loses the capacity to retain control of the attention structure of a society will lose power.

2. Like the dominant male of a primate band, a politician's perceived dominance is correlated with the frequency of his display of dominance gestures and non-verbal cues.

a) If citizens or voters have strong party identification or are strongly committed supporters—or opponents—of candidates or political leaders, gestural cues and other features of dominance behavior will reinforce cognitive attitudes or be screened out of perception (but will have only a marginal influence in changing attitudes).

b) In electoral campaigns, if voters have relatively little cognitive information, are cross-pressured, or lack strong party identification, perceptions of dominance will be a more salient determinant of preference for a candidate.

c) Perception of personal dominance is a more important variable in determining voter preferences in Presidential elections than in campaigns for Congressional seats.

d) Presidential candidates are more likely to use highly dominant gestures in "public" settings than Congressional candidates, who will more frequently exhibit appeasement gestures in "social" settings.

e) The frequency of dominance gestures will be correlated with a politician's success in attracting public attention and support, and will decline in the course of a manifestly unsuccessful electoral campaign.

3. Like other primates, most humans feel a sensation of physical pleasure when in close proximity (Hall's "personal" space) to a highly dominant leader.

a) Leaders who enter crowds of supporters, exhibiting appeasement as well as dominance gestures, thereby reinforce their dominant status.

b) The more isolated a leader from public settings in which he can exhibit dominance and appeasement gestures, the more he must rely on bureaucratic, legal, or purely military devices for assuring his dominance.

c) The more radical the social changes sought by a leader, the more likely his attempts to engineer massive public manifestations of submissive behavior and support—and/or to maximize bureaucratic or military structures of power.

d) The closer the physical proximity of an individual to a dominant leader, the less extreme the contradiction between their cognitive attitudes or opinions.

e) The greater the concession made by the representative of a group in bargaining or negotiating with rivals, the more he must either increase his own dominance within his client group or lose its support.

4. Like primate bands (and other animals groups), "mood convection" can rapidly spread the disposition to act in human societies.
   a) Massive popular manifestations of political activism are triggered by different releasers than those in everyday behavior; hence such "firestorms" cannot normally be predicted by merely extrapolating prior rates of change in public opinion.
   b) Explosions of public discontent, triggered by "mood" convection, only produce lasting political change when they alter the dominance hierarchy of a society.

The above hypotheses suggest that an ethological perspective, when fully integrated in modern empirical political science, is capable of improving our understanding of politics. It will be noted, however, that these hypotheses are stated *solely* as correlations; they do not pretend to explain fully the phenomena in question, and still less to "reduce" them to merely animal behavior. Rather, they suggest that ethology can direct political scientists to new aspects of political behavior which need better study.

Nothing has been said, moreover, concerning the relevance of ethology to that area of political science variously known as traditional political theory, political philosophy, or normative theory. Suggestive comments on this level have been made elsewhere (Masters, 1975; Hummel, 1971; Davies, 1973; Schubert, 1973; Tiger, 1973). In the long run, it is not unlikely that the impact of ethology on political science will be greater in this area than in any other. But before ethological research can establish the nature of the political animal with sufficient precision that our understanding of the ends of the just human society can be improved, it is perhaps necessary to devote more careful attention to the description of what actually happens in politics.

Notes

1. For a review, see Somit (1972) as well as the works cited below.
2. For example, in 1960 Gabriel Almond emphasized the element of "legitimate *physical* compulsion" defining authority, the importance of the concept of the "political system," and the utility of a "probabilistic theory of the polity" (Almond, 1960, pp. 6-9, 62, et passim); while referring in passing to the biological model of "functioning, living organisms" (p. 13), most of the implicit epistemological references are to physics and mathematics (cf. also Almond, 1956; Almond / Powell, 1966, p. 19). Without abandoning these concepts, Almond has recently used more explicitly biological terminology in presenting his approach to comparative political science, speaking of "system" as an "ecological concept" (1974, p. 4).
3. Ignorance of this property of biological research has perhaps been important in the criticisms of Konrad Lorenz by American social scientists. The intervention of the biologist is more likely to alter the "natural" relation of an organism to its environment than is a physicist's experiment—even if a biologist hasn't killed a living organism to dissect it. The biological equivalent of Heisenberg's uncertainty principle (in which apparently passive observation modifies the system being studied) thus intervenes at a grosser level of analysis than in physics, thereby suggesting the parallel with the social science (cf. Mannheim's paradox).
4. On teleological or functional analysis in biology, see Simpson (1969), Pittendrigh (1958), and—for an extreme view—Grassé (1971). On the role of ends or teleology in political science, see Strauss (1953).
5. Cf. Stent (1969), Morin (1973), Pattee (1973). On the analogies between genetic material

and human language as information stores essential to the organization, functioning, and evolution of complex physical systems, see Masters (1970).

6. I have elsewhere developed the logic of analogical comparisons between humans and other animals in detail, suggesting both the dangers to which it is subject in practice and its legitimacy in principle (Masters 1973a, 1973b). In addition, see Lorenz (1974), Larsen (1974), and Harlow / *et al.* (1972).
7. In addition to my own statement of the implications of treating politics as a biological phenomenon (Masters, 1975), see Corning (1971), Hummel (1971), Schubert (1973), Somit (1972), and Tiger / Fox (1971).
8. Needless to say, "behaviorism" in political science is not simply identical to behaviorism in experimental psychology, merely because the name's the same. But while Dahl (1961) spoke of the objective of the "behavioral mood" in political science as a "scientific outlook," without specifying experimental psychology as a model, the social science in the U.S. which came closest to promising empirically verifiable, general "laws" of behavior was evidently psychology, if only because of American psychologists came closest to approximating classical physics in their methods and epistemology.
9. For a representative as well as influential statement, see Skinner (1965). In their reader representing the field, Ratner / Denny (1964) admit that "No special theory has been followed" although their "book is organized around a stimulus-response (S-R) approach, which can be briefly characterized as systematic, common sense" (p. 7). In a philosophical sense, one could almost say that American behaviorist psychology is a-theoretical.
10. The exposition here is derived primarily from Lorenz's scientific papers (Lorenz, 1956, 1970, 1971; Lorenz / Leyhausen, 1973) and his explicit rebuttal of American behaviorism (Lorenz, 1965). For a convenient summary, see Hass (1970).
11. This problem has now been clearly recognized as crucial by American experimental psychologists (e.g., Garcia / *et al.*, 1974). Earlier, behaviorists had asserted "that just about any stimulus can be used as a CS (Conditioned Stimulus) so long as it is above threshold and not too intense" (Ratner / Denny, 1964: 529). The epistemological implications of Lorenz's view are, of course, far reaching (cf. Stent, 1972, pp. 51-52).
12. "The organism exhibits an 'appetite' for performing its own instinctive behavior patterns. Disinhibition and performance of the instinctive behaviour pattern are accompanied by certain subjective phenomena, and both man and animals actively *strive* to enter stimulus situations in which these processes will occur." (Lorenz, 1970, p. 325.) Need it be added that Lorenz's approach, particularly as modified by recent research in neurophysiology, points to the possibility that Kant was correct when he asserted (in the *Critique of Pure Reason*) that "we are in possession of certain modes of *a priori* knowledge ?"
13. The best known empirical example of such behavior chains is probably the mating behavior of the stickleback fish (Tinbergen, 1967, pp. 20-27; Morris, 1956).
14. It follows, on the basis of Lorenz's own work, that "aggression" is not a single instinct, but a functional category of behavior (cf. Moyer, 1969). The same can be said, of course, of "territory" (Vine, 1973).
15. In particular, Lorenz's emphasis on how an organism recognizes a stimulus as a stimulus (e.g., 1970, pp. 101-258) frees the study of comparative animal behavior from the anthropomorphism of assuming that the experimentor's definition of a stimulus or a drive is identical to the perception of the organism being studied. It is thus ironic that Lorenz has been charged with anthropomorphism (e.g., Montagu, 1968), since on epistemological grounds his work opened the way to approaches that minimize this danger.
16. Consider the following account of the primary campaign: "Early in a Presidential election year [public opinion polls] are relatively changeable, and more than any other thing what they measure is name recognition, the *relative visibility* of candidates, and not the stable or well-thought-out preferences of committed voters. This explains why George McGovern has been moving up in the affections of voters and Edmund Muskie has been moving down. As Muskie has dropped by the wayside the *coverage* he has received in the media has subsided, and this has immediately been translated into losses of first-choice support among Democratic voters. Meanwhile, as McGovern has picked up a few important primary victories, he has gotten *more and more favorable attention* from the media. *As he has received more attention and become better known, his popularity has increased in the opinion surveys.*" (Polsby, 1974, pp. 31-32, italics added.) This assessment was originally published in the *Wall Street Journal,* May 25, 1972. Note that McGovern's first so-called "victory" in the Primaries—namely his strong showing in New Hampshire—was not a victory in the strict sense of the term. Muskie "won" the New Hampshire Primary in the sense of get-

ting the most votes, but McGovern got the attention by a better showing than had been predicted.

17. For evidence of such rituals in past civilizations, artistic presentations of *mythological* scenes often reveal the gestures which were probably characteristic of *human* behavior (e.g., for ancient Egypt, see Viaud, 1968, p. 42 et passim).
18. The facial components of this gesture, which incidentally are also exhibited without the raised arms (especially by campaigning politicians entering cheering crowds), are strikingly similar to the "staring bared-teeth scream face" and "silent bared-teeth face" display gestures, which are found in most primates (for a detailed description, see Van Hoof, 1969, pp. 32-45).
19. As Lorenz has pointed out, the "triumph gesture" in greylag geese involves threatening motor coordinations which have been redirected into an appeasement ritual that produces a social bond (Lorenz, 1966, pp. 175-183). More broadly, a behavior which seems superficially to be entirely aggressive—or sexual—is often found by ethologists to combine elements linked to flight and social facilitation as well as threat (e.g., Morris, 1956; Van Hoof, 1969).
20. In principle, one might therefore develop a 3 × 3 matrix as folows:

		Nature of Setting		
		Public Space	Social Space	Private Space
Performance of Dominance Gestures	Ease and High Frequency of Displays			
	Minimal Performance			
	Inhibition and Low Frequency of Displays			

Systematic observation could then be used to see whether different political men exhibit characteristic profiles on this matrix, so that an objective classification of political styles would emerge.

21. E.g., "Your endorsement of McGovern appears a reasoned and realistic statement of the case. It comes down to supporting a man who at the very least seems to be stumbling in the right direction, as against a President striding briskly in the wrong" (Letter from Peter DeVries, *New York Times,* October 1, 1972, p. 14E).
22. It is worth adding that at the time of this conversation, the editor in question was vigorously defending Nixon against the charge of a possible impeachment.
23. An anecdote from personal experience is striking. I once knew a girl who in many respects shared the opinions of the "counter-culture." She cordially detested Richard Nixon (as might be expected). During Nixon's first term, however, she had the occasion to go to a non-political meeting at which she was but a few feet from him—and came away saying that "he was not as bad as she had thought." This attitude change lasted, however, only a couple of days. . . Similarly, Robert Sheer—former editor of *Ramparts* and an avowed Marxist—recently wrote a scathing analysis of Nelson Rockefeller, based on a month of personal interviews; despite his hostility to Rockefeller, "Sheer even found himself slipping from time to time. 'Hell, it's hard *not* to like the man when you're around him. . .'," *Playboy,* November 1975, p. 3.

Bibliography

Almond, G. A.
1956 "Comparative Political Systems," *Journal of Politics* 18, pp. 391-409.

Almond, G. A. (ed.)
1974 *Comparative Politics Today,* Boston, Mass., Little, Brown.

Almond, G. A. / Powell, G. B., Jr.
1966 *Comparative Politics: A Developmental Approach,* Boston, Mass., Little, Brown.

Almond, G. A. / Coleman, J. S. (eds.)
1960 *The Politics of the Developing Areas,* Princeton, NJ, Princeton University Press.

Altmann, S. A. (ed.)
1967 *Social Communication among Primates,* Chicago, Ill., University of Chicago Press.

Autuori, M. / *et al.*
1965 *L'instinct dans le comportement des animaux et de l'homme,* Paris, Masson.

Barash, D. P.
1974 "The Evolution of Marmot Societies: A General Theory," *Science* 185, August 2, pp. 415-420.

Bentley, A. F.
1908 *The Process of Government,* Chicago, Ill., University of Chicago Press.

Bischof, N.
1972 "Biological Foundations of the Incest Taboo," *Social Science Information* 11 (6), pp. 7-36.

Bohr, N.
1958 *Atomic Physics and Human Knowledge,* New York, Harper and Row.

Brinton, C.
1957 *The Anatomy of Revolution,* New York, Vintage.

Callan, H. M. W. / Chance, M. R. A. / Pitcairn, T. K.
1973 "Attention and Advertance in Human Groups," *Social Science Information* 12 (2), April, pp. 27-41.

Campbell, A. / Converse, P. E. / Miller, W. E. / Stokes, D. E.
1960 *The American Voter,* New York, Wiley.

Chance, M. R. A.
1967 "Attention Structure as the Basis of Primate Rank Orders," *Man* 2, November, pp. 503-518.

Corning, P. A.
1971 "The Biological Bases of Behavior and some Implications for Political Science," *World Politics* 23, April, pp. 321-370.

Dahl, R. A.
1961 "The Behavioral Approach in Political Science," *American Political Science Review* 55, December, pp. 763-772.

Davis, M.
1972 *Understanding Body Movement,* New York, Arno Press.

DeVore, I. (ed.)
1965 *Primate Behavior,* New York, Holt, Rinehart and Winston.

DeVore, I. / Hall, K. R. L.
1965 "Baboon Ecology," in: DeVore, (ed.), *Primate Behavior, op. cit.,* pp. 20-52.

DeVore, I. / Washburn, S. L.
1967 "Baboon Ecology and Human Evolution," in: Korn, N. / Thomson, F. (eds.), *Human Evolution,* pp. 137-160.

Deleurance, E. P.
1956 "Analyse du comportement bâtisseur chez les 'polistes' (Hyménoptères Vespides)," in: Autuori / *et al., L'instinct..., op. cit.,* pp. 105-141.

Dethier, V. G. / Stellar, E.
1970 *Animal Behavior,* 3rd ed., Englewood Cliffs, NJ, Prentice-Hall.

Easton, D.
1953 *The Political System,* New York, A. A. Knopf.
1965a *A Framework for Political Analysis,* Englewood Cliffs, NJ, Prentice-Hall.
1965b *A Systems Analysis of Political Life,* New York, Wiley.

Fast, J.
1972 *Body Language,* New York, Harper Colophon.

Fontaine, M.
1956 "Analyse expérientale de l'instinct migrateur des Poissons," in: Autuori / *et al., L'instinct..., op. cit.,* pp. 151-167.

Fox, R.
1967 "In the beginning: Aspects of Hominid Behavioural Evolution," *Man* 2, September, pp. 415-433.

Galle, O. R. / Gove, W. R. / McPherson, J. M.
1972 "Population Density and Pathology: What Are the Relations for Man?," *Science* 176, April 7, pp. 23-30.

Garcia, J. / Hankins, W. G. / Rusiniak, K. W.
1974 "Behavioral Regulation of the Milieu Interne in Man and Rat," *Science* 185, September 4, pp. 824-831.

Goffman, E.
1972 *Relations in public*, New York, Harper Colophon.

Goldberg, A. S.
1963 "Political Science as Science," in: Polsby, N. W. / Dentler, R. A. / Smith, P. A. (eds.), *Politics and Social Life*, Boston, Mass., Houghton Mifflin.

Grassé, P. P.
1971 *Toi, ce Petit Dieu*, Paris, Albin Michel.

Haldane, J. B. S.
1965 "Les Aspects physico-chimiques des Instincts," in: Autuori / *et al.*, *L'instinct...*, *op. cit.*, pp. 545-557.

Hall, E. T.
1959 *The Silent Language*, Greenwich, Conn., Fawcett.
1969 *The Hidden Dimension*, Garden City, NY, Doubleday Anchor.

Hall, K. R. L.
1967 "Social Interactions of the Adult Male and Adult Females of a Patas Monkey Group," in: Altmann (ed.), *Social Communication among Primates*, *op. cit.*, pp. 261-280.

Hall, K. R. L. / DeVore, I.
1965 "Baboon Social Behavior," in: DeVore (ed.), *Primate Behavior*, *op. cit.*, pp. 53-110.

Handler, P. (ed.)
1970 *Biology and the Future of Man*, New York, Oxford University Press.

Harlow, H. F.
1939 "Forward Conditioning, Backward Conditioning, and Pseudo-Conditioning," *Journal of Genetic Psychology* 55, pp. 49-58; also in: Ratner / Denny (eds.), *Comparative Psychology*, *op. cit.*, pp. 534-543.
1958 "The Evolution of Learning," in: Roe, A / Simpson, G. G. (eds.), *Behavior and Evolution*, New Haven, Conn., Yale University Press, pp. 269-290.
1971 *Learning to Love*, New York, Ballantine.

Harlow, H. F. / Harlow, M. K.
1963 "A Study of Animal Affection," in: Southwick, C. H. (ed.), *Primate Social Behavior*, New York, Van Nostrand, pp. 174-184.

Harlow, H. F. / Gluck, J. P. / Suomi, S. J.
1972 "Generalization of Behavioral Data between Nonhuman and Human Animals," *American Psychologist* 27, August, pp. 709-716.

Hass, H.
1970 *The Human Animal*, London, Hodder and Stoughton.

Heisenburg, W.
1958 *Physics and Philosophy*, New York, Harper and Row.

Hinde, R. A. / Tinbergen, N.
1958 "The Comparative Study of Species-specific Behavior," in: Roe, A. / Simpson, G. G. (eds.), *Behavior and Evolution*, New Haven, Conn., Yale University Press, pp. 251-268.

Hummel, R. P.
1971 "Teaching Political Theory: the Impact of Biopolitics," Paper at the American Political Science Association Convention, Chicago, Ill., September 7-11.
1973 "A Psychology of Charisma," Paper presented to the 9th Congress of the International Political Science Association, Montreal, August 21.

Jay, P.
1965 "The Common Langur of North India," in: DeVore (ed.), *Primate Behavior*, *op. cit.*, pp. 197-249.

John, E. R.
1972 "Switchboard vs Statistical Theories of Learning and Memory," *Science* 177, September 8, pp. 850-864.

Koford, C. B.
1963 "Group Relations in an Island Colony of Rhesus Monkeys," in: Southwick, C. H. (ed.), *Primate Social Behavior*, New York, Van Nostrand, pp. 136-152.

Kolata, G. B.
1975 "Sociobiology, 2: The Evolution of Social Systems," *Science* 187, January 17, pp. 156-157.

Kummer, H.
1971 *Primate Societies*, Chicago, Ill., Aldine Atherton.

Kuo, Zing-Yang
1967 *The Dynamics of Behavior Development*, New York, Random House.

Larsen, R. R. / Coffin, R.
1974 "Rules of Inference when arguing from Animal to Man," Paper at the Animal Behavior Society, University of Illinois, May.

Lehrman, D. S.
1956 "On the Organization of Maternal Behavior and the Problem of Instinct," in: Autuori / *et al., L'instinct..., op. cit.*, pp. 475-514.

Lodge, M. / Tursky, B.
1975 "The Analysis of Political Behavior: a Bio-behavioral Approach," Paper at IPSA Colloquium on Biopolitics, Paris, January.

Lodge, M. / Tursky, B. / Tanenhaus, J.
1973 "A Bio-behavioral Approach to Political Behavior," Paper at the 9th Congress of International Political Science Association, Montreal, August.

Lorenz, K. Z.
1956 "The Objectivistic Theory of Instinct," in: Autuori / *et al., L'instinct..., op. cit.*, pp. 51-64.
1961 *King Solomon's Ring*, New York, Crowell (first ed. 1952.)
1965 *Evolution and Modification of Behavior*, Chicago, Ill., University of Chicago Press (first ed. 1961).
1966 *On Aggression*, New York, Harcourt, Brace and World (first ed. 1963).
1970 *Studies in Animal and Human Behavior*, Cambridge, Mass., Harvard University Press, vol. 1 (first ed. 1931-1942).
1971 *Studies in Animal and Human Behavior, op. cit.*, vol. 2 (first ed. 1941-1963).
1974 "Analogy as a Source of Knowledge," *Science* 185, July 19, pp. 229-234.

Lorenz, K. Z. / Leyhausen, P.
1973 *Motivation of Human and Animal Behavior*, New York, Van Nostrand Reinhold.

Masters, R. D.
1964 "World Politics as a Primitive Political System," *World Politics* 16, July, pp. 595-619.
1969 "Les Racines biologiques d'un Révolté," *Preuves*, February, pp. 74-81.
1970 "Genes, Language and Evolution," *Semiotica* 2, pp. 295-320.
1972 "Political Behavior as a Biological Phenomenon," Paper presented to the Meeting of the American Political Science Association, Washington, DC, September 7-11.
1973a "Functional Approaches to Analogical Comparison between Species," *Social Science Information* 12 (4), August, pp. 7-28.
1973b "On comparing Humans—and Human Politics—with Animal Behavior," Paper presented to the 9th Congress of the International Political Science Association, Montreal, August.
1975 "Politics as a Biological Phenomenon," *Social Science Information*, 14 (2), April, pp. 7-63.
in press "Human Nature, Nature and Political Thought," in: Pennock, J. R. / Chapman, J. (eds.), *Human Nature and Politics*, Nomos 17, New York, Lieber-Atherton.

Montagu, M. F. Ashley (ed.)
1968 *Man and Aggression*, New York, Oxford University Press.

Morin, E.
1973 *Le Paradigme perdu*, Paris, Éd. du Seuil.

Morris, D.
1956 "The Function and Causation of Courtship ceremonies," in: Autuori / *et al., L'instinct..., op. cit.*, pp. 261-284.

Moyer, K. E.
1968 "Kinds of Aggression and their Physiological Basis," *Behavioral Biology* 2, pp. 65-87.
1969 "Internal Impulses to Aggression," Trans. New York Academy of Sciences, Series 2, vol. 31, February, pp. 104-114.

Parsons, T.
1959 "Power, Party and System," in: Burdick, E. / Brodbeck, A. J. (eds.), *American Voting Behavior*, Glencoe, Ill., Free Press, pp. 81-93.

Pattee, H. H. (ed.)
1973 *Hierarchy Theory*, New York, Braziller.

Piéron, H.
1956 "L'évolution du comportement dans ses rapports avec l'instinct," in: Autuori / *et al., L'instinct..., op. cit.*, pp. 677-695.

Pittendrigh, C.
1958 "Adaptation, Natural Selection and Behavior," in: Roe, A. / Simpson, G. (eds.), *Behaviour and Evolution*, New Haven, Conn., Yale University Press.

Ploog, D. W.
1967 "The Behavior of Squirrel Monkeys *(Saimiri sciureus)* as revealed by Sociometry, Bioacoustics and Brain Stimulation," in: Altmann (ed.), *Social Communication among Primates, op. cit.*, pp. 149-184.
1970 "Neurological Aspects of Social Behavior," *Social Science Information* 9 (3), June, pp. 71-97.

Polsby, N.
1974 *Political Promises*, New York, Oxford University Press.

Ratner, S. C. / Denny, M. R. (eds.)
1964 *Comparative Psychology*, Homehood, Ill., Dorsey.

Ripley, S.
1967 "Intertroop Encounters among Ceylon Grey Langurs *(Presbytis Entellus)*," in: Altmann (ed.), *Social Communication among Primates, op. cit.*, pp. 237-253.

Rousseau, Jean-Jacques
1964 *Firt and Second Discourses*, Masters, R. (ed.), New York, St Martin's (first ed. 1755).

Rowell, T. E.
1969 "Variability in the Social Organization of Primates," in: Desmond, M. (ed.), *Primate Ethology*, Garden City, NY, Doubleday Anchor.

Schaller, G. B.
1965 *The Year of the Gorilla*, New York, Ballantine.

Schneirla, T. C.
1956 "Interrelationships of the "Innate" and the "Acquired" in Instinctive Behavior," in: Autuori / *et al., L'instinct..., op. cit.*, pp. 387-439.
1968 "Instinct and Aggression," in: Montagu (ed.), *Man and Aggression, op. cit.*, pp. 59-64.

Schubert, G.
1973 "Biopolitical Behavior: the Nature of the Political Animal," *Polity* 6, Winter, pp. 240-275.

Simpson, G. G.
1969 *Biology and Man*, New York, Harcourt, Brace and World.

Skinner, B. F.
1965 *Science and Human Behavior*, New York, Free Press.

Somit, A.
1968 "Toward a More Biologically-oriented Political Science," *Midwest Journal of Political Science* 12, November, pp. 550-567.
1972 "Review article: Biopolitics," *British Journal of Political Science* 2, April, pp. 209-238.

Stent, G. S.
1969 *The Coming of the Golden Age*, New York, Natural History Press.
1972 "Cellular Communication," *Scientific American* 227, September, pp. 49-51.

Strauss, L.
1953 *Natural Right and History,* Chicago, Ill., University of Chicago Press.
Sugiyama, Yokimaru
1967 "Social Organization of Hanuman Langurs," in: Altmann (ed.), *Social Communication among Primates, op. cit.,* pp. 221-236.
Tiger, L.
1969 *Men in Groups,* New York, Random House.
Tiger, L. / Fox, R.
1966 "The Zoological Perspective in Social Science," *Man* 1, March, pp. 75-81.
1971 *The Imperial Animal,* New York, Holt, Rinehart and Winston.
Tinbergen, N.
1967 *La vie sociale des animaux,* Paris, Payot.
1974 "Ethology and Stress Diseases," *Science* 185, July 5, pp. 20-27.
Truman, D. B.
1951 *The Governmental Process,* New York, Knopf.
Tsumori, Atsuo
1967 "Newly acquired Behavior and Social Interactions of Japanese Monkeys," in: Altman (ed.), *Social Communciation among Primates, op. cit.,* pp. 207-219.
Van Hoof, J. A. R. A. M.
1969 "The Facial Displays of the Catarrhine Monkeys and Apes," in: Desmond, M. (ed.), *Primate Ethology,* Garden City, NY, Doubleday Anchor, pp. 9-88.
1973 "The Comparison of Facial Expression in Man and Higher Primates," Paper at the International Social Science Council on "Logic of Inference from Animal to Human Behavior," Murten, Switzerland, March.
Van Lawick-Goodall, J.
1971 *In the Shadow of Man,* New York, Dell.
Van Lawick-Goodall, H. / Van Lawick-Goodall, J.
1970 *Innocent Killers,* New York, Ballantine.
Viaud, J.
1968 "Egyptian Mythology," in: *New Larousse Encyclopedia of Mythology,* New York, Prometheus Press, pp. 9-48.
Vine, I.
1973 "Social Spacing in Animals and Man," *Social Science Information* 12 (5), October, pp. 7-50.
Wickler, W.
1972 *The Biology of the Ten Commandments,* New York, McGraw-Hill.
1972 *The Sexual Code,* Garden City, NY, Doubleday Anchor.
Wilson, E.
1975 *Sociobiology: The New Synthesis,* Cambridge, Mass., Harvard University Press.

Commentary and Discussion

DAVID EASTON

The Relevance of Biopolitics to Political Theory

OUR CHAIRMAN HAS PUT the question to us "*quo vadimus?*". By this I assume us he means to direct our attention both to where we are going and to where we ought to go in the field of biopolitics. Among the three discussants I am undoubtedly the one least able to proffer very useful substantive advice. My knowledge of research in this area is second-hand and rather casual, even though my concern, for reasons that must have become clear in earlier discussion, is deep and serious. From the point of view of the continuing development of general political theory, the biological basis of political behavior has a significant contribution still to be made. This is a consideration, then, in addition to those of a strictly substantive sort that must be taken into account in assessing the direction in which this new field ought to move.

We can perhaps sharpen our own awareness of what we are about and put our expectations in a realistic perspective if we begin with a glance at the development of political science itself, particularly as it relates to the reception of new subspecialties into the discipline. With this in mind my first observation must be that if there is one outstanding characteristic of our discipline, it has been the way in which political science has proliferated new subfields in the last half century or so. And these new specialties have for the most part been constructed out of knowledge drawn initially from adjacent disciplines. Bentley borrowed massively from the sociology of his day to persuade us of the vital significance of groups in political processes. In due course group politics formed a central specialty within the discipline. Lasswell alerted us to the potentialities of psychoanalysis and this ultimately led to the emergence of political psychology as a new field. Voting behavior had its origins in the work of a sociologist and communications expert, Lazarsfeld and Berelson. Political socialization drew heavily from anthropology and psychology. Formal modelling has taken its lead from economics and the mixed field of decision-making. And from the cybernetic and system sciences, substantive contributions have been borrowed in the hope of obtaining a richer understanding of the operations of political systems.

There is little new, therefore, in political science continuing to expand its horizons by adopting substantive ideas as well as methods of research from other disciplines. This seems to have been the source of innumerable if not most innovations in political research. The emergence of biopolitics in the last few years is not extraordinary as a procedure; rather, we might be inclined to ask why it has been so late in arriving. Now that it is here, however, we must recognize that it continues the long tradition of expanding our explanatory horizons by adding new substantive determinants for political phenomena, determinants drawn from external disciplines. Because of the specialization of labor, each addition tends to take institutional form within political science as a recognized subdiscipline, ultimately with its own special sub-vocabulary and network of specialists or invisible college. Insofar as this now seems to be happening with respect to biopolitics it is not a unique development but only a reaffirmation of the way in which explanatory changes have typically occurred in our discipline. Biopolitics is less an upstart whose presence has to be justified than the start up of yet another externally generated source of understanding.

The second observation that might occur to us as we look at the way the discipline has grown is that as new subfields emerge, their content usually takes shape in a hit-or-miss way. At least to the onlooker it appears highly random. In the end there may well prove to be some controlling inner logic to the reception process but at the outset it is seldom apparent to those who are most involved. Initially the innovations tend to look very disorganized, if not fragmented, with little apparent coherence in the selection of ideas brought in from the adjacent discipline, in this case, biology.

We ought not to be too surprised or alarmed at this unsystematic character of our borrowings. It is not only understandable but also very appropriate if only because in the initial stages of new intellectual developments we often find a kind of probing procedure. In exploring a new territory it is only natural to scan the landscape broadly looking for the easiest points of entry. Hence in commenting during our earlier discussions that in reviewing the existing hundred papers or so in biopolitics, my first impression was one of lack of intellectual coherence, this was by no means intended as a negative statement. Instead it describes what, on the basis of our experience with innovations in other periods in the history of political science, we might expect to find at this stage as well.

On the other hand, this does not mean to imply that, given the fragmentary nature of the literature, it is a state we would wish to prolong indefinitely or unnecessarily. Enough research would appear to have been undertaken by this time to impress on us the need to draw our thoughts together in some more systematic fashion. The sooner special efforts are made to demonstrate the feasibility of the task, the more valuable the whole new specialty of biopolitics will be for political science. It will accelerate the process whereby those of us on the outside who see the need for enriching our understanding of politics by absorbing the relevant knowledge from biology will be able to do so. It will provide us with a manageable and economical conceptual apparatus for application to our own special problem areas.

One final observation might be made about the introduction of materials from other disciplines. The rate of acceptance of innovation in political science has increased remarkably. A logarithmic curve probably best fits the facts. If you look at the history of the discipline you find that just as in society as a whole the rate of change has been increasing at an increasing rate, so the rate of reception and expansion of innovations in the substantive subfields of political science has been following a similar course. This phenomenon can be readily demonstrated. Bentley had written *The Process of Government* at the turn of this century. His book lay dormant for a number of years, until the 1920s. Then, largely through the efforts of Herring, the group approach began to take root without reference to Bentley. It was not until the 1940s that group analysis broke into full bloom, flowering finally in Truman's text of the fifties and in the belated discovery of Bentley. In the area of group politics we find a lag of almost half a century between one scholar's awareness of the apparent political importance of groups and its ultimate and complete legitimation through incorporation into a systematic text on the subject.

In contrast to this slow reception, we might consider a field in which I have had more than a passing interest, that of political socialization. At the time I began my work in this area, about 1959, I do not think that there was even one article written by a political scientist. There were of course many publications in other disciplines but few touched on politics. By the time some of my own publications in the subject began to appear in the sixties, research by others was already under way. When colleagues and I presented the analysis of our data in book form in the late sixties, one or two other books on political socialization had been published as well as a number of journal articles. Today, no more than about fifteen years after my initial interest in what appeared to be virgin territory at the time, political socialization has been accepted as a legitimate specialty within our discipline, hundreds of articles and numerous books have appeared, and in *Youth and Society* we have a journal devoted to the subject. No one has to take the time to persuade the discipline that socialization is a matter worth pursuing. And this whole introduction, reception, and legitimation took place in the short period between 1960 and 1975. The same accelerated rate of absorption of new substantive fields within the mainstream of our discipline could be illustrated in numerous other specialized areas as well.

If the rate of change in the discipline has now acquired this kind of exponential characteristic, we might expect that it will apply to the new field of biopolitics as well. If there is real substance and significance for political understanding and explanation in the addition of knowledge from biology, as I firmly believe there is, the special advocacy with which Schwartz appropriately began our sessions the other day will be considered out of place within a few years.

*

So much for biopolitics considered from the point of view of the sociology of knowledge of political science. I will now turn to a major concern I have about some of the assumptions that are accompanying the growth of this new subfield. There seems to be little clarity about and virtual neglect of the varying modes of explanation used not only at this conference but in the literature in general. In fact I have been able to identify at least four kinds of latent approaches towards explaining the relationship of biology to politics. Depending upon the mode used we would be attributing a different kind of significance to biology for helping us to understand political phenomena. Unless we do sort them out clearly, we may well think that we are all talking about similar kinds of linkages between biology and politics when in fact we may be miles apart.

The first mode of explanation to which I would draw your attention I shall call *biological reductionism*. I know that in some circles reductionism is a dirty word, and many scholars in the social sciences would disown it on sight. Yet this aversion to the idea cannot be allowed to conceal the fact that statements are frequently made, both in the literature and in discussion, that are strongly reductionist in character. Even though it may seem somewhat pedantic, I want to take a moment to explain what I mean by reductionism. It is important to do this so that I may distinguish it as clearly as possible from other closely related modes of explanation.

By reductionism I mean the conviction that we cannot understand an entity without decomposing it into its parts and reducing these parts in turn into their components until we have reached some fundamental, irreducible unit of analysis. By reconstituting the knowledge of the parts so obtained we shall then be in a position to understand the functioning of the whole entity with which we began. The alternative, so-called holist position would maintain that even if we were able to break social behavior down into all its constituent elements, we would never be able to understand fully how groups of human beings behave. Each new combination of human beings may have properties that cannot be inferred from knowledge about the parts of which the collectivity is composed. The whole is other than simply the "sum" of the properties of the parts.

In the area of biopolitics the implication seems to be that it is insufficient to restrict our research to those perspectives manifested by actors as they relate themselves in parties, interest groups, legislatures and the like. We need to understand the way in which the behavior of each individual actor is made up of properties that derive from his species' specific biological make-up. Ultimately we may be driven back to explanation based on the properties associated with the genes if not the molecular structure of the DNA molecule itself and the way it regulates the growth and predispositions of the human organism. From knowledge about these elemental parts we would then be in a position to deduce the impact of biological elements on various aspects of political behavior.

I am not about to engage in a discussion around the complex issues raised by reductionism, given the present state of knowledge in both the natural and social sciences, although I would caution that reductionism and emergence or holism are not necessarily irreconcilable. The only point I wish to press here is that as I have read the literature in biopolitics and listened to the language used in our discussions, reductionism appears in an ambiguous light. No one seems to wish to enunciate it forthrightly, let alone define or defend it. Nonetheless its presence is constantly felt. Time and again we hear it implied that we really cannot understand various political actions—political participation, conflict, aggression, leadership, and the like—unless we penetrate to their biological roots, whether in the brain, the genetic code, or the nervous organization of the human species. That is to say, at times we seem to be saying that a final understanding of such behaviors will depend on our capacity to see them as manifestations of the biological constitution of men. In the end social behavior can be reduced to and explained by "human nature" as defined in contemporary biology.

Whether or not we agree with this conclusion, we ought at least to be aware of its unavowed presence. Even more importantly, we ought to be able to fit it into context so as to be able to show just how far biological explanation can carry us, if it is at all relevant, and how far we would continue to depend upon socio-political relationships to explain political phenomena. Enthusiasm for what look like newly discovered biological roots of some kinds of political behavior can easily lead us to neglect a more balanced inquiry. We might only too readily conclude that through ethology and genetic analysis alone we have discovered the fundamental units out of which a virtually complete understanding of our political behavior and the operations of our political institutions can ultimately be reconstructed.

A second type of explanation is represented in what I shall call *biological determinism*. This is different from biological reductionism although in discussion both positions may become so interwoven as to be almost indistinguishable. For the determinist, biology is also fundamental to the understanding of all human behavior but in a very special sense. The reductionist seems to be asking us to deduce observed behavior from its genetic basis, leaving it an open question as to how we can connect the biologically given internal states of the organism with institutionalized political behavior. The reductionist need not believe that the soma determines political behavior in the special sense in which the concept of determination will be used here.

The determinist seems to be saying that it is in the very nature of the human organism, that is, because of its genetic structure, that motivations and other predispositions arise, and these compel or propel the human being in a given direction. Biology does more than impose passive constraints or limits. It operates actively to direct the behavior of the organism. Since this compelling behavior is genetic it cannot be expugned. We act in a certain way whether we like it or not. It is this insurmountable character of our genetic heritage that reveals the continuity of evolution among the varied species of which man is only one. Among lower animals these genetic givens may lead to imprinting, nesting, hunting, care of the young, courting rituals, social cueing and the like. The activities involved are avoidable only with the extinction of the individual or at overwhelming somatic and psychic costs. In the human species these irresistible and propelling drives are seen in the need for territorial control, aggression, perhaps male bonding, typical responses to crowding, and so forth.

These biologically based needs are to be viewed as institutional hazards, as it were. If the needs are not met, social institutions that stifle or divert them will operate ineffectively or crumble. Not that a certain amount of accommodation between needs and institutions is impossible. Constant needs may find their expression in variable institutions. But whatever form the political institutions take, they must provide for the expression of these irreducible needs. In this sense our genetic inheritance determines, shapes, or controls in some ultimate sense the nature of social relationships.

In a way biological determinism resembles forms of Marxist explanation and if for that reason alone it cannot be lightly dismissed. In fact, for our purposes a useful illustration is to be found in the writings of the structural Marxists where Althusser speaks about "determination in the last instance," a much discussed phrase. This does not mean to say that for this school of Marxism, ideology or politics, for example, are without effect and may not even be "dominant" under the appropriate circumstances. In the end, however, it is the mode of production that forms and shapes the fundamental social relationships and compels the functioning of social organizations and the movement of change in certain directions. There is a certain inescapability to these effects aside from the utter destruction of society itself. In the last instance, therefore, what determines the character of society is the mode of production. It is as though the economy, so described, were the genetic code of society, a code that permits considerable variability but one that reproduces existing social relationships or channels change in determinable directions. Biological determinism seems to bear the same implications.

We are likely to find this kind of meaning for determinism in the discussions around the subject of ethology. Even though tendencies towards dominance, territoriality, bonding, aggression and so on seem to have their place in the genetic structure of certain species, analogically or though the assumption of evolutionary continuity, they have been applied to human beings in their social relationships. Our discussions during this conference, however, did not address themselves significantly to questioning the validity of this mode of explanation even though a protest or two was heard.

Indeed, there is much to be said for the other side, and this has yet to be brought out or at least compared with the ethological arguments for continuity among the species and their genetic traits. Certainly no attention was devoted to specifying the linking mechanisms whereby presumed genetic compulsions transformed themselves into human institutions or influenced the form and operation of these institutions. Even if the ethologists were correct, we might still wish to consider the possibility that a basic genetic trait of the human species, distinguishing it from others, is its susceptibility at various stages to behavior modification through the environment. For example, the biologist Vandenberg has questioned the validity of the studies of intelligence among identical twins in order to test for the compulsory results of identical genetic characteristics (see 21 *Items,* 1967). And the implications of Dobzhansky's thinking would seem to be that there is no one human nature but many human natures—biological human nature is not uniform but multiform. In that case each person would have his own combination of genes, and depending upon the variability of the environment, a person would have free rein to express his full personality.

Once again, however, I would remind you that I am only an amateur in the substantive problems relating social behavior to its biological base and would not pretend to argue seriously for the validity of any position. My only point is that determination in the sense of leaving the human being no option but the one that is genetically based—aside from the destruction or basic impairment of the organism—is an assumption that underlies much of our discussion, especially about ethology and its relationship to political behavior. This position not only must be recognized in relation to other alternatives but it needs further clarification especially about its implications for the functioning of political institutions.

There is a third mode of explanation. It seeks to demonstrate the biological impact by describing what I shall call the *organic limits* to human behavior. Here we have two sub-categories: one we may call fixed limits, the other, variable limits.

There are those who seem to be saying that the soma—the organic structure of the human being—imposes limits on what society can expect from persons in the political sphere, and these may be unchangeable. Insofar as these may be genetic, these kinds of limits presumably would be included. But the emphasis is not on the genes. Instead it is on the physiological and anatomical organization of men. The soma in this sense imposes virtually absolute organic limits on the way people can be expected to interact politically. Man cannot fly. That is a fixed boundary on his behavior. If he is going to fly he has to supplement his organic capabilities in some way. Similarly, it would appear from what Schwartz has so effectively argued, persons of a certain health status or with nutritional deficiencies are less likely to participate politically in given ways. Institutional modifications, ideological exhortations, and laws notwithstanding, hunger is likely to reduce the quantity and quality of political participation. A person weakened by disease is less likely to drag himself to the ballot box or even to the barricades than to the nearest source of medical care, if available. It is difficult for society to change this relationship between somatic states and political participation. It was even suggested by Tiger that among these fixed limits there may be design defects in the human brain that prevent it from seeing into parts of its reality. We may therefore be unable to cope with all parts of the environment, an interesting if frightening thought. In other words, regardless for the moment of the direction in which any presumed genetic needs may impel us, variations in organic states impose major barriers to discernible kinds of political behavior.

The second sub-category of these biological limits are variable in nature. No paper at this conference devoted its attention to this aspect although Somit's review of the literature did touch on the subject. Here I refer to the extent to which human biology is of such a character that it permits control or manipulation through psycho-pharmacological means. There is little question about the potential here for behavior modification, voluntarily accepted or otherwise. On a mass scale biological warfare is certainly the most discussed of these variable limits. The potential for exposing whole or selected parts of a population to a large arsenal of biochemicals dramatically calls attention to the need to consider not only the policies about and consequences of such action but the underlying mode of explanation of biological effects.

This area deals with the enormous and growing potential for control through direct chemical assault on the organism. It ranges all the way from the 1984 type of control through test tube reproduction of selected species types, to limited regulation of behavior of specific parts of the population through pharmacologically induced states of mind. But without letting our imagination roam too wildly here, the emphasis of this mode of explanation about the relationship between biology and politics is on the indeterminate but undeniable variable limits to the modifications that can be induced in political relationships through direct action on the soma.

The fourth and final kind of explanation of biological linkages to politics can be found in the work that has been going on at Stony Brook. Here there is a subtle shift in emphasis of which we have failed to take due notice. Our interest in biology arises not out of the consequences that varying states of the body will have for political behavior but because these varying states may provide us with better indicators and measures of observable political attitudes and actions. The basic assumption here is that for many social interactions and attitudes there are corresponding biological events. If we can obtain precise measures of these organic events through the use of well-developed devices already in use, as in psychometrics, then we might have far more reliable indicators than are currently available either through self-reporting attitude surveys or the uneven data of objective observers. As Tursky and Lodge have reported, we now have the beginning of the development of physiological and psychophysical devices designed to measure political attitudes more accurately.

This fourth kind of explanation—*bio-polimetrics* if I may dare call it that—seems to fall into a category substantially different from the other three modes of explanation. In all the latter, the objective was to use biological knowledge as a way of adding to our substantive understanding of how or why individuals engage in politics. Bio-polimetrics turns to the soma only because it gives us better ways of measuring political phenomena. But the fact that this use of the organism does not seem to stand apart from the rest of the discussion at this conference only emphasizes what we have already recognized: We have no biological theory which can instruct us in advance of the likely and different points of connection between politics and biology. If we had such a theory or set of conceptual guidelines, perhaps such measuring devices as the hand dynanometer or galvanic gauges might not seem so alien to the field of biopolitics.

I am not suggesting either that any single paper at this conference or elsewhere takes advantage of all these four modes of explanation or even, for that matter, confines itself to one alone. In all probability each confrontation between biology and politics involves some unwitting combination of two or more of these ways of explaining the linkage. If this does indeed reflect current practice, it points up the desirability of explicitly recognizing the alternatives that are available and justifying the utilization of one or another. Certainly, if we are to produce valid knowledge in this new field, the difficult problems raised by such premises as reductionism, determinism, organic limits and physiological or psychophysical measurement merit the fullest exploration. Since at best we have only touched on these subjects, any future conference might well take modes of explanation in biopolitics as one of its central themes.

*

I would like to say a word or two about a final point. It goes to the heart of my concern with biopolitics as a developing subfield. My observation here deals not with the substantive findings or speculations to which we have been exposed or even with the kinds of explanatory assumptions but with the political objects to which biological knowledge ought to be related. After all, we say very little to our colleagues in the profession when we point out that our subjects are all human beings and, as such, we might expect that their organic and conspecific properties ought to have some significance for their political behavior. Every freshman knows that Plato made a point about human nature and politics some little time ago. Rather, we lack two things: some systematic, even if only preliminary ideas about those aspects of man's biological make-up that could possibly be relevant to his political behavior—and this I have already mentioned; but also a theoretically coherent way for identifying those areas of political life with which we might most fruitfully connect biological knowledge.

In the existing literature it would appear that the political aspects which are presumed to be influenced in some significant way by the organic nature of the human being have been selected on an *ad hoc* or intuitive basis. I would not want to press this point too vigorously since, as I have already suggested, in any innovative area the pathbreakers are compelled to take advantage of targets of opportunity, and this should be no less true for biopolitics. Hence we have had to work with what we have at hand. Ethology, data on the health and nutrition of populations, psychoneurological studies of aggression, the effects of animal crowding, the uncontrolled exploitation of one's biological niche, biochemical effects on individual and group behavior, and so forth suggest immediate linkages to interesting political problems, especially of currently topical kinds.

Now that we have made this beginning, however, we might begin to ask whether we ought to proceed very much further without asking what must always be the logically prior question even though empirically it follows after the facts: Are we really posing theoretically relevant and interesting questions if some day we hope to be able to produce a special theory about biopolitical effects? Even while our hit-and-miss research is under way, ought we not to be asking ourselves: What are the important areas and issues in politics for which we should be seeking additional understanding potentially available from a biological perspective?

We can learn here from the history of the voting studies. As we all know, they initially got their strongest push not from political scientists but from Lazarsfeld and Berelson, a sociologist and a communications expert. The disciplinary affiliation of these scholars need have little meaning for us in the abstract. Disciplines are only arbitrary classifications to meet the needs of an enforced specialization of intellectual labor. Nevertheless Lazarsfeld and Berelson came to voting behavior in pursuit of an understanding about the way in which people make choices. Research in the economic market place had already thrown some light on this, and Lazarsfeld hoped to be able to get at the matter from another direction by examining the way in which choices are made in the political market place. No one would want to detract even in a small way from the credit due to both these social scientists for breaking open the whole area of voting research. It is now clear, however, that their *optique* derived from and pointed towards the social psychology of choice. Political scientists of similar stature and wisdom at the time would surely have immediately sought to build the study of voting into the institutional context in which political systems operate. Only in recent years have efforts been made to reorient the theoretical apparatus within which voting research takes place so as to make it more useful for solving the institutional problems of political systems.

I mention the voting field only as a warning. Our experience there should sensitize us to the need to draw up short, while there is still time, rather than to rush headlong after our intuitions about the political areas significant for biological inquiry. At the very least some of our thinking ought to go into considering what might be the most important political areas that merit investigation. To this point we seem to be asking: Given the kind of biological knowledge that we now have, how can we bring it to bear on one or another aspect of political behavior. It may not seem appropriate to say that this is putting the cart before the horse but it is certainly putting the body biologic before the body politic.

It is clear that I do not come to this conclusion from the point of view of a disinterested bystander. I have already gone to some pains to suggest that there are certain kinds of political variables—such as the inputs of demands and support, conversion processes, and policy outputs—that in systematic terms ought to command our prior attention. To me these seem critical for an understanding of the way any and all kinds of political systems function. They are influenced by the environment in which they operate, and as part of that environment the biological nature of man has a significant place. From this perspective the soma and genetic structure constitute two among a number of major determinants including the economic, social structural, cultural, and the like. From a systems perspective we have some guidelines for helping us to sort out from the whole body of biological knowledge those parts that might be most useful in adding to our understanding of the way in which political systems function. A framework for political analysis also helps us to detect those lacunae in the field of biology that, as political scientists, we would like to see filled. Thereby we could act not only as passive consumers of biological information but as stimulators to the biologists. In this way we might hope to generate more rapidly a true give and take in this interstitial field of inquiry.

I hasten to add that I do not intend to say that biology would be of value only if it could be used to help us address ourselves to those political issues suggested by systems analysis. I have little doubt that there are biological elements involved, say, in the fluctuations of support for political authorities or a regime, or for the feedback implications of outputs insofar as they depend upon the energy levels of a population. I am, however, using systems analysis only for illustrative purposes. What I am suggesting is that some broader conceptualization of politics might appropriately be in the minds of our biopolitical scientists so that they can do more than intituitively justify the political problems selected for explanation or seek refuge in the fact that they are dealing with political areas normally considered important. Here I return to one of the opening remarks to the conference by Wahlke. Not only is there a shortage of conceptual analysis on the biological side. It exists on the political side as well. In one way or another we need to devote specific and special attention to bringing the two together.

I would conclude then by saying: first, if the history of the addition of past subfields to political science is any guide, we can expect a rapid acceleration of interest in biopolitics; second, we must be far clearer about our explanatory assumptions when we speak about the biological bases of political behavior; third, we are already at the stage where we can begin to contemplate our biological interests in more systematic ways; and finally, it is imperative that we lock our new-found biological knowledge into those aspects of political systems which are theoretically viewed as most significant. Under these conditions a profitable exchange between biology and political science may begin to take place, one in which we not only learn from biology but in which biologists modify their own research so as to help us to answer issues of significance to ourselves as well. Each of these points might well form a central theme in one of a series of succeeding conferences on biopolitics as research in this field continues to develop.

WILLIAM MACKENZIE

Biopolitics: A Minority Viewpoint

I WILL SPEAK OF SOME QUITE SPECIFIC THINGS of which I'm bound to have a minority point of view and it may be useful to get those in at this stage. But before saying that I would say two things from my notes. First, David Easton illustrated perfectly how you can take the bio-politics focus and build into it a sophisticated discussion of the "great issues," and that this could be used like other focuses as a means of teaching students about the tradition of political study and would "click" in with many different frameworks for general, non-specialized, courses. That's point one. Point two is that some of what I'm saying may sound skeptical, but I am for various reasons very much committed to this whole approach, partly due to an accidental series of encounters in which I have been personally involved, I'm probably earlier than anybody else here, for instance, in my encounter with the first *Territory in Bird Life* book* which appeared in 1920, and which my father had. I'm quite sure I'd read that before I'd left school.

Now the minority points of view. The first is about the name. The name "bio-politics" is questionable on two grounds. First, and this may again be a unique point of view. I may not be the only person who actually owns a copy of Morley Roberts' *Bio-Politics,* which was published in 1938,** but I'm quite certain I'm the only person here who ever met Morley Roberts. Roberts was a very interesting man and I think he is entitled to some proprietory rights. He's entitled perhaps not to a Ph. D. dissertation but maybe an article by somebody who is interested in him, because he was one of these wandering Englishmen of the late nineteenth century like Cunninghame Graham and W. H. Hudson. He travelled for twenty years, earning his way; then he wrote minor books for twenty years. Then possibly from some personal crisis when he was over fifty he "clicked" in on cancer research, whether or not cancer was his own fate. He had

* By H. Eliot Howard (London, John Murray, 1920).

** London, Dent, 1938. He was born in 1857, was one of the first graduates of Manchester University, and I reckon sixty-one books in his *Who was Who* entry. Most of them seem to be travel, adventure, "boy's books." But by a biography and an edition of the novels he did much to re-build the literary reputation of his Manchester contemporary George Gissing.

sufficient spirit of inquiry and sufficient plausibility in status and access to meet the key authorities of this period, and he did this on the level of the best amateurs around this table. He produced the first book in this field; produced it when he was nearly seventy. It was on "malignancy," in a physiological sense, and was introduced by Arthur Keith who will of course be known to all as a physical anthropologist.

I think the title should not be pinched without some reference to him. About his position historically in this field, I personally think that because of his peculiar wandering life he cut in on the intellectual discussion at a fresh period, and that one could claim that he did give us a fresh start. The Herbert Spencer type of bio-politics had been slaughtered on the one hand by the first generation of the positivists,* on the other hand by the idealists. Roberts was quite outside these academic contests; he began again on bio-politics with a better training in physiology and ecology than perhaps the Spencerites ever had. It's not quite a discovery like that of Bentley, but there are some analogies with Bentley and the way he cut into a new field which nobody else looked at again for thirty years.

I suppose there were two interesting things, that he said at the top level. First, the organic analogy, and he's very clear about analogies and he knows very well that he's talking about analogies and he knows how to use them. First, the organic analogy. The body is not a peaceful, well organized, static equilibrium kind of thing. The body itself is a center for perpetual struggle, and if you are taking an organic analogy (and he, mind you, was taking it at a cell level), it is not one of authoritarian government; it's one of continual internal strife. The second point in his analogy is this. He said that if you take the political system and give it an organic analogy, the true analogy is more like that of a jelly fish, than that of a mammal. It's a very primitive sort of organism, but it is an organism and it is capable of reactions. You can play the analogies on from there.

Well, that's all about the founding father. Leading from that to the second point, another in which I'm bound to be in a minority, I don't much like the term "bio-politics" because at least in our context it tends to mean, as David Easton said "bio-political science" and I'm not terribly interested in "bio-political science," I'm interested in biology and politics. And I'd like to hold fast to the ambiguity of the word "politics" as meaning both a practical activity and a meta-activity. So I propose calling it biology and politics, and not bio-politics.

My second main heading is about tactics. And I do read P.S. and I do regard the APSA as a remarkable forum of micro politics, and I study its workings with gigantic interest and I realize how they affect us all. But, of course, I stand outside this and perhaps could put this in terms of two dilemmas. The first one: Is this a new sector of political science or is it a possible focus of all existing sectors? I think I realize something of the dynamics of APSA and the practical impossibility of approaching the problem like this. But I see this as a field which is now concentrating peoples' minds, as the H-bomb and the cold war concentrated peoples'

* See in particular G. E. Moore, *Principia Ethica* (Cambridge University Press, 1903) on "the naturalistic fallacy," and on Spencer in particular (§§ 29-34).

minds in the 1950s; the themes of Doom Watch, Space Ship Earth, planetary ecology have generated a good deal of new thinking, so that now practically everybody is thinking about biology and politics in some sense. Of course, we may be moving from a biological interest into an energy interest, in terms of sheer saleability to the public, and I'm quite prepared to take biology and treat it as a subset of energy politics in terms of energy/information systems etc. There may have been a special biology vogue floated off by the saleability of not only Lorenz but also Jane Lawick-Goodall and others, and this may now be pushed into the back pages. But I don't very much mind; I think biology is now an enormously stimulating focus of thought, which can be operated over the whole range of first-level politics, and will stimulate greatly in that sense the development of the discipline of meta-politics.

But this is my dilemma as between these two. Should one think of a sub-discipline of meta-politics ? or should one think in terms of the following ? I think it is still possible, in the United Kingdom at least, to consider the possibility that the professional associations have developed on quite the wrong footing. This is a subject where one ought to read bio-sociology and bio-psychology at the same time as one handles bio-politics. That's to say, I think this field of biology and politics cuts across meta-politics, and that it cuts all the social sciences too. And the other social sciences are undoubtedly in the same condition. This leads me to realize how badly briefed I am about bio-sociology and bio-psychology. I think that I know bio-economics does not exist; but I'm perfectly convinced that the others do and that it's a rather dangerous enterprise, in terms of the development of knowledge about the working of human society, to handle this as politics only and not as a general social science enterprise.

I've pages more stuff here, which I must cut—except to say that I owe enormous gratitude to the whole of this meeting in terms of new ideas, new bibliographies, new people in a personal sense. I'd like to make two points only to end the discussion. One of them: Have you picked the right name ? The other: Have you picked the right tactics ? And that I know has to be looked at in the American context. My tactics are different, but this is not to condemn those who are experts in the bio-politics of APSA; the nutritional politics of APSA, one might say, or even the energy politics of APSA.

JOHN C. WAHLKE

Observations on Biopolitical Study

As an anticipatory note, let me add briefly to David Easton's comment about the extent to which diffusion, fragmentation, and general lack of logical interconnectedness seem to be characteristic of the field with which we are dealing. Without denying the frequent aptness of such characterizations, I suggest that it is still possible to detect some widespread, though admittedly tacit, consensus on a number of important scientific premises among workers in what is broadly and vaguely called "biopolitics." It is useful to bring these points of agreement to the surface, state them with precision, and recognize them for what they are, as has been done, for example, by Peter Corning in his list of sixteen fundamental propositions, and by David Schwartz in explication of his three basic assumptions. Epistemological and methodological clarity will surely not hamper our search for increased substantive understanding of political behavior.

*

More directly to the point of that main purpose, now, let me first consider in this paper* our chairman's futuristic question "where are we or ought we to be going?" from the standpoint of his particular formulation of it: *How much and what kind of biology should we bring to political study and how should we do that?*

To begin with, I agree with Roger Masters that biological thinking will and should have a powerful "paradigmatic effect" on our formulation of questions and problems for political behavior research, although the term "heuristic" may describe such effects more appropriately than the term "paradigmatic." But I think it is even more important to recognize, as Lionel Tiger has reminded us, that the "biological thinking" we are talking about is not a matter of figures of speech, purely literary analogies,

* I am indebted beyond measure to the Netherlands Institute for Advanced Study in the Humanities and Social Sciences for enabling me to pursue my studies of the matters touched on in this volume as a Resident Fellow for the year 1974-1975. It provided not only the opportunity and the facilities for that study, but the most congenial and favorable climate, intellectually, socially, and culturally, imaginable.

"as-if" comparisons, and other non-verifiable statements, but of propositions about important "facts of life," genuine *knowledge* of human behavior, so far as this is possessed and communicated to us by colleagues in the various biological and biologically related sciences. To ignore these facts about how human beings are constructed and how they behave (both individually and in collectivities), or to pretend that the facts are otherwise, is to advance the cause of error and ignorance. I would therefore emphasize the importance of acquainting ourselves with such biologically based knowledge, and heavily underscore Glen Schubert's critical estimate of our performance as a discipline in this respect:

> Political science has books that pretend to be biological in approach but that in fact are not so at all; and this is almost certainly worse than to have no biopolitical books at all.[1]

What are the main elements of this knowledge which genuinely "biopolitical books" should incorporate? Let me, without claiming to have access to some encyclopedic classification of such knowledge, suggest one or two points that seem to me directly relevant to the study of political behavior.

First is the notion, well developed in works on ethology and on evolution, that there is what might be (indeed, has been) called "morphology of behavior" as well as morphology of physical structure or anatomy. Broad patterns of behavior, among humans as among other species, are developed and transmitted through evolutionary processes. These "evolutionary processes" must be understood, of course, to comprehend both culturally learned behavior patterns and behaviors which are physiologically programmed through genetic inheritance. Examples are legion of physiologically programmed behavior patterns shaping and channeling culturally learned behaviors, setting limits within which these may be altered, and governing the mechanisms for such alteration.[2] Investigation of these types of behavior pattern and their development is the business of behavioral genetics and ethology, to be carried out through research species-by-species, including the human species. It is certainly the business of political scientists to know what such researchers find out about the bases of human behavior; it is perhaps not inappropriate for political scientists to suggest to biological or ethological or other researchers fairly specific questions about human behavior patterns relevant to or constituent of politics and government.

Second, at a different level of analysis, we have some fundamental principles of the mechanics, the physical and physiological processes within the individual organism, by which it engages in actions manifesting the kinds of phylogenetic behavior patterns, as well as the more idiosyncratic day-to-day actions, which concern students of social and political behavior. Without detailing or systematically characterizing main propositions found in this particular body of knowledge, we can note that this conference heard, for example, of the relevance of certain classes of neurophysiological knowledge to certain classes of political behavior.[3] The powerful methodological application of fundamental principles of psychobiology, psycho-physics, and psycho-physiology, has been well demonstrated here (as well as in other works) by Milton Lodge and Bernard Tursky.

Both Milton Lodge and Roger Masters have pointed out to the conference the importance of basic principles from behavioral psychology for our conceptions of the behavioral dynamics of individual human organisms. Particularly worth noting are their warnings about the obsolescence of the simplistic "S-R" conceptions which some critics seem to think constitute "behavioristic psychology," and their recognition of the potential explanatory power of more realistic "S-O-R" models which take into account "endogenous" stimuli occurring within the organism and view certain physiological, neurophysiological, and psychophysiological processes as chains of internal "S-R" linkages embedded within the overall "S-O-R" process. Not represented in papers or discussions here but surely potentially an important contributor to our knowledge in this area is the discipline of behavioral anthropology, as pursued by such scholars as Eliot Chapple.[4] To repeat, this brief paragraph in no wise purports to list or summarize the substantive principles of behavior taught us by biologically oriented disciplines like physiology, neurophysiology, or others. The point is simply that students of political behavior must know and incorporate those principles in the "models of behavior" and "conceptions of human nature" which shape and guide their formulation of questions and manufacture of research designs, however consciously or unconsciously.

One main thrust of all the knowledge coming at us out of the biological sciences seems to me to be an irresistible compulsion toward drastic revision of the long-prevailing, simplistic paradigm of political behavior which has been drawn almost wholly, as Glen Schubert rightly noted, from inadequately digested social psychology. I refer to the notion of a dichotomous set of political actions, *behavior,* which is presumably motivated (caused) by usually self-conscious and articulatable *attitudes,* variously defined or analyzed as comprising cognitive, affective, and emotive dimensions or elements; opinions, beliefs, orientations, and other "types," and so on. As Milton Lodge in particular has pointed out, so much political behavior research consists of "explaining" one attitude by another attitude which presumably lies closer to the behavior of interest to the research that a scandalous proportion of political behavior research findings ultimately boils down to explanations that account for hardly more than one per cent of the variance in the behavior in question.[5] It is therefore painfully clear that radical reconceptualization of political behavior in light of available "hard" knowledge about the working of human organisms must have high priority on the agenda for biopolitical research. We shall make precious little contribution to knowledge if we do nothing more than use some newfangled "biological variables" to explain another two or three per cent of the variance in a few of the same old attitudinal variables that have all along been the weak links in political behavior research.

*

Now let me come at the question of where we are or ought to be going from the other direction, not from the question of how much or what biology we might apply to political study, but from the question: *What political science questions should we bring biology to?* This question grows out of but goes beyond my preceding remarks about conceptions of attitude and behavior in political research. It relates to David Easton's concern with the "linkage problem," i.e., the relevance of biological knowledge to macro-questions and system-concerns, and to John Meisel's plea for some kind of "theoretical framework" to guide our applications of biology to political study.

A few years ago it was suggested that "what is necessary [for progress in biopolitical research] is a biobehavioral model of man susceptible of being linked with models of sociopolitical processes."[6] I think the order of need is the reverse: we now have within reach a far better "biobehavioral model of man" than the prevailing outmoded social-psychological model, but we do not really have a justifiable and applicable "model of sociopolitical processes." Particularly do we lack a comprehensive and orderly conception of the generic character and role of government within or without such a model. I think I am here echoing the plea made earlier by John Meisel for some kind of "theoretical framework" to guide our applications of biology to political study.[7] And I suspect that some of the primordial hostility expressed in several discussions toward ethological perspectives—or rather, as John Crook properly reminded us, toward the bowdlerized Lorenzianism that passes for ethology among some political scientists—stems from legitimate concern that biopolitical scientists may too readily go off half-cocked in precisely this sense, i.e., plunge helter-skelter into fragmentary research projects, operationalizing miscellaneous dependent variables more or less randomly *ad hoc,* and leaving it to others (or to nobody) to find out to what (if any) fundamental theoretical concepts, out of which (if any) broad conceptions of the polity, they might correspond.

Critics of biopolitics are, of course, correct when they insist that there is little theoretical or conceptual value in merely relabeling assorted political phenomena with biopolitical names, for example calling politics "dominance behavior" or obedience to law or authority "operant-conditioned responses." Whatever the words used, what matters is to find out (for example) the consequences of different forms of dominance behavior for different forms of political society and for the species, which inquiry can be properly oriented and guided only by some conception of the generic character of human society and government, some basic notion of how such phenomena as obedience or dominance behavior relate to our other concerns as political scientists with the behavior of humans in the social aggregates we call polities or political communities.

It is in developing a more fundamental and unifying grasp on the generic character and function of government among humans that the paradigmatic (heuristic) influence of biological knowledge is potentially most helpful. The conference has several times been given hints about how such a perspective can force us back toward base-line, axiomatic, but empirically defensible starting points for our theoretical and research activity. Note, for example, Roger Master's definition of political life as "the arena of 'agonistic' behavior directed to the establishment, maintenance or change of social rules,"[8] or Peter Corning's proposition that, "politics [. . .] is the process by which man attempts to cope with public problems, make authoritative decisions for the group, and organize and coordinate the behavior of his fellows."[9] Their formulations do more than just rephrase those almost-ancient definitions which revolve around "monopolistic control of instruments of coercion" or "authoritative allocation of values for society," although they are by no means incompatible with them. They carry such formal definitions, which more often than not become sterile, abstract, and hopelessly "academic" in the hands of modern researchers, back to the kinds of macro-concern which agitated the earliest and the greatest of inquirers into political affairs of men. More important, they appear clearly as suppositional points from which to begin study, not as embodiments of substantive knowledge resulting from it; they require us to go on in search of verifiable existential and synthetic statements about uniformities of behavior common throughout the human species, as well as patterns and uniformities differentiating among sub-sets within it, so that we can better understand the provision and maintenance of "government" as a functional requisite for the cohabitation of men in groups.

Formulation of inter-related "big" questions about the polity and equally "big" questions about the behavioral mechanics relating to them is better done by long, intensive effort of cooperating individual scholars than in conferences or committees. But a few tentative, crudely formulated examples here may help illustrate the character of the problem as I see it:

- What are the bases and the dynamics of communal sub-grouping (pseudo-speciation) in observable human societies? Of change in given groupings?
- What are the consequences of different types of sub-grouping: for inter-group conflict? for ecology (man's "niche" at a given moment)? for human survival?
- What behavioral patterns are amenable to shaping *by* governmental processes in different kinds of communities and societies? I.e., what are the phylogenetic limits within which societal and governmental processes operate in patterning human behavior?
- What kinds of individual play what kinds of roles (functions) in the governmental processes of different kinds of community and society? I.e., how much is the governing of human groups determined by phylogenetically programmed factors?

To repeat, these unsystematic examples are offered only to suggest the kinds of macro-concerns that might profitably guide biopolitical research on human behavior. They may also suggest how far biopolitical approaches take us from the usual topics of political behavior research, which tend to center on such ritualized individual behaviors as voting. They may well suggest, finally, that it will be at least as difficult to formulate such questions as to design research to get answers to them.

*

About the implications of biopolitics for more mundane, day-to-day aspects of our activity as professional scholars, I have only a few words to say.

With respect to the practical objectives of biopolitical research and the uses of biopolitical knowledge obtained by it, some critics seemed to suggest that the kind of "policy-science" effort represented by Peter Corning's paper pose in quantitatively and qualitatively different measure than other policy research the dangers and difficulties of "premature postbehavioral policy science" (apologies to David Easton for bastardizing that concept from two of his brain-children!). It is certainly legitimate to examine, question, and reject the premises and the criteria he offers as bases for policy decision on the matters with which he is concerned. But it is hardly legitimate to refuse to consider the question of making policy decisions on grounds of their awesome possible consequences unless it can be shown convincingly that there will be no "policy" or consequences emerging by non-decision if we do refuse to think about such problems. Nor is it legitimate to consider the problem important but to reject Corning's criteria and premises unless one can show that more desirable consequences will follow from basing policy decisions on premises and criteria which he poses as alternatives to Corning's. In principle, the intellectual, ethical, and scientific problems and issues raised when we move from the "pure-science" to the "applied-science" or policy framework in biopolitics are those encountered when we make the same shift in any other field of interest. That is hardly to say, of course, that the problems are unimportant or easy to solve.

With respect to the still more concrete problems of research strategy and tactics, it is difficult in a conference like this to proceed much beyond the most general sort of prescription. It is easy (and I think correct) to say that increasing use of experimental methods, both laboratory and natural-setting, seems likely and desirable. So does wider and more sophisticated use of field-observation methods borrowed from behavioral anthropologists and ethologists. By the same token, it seems both probable and desirable that we shall become much less dependent upon the "classical" contemporary version of the sample opinion survey. Beyond this, we can hardly venture here. The most hopeful sign of progress in biopolitical study would probably be demonstration of the necessity for another conference to deal technically and in explicit detail with problems of research strategy and design relative to specified research topics.

Finally, under the heading of implications for our role as professional scholars communicating to less biologically oriented students and colleagues, I think the controlling principle is recognition that the time has come to "put up or shut up." Our prospective audience probably does not want more pious sermons about the desirability of somebody, some day, doing some far-reaching biological research and study, or more "neat ideas" about what other people might profitably do. What is desired is some concrete additions to our knowledge of human government and behavior.

There is, of course, the related problem of finding enough properly and adequately prepared, biologically oriented political scientists to do that kind of work. What is called for is not just ability to recognize what "biopolitics" is, but internalization of at least enough substantial biobehavioral knowledge to enable the scholar to design hard, scientifically defensible research about important questions relating to government and politics. For new academic generations of graduate students, this may entail new courses and seminars and new cross-disciplinary programs. But for even slightly advanced students, and for most faculty who might think of devising the new programs and courses or of themselves pursuing bio-behavioral and biopolitical research, it may also entail "going back to school" to acquire some knowledge of elementary biology, chemistry, physiology, and other basic biobehavioral subjects, much as many of us went back to school in the 1940s and 1950s to acquire or re-tool our knowledge of mathematics and statistics.

*

Since these remarks opened with an observation concerning the desirable consensus about scientific premises, they might well close by observing that premature consensus going much beyond that would be undesirable. We are still far from agreeing on what one theoretical problem or approach holds most promise. We do not know any one research strategy or technique that better guarantees results than others. And we hold diverse views about how best to equip ourselves and our students for coping with these matters. But we do all know, to close on an appropriately biological note, that there are usually many ways to skin a cat.

Notes

1. Glendon Schubert, "Politics as a Life Science," *supra*, p. 188, note 5.
2. Acceptance of the general proposition here was indicated most clearly in discussions of Jean Laponce's findings concerning "handedness" and of Konrad Lorenz's concept of "fixed action patterns."
3. See, for example, the Conference paper by James C. Davies and related comments.
4. E. D. Chapple, *Culture and Biological Man*, New York, Holt, Rinehart and Winston, 1970.
5. Among the many examples given, one must suffice. Explanations of participation in voting frequently involve explaining one attitudinal variable, such as "intention to vote," by another attitudinal variable, such as "sense of political efficacy," "political competence," "political alienation," etc. Rarely is there data about the act of motor behavior (voting) which is the only variable of real ultimate concern; instead there is only the assumption, usually not even mentioned, that the correlation between the act of voting and the attitude of efficacy, competence, etc., is perfectly indexed by the correlation of the latter attitude with "intention to vote," i.e., that the correspondence between voting and "intention to vote" is nearly perfect. For an interesting discussion and set of articles suggesting the complexities of determining what that correlation actually is, and offering reasons for suspecting it to be extremely low, see I. Deutscher, *What We Say/What We Do*, Glenview, Ill., Scott Foresman, 1973. For methodological problems concerning correspondence between overt motor actions and verbal behavior presuming to report subjects' attitudes, see John C. Wahlke and Milton Lodge, "Psychophysiological Measures of Political Attitudes and Behavior," *Midwest Journal of Political Science* 16 (4), November 1972, 505-537.
6. H. Beck, "The Rationality of Redundancy: Neurocybernetic Foundations for a Theory of Biopolitics," *Comparative Political Studies* 3 (4), January 1971, pp. 469-478, at p. 469.
7. Glendon Schubert's survey (*supra*, pp. 187-195) of reviews, articles, and book advertisements in our principal professional journal offers a not unfair measure of the current state of our guiding conceptions.
8. Roger D. Masters, "The Impact of Ethology on Political Science," *supra*, p. 207.
9. Peter Corning, "Toward A Survival Oriented Policy Science," *supra*, p. 131.

LIONEL TIGER

Ions of Emotion and Political Behavior: Notes on a Prototheory

I WOULD LIKE TO MAKE ONE GENERAL REMARK, then a number of specific comments on particular points, then several very brief general remarks. My own background is neither in physiology nor medicine, but rather anthropology by way of political sociology—anthropology because it has retained a slightly more than lingering interest in "human nature" and hence is more hospitable than sociology to work at the biology/social science interface. So my competence is decidedly questionable to cope with the specific brain materials which are dealt with in Dr. Davies' paper. However, I am reasonably sure that studying the neurological component of political behavior is, perhaps, the most economical high-gain strategy in approaching this whole matter. One can only restate on this occasion that admiration which one happily felt for Davies' exceptionally adventurous—in fact, at the time, intrepid synthesis of the available material he published in 1963 in his *Human Nature in Politics*. Perhaps it's one measure of the long-run success of that study, if not only as a fulsome statement, but also a troublesome goad to Davies' colleagues, that a paper as biologically specific and fundamental as his present one, can be displayed in polite company.

Now some specific points about the paper. I'm compelled to restate something said at a meeting of the United States Political Science Association in Washington which has to do with Darwin and Marx, their relationship, and the implications of their work for each other's systems (Tiger, 1973). It is widely unknown that Marx offered to dedicate *Das Kapital* to Darwin and it is to be recalled that at Marx's graveside, Engels reiterated the connection between Marx's thought and the scientific work and perspective of Darwin. As Marx wrote in the Paris Manuscripts, "history itself is a *real* part of natural history—of nature developing into man. Natural science will in time incorporate into itself the science of man, just as the science of man will incorporate into itself natural science. There will be one science" (1972, p. 143). Both Marx and Engels turned their hands to pure science: ". . . although Engels' efforts are acknowledged and readily available, Marx's have remained unknown. During the latter part of his life he drafted about 900 pages of mathematical manuscript which today are accessible only in Russian. . . Perhaps Marx saw in Darwin's theory of natural selection the potential for mathematical precision which contemporary evolutionists have since developed." (Heyer, 1975, pp. 10-11.) Paul Heyer's recent study of the matter freshly and persuasively presents a host of new data about the Marx-Darwin relationship and the similarity of their approaches to living systems; it underlines the importance of their similarity as systematizers and the even greater importance of the lack of any cohesion between their followers in the respective systems of biology and Marxism. Both thinkers were interested in interdependence, both were secular thinkers and both saw distinctions between organisms as having real implications for their life chances—both immediately and reproductively; and neither was prepared to assign value to organisms on some abstract ground of being pagan, middle class, aristocratic, etc., etc. (Marx's antipathy to the bourgeois classes was clearly a comment on their historical situation and not on their organic constitution, unlike racist thinkers who used biology to draw such implications). Marx was, in fact, of course, very interested in the human nature problem as Engels was and both were convinced that biology was a realistic contributor to political thought.

I make this point simply because both these thinkers have been rejected by the two great traditions which they either stimulated or shared: in the West, the psychological tradition culminating in Skinnerian psychology and in the Communist world first with Pavlov and then Lysenko. In both cases an environmentalist meliorism overcame any concern for the endogenous contribution of organisms to social behavior as opposed to the effect of social and material environments on behavior. I stress this because this was an extremely interesting and important *political* decision about the nature of science or at least about what kind of science could be permitted to seek demystification of social structure. In a word, I suggest that the ideological leadership or the political leadership of both the Communist and Non-Communist traditions were content with and supported a psychology which focused on their rights and options as manipulators rather than on the proclivities, needs, enthusiasms and "biological rights" of the manipulated. Parenthetically, it is also no accident that both the American and the Russian traditions arose in large communities marked by immense regional diversity which could not be tolerated by elites if they were to truly benefit successfully from the extensiveness of their nations. The New Soviet Man, the great American Melting Pot and sundry similar notions all reflected a psychological theory content with a *tabula rasa* conception of individuals and intrigued by the mechanisms of control deriving from psychological, and in the case of Lysenko, genetic control. The issues arising from so-called "behavior modifications" are a relatively extreme rendering of the same problem. Incidentally, I do not mean to imply that other cultural systems may have a more generous attitude or welcoming attitude to the question of human biology and its political implications. Certainly the ascetics of Asia or the spiritualists of Africa could be said to have equivalent antibiological social theories. But I'm here concerned with the two major economic systems which have generated the predominating scientific schema. There may even be an anterior problem underlying this, such as a "design defect" in the brain rendering human beings unable to see themselves as they are—a technical difficulty which may underly the universal distinction made between nature and culture which Lévi-Strauss has so extensively analyzed in buildings not very far from the one in which we are meeting.

The Expression of Potential

Davies discusses the expression of potential almost as if potential were a secretion. I confess that the introduction of this conception is unsettling, at least to me, if only because really or potentially, it introduces an idealistic *as if* quality to the discussion. What on earth is self-actualization? Is this a political perception or a Ph. D. psychologist's updating of the neo-Protestant credo? Where does the therapy start and a theory end? How does one "find and fulfill one's individual self" in Maslow's terms? What would Maslow tell an Eskimo or Bushman or a long distance runner or a Chinese peasant? This concept, breathtakingly generous though it may appear to be about the opportunities for life of one's fellow man, may well be an extremely culture bound concept implying certainties about psychological superiority of particular individuals who operate their lives in particular ways. This can easily (and I think has already) conduce to a psychological imperialism of those who decide that they know who they are and are fulfilled over those who decide they cannot make this claim. Perhaps this conception can be valid cross-culturally but, on the face of it, it would appear to me difficult to demonstrate its validity given the clumsy benignity of the conception as it presently stands. Far more rigorous and specific notions of what constitute individual fulfillment would have to be required before any class of persons, called psychologists or anything else, should be permitted to pronounce on the final value of other people's experience of life in terms of a set of norms of psychological activity allegedly marking the most successful members of a community and, in turn, of the species. And, to put the matter at its most bitterly simple: in what possible sense can a highly skilled gregarious animal not fulfill himself? And, on what grounds again is it possible to create an orthodoxy about psychological state given that if we know anything about biology it is that variation in complicated social species is structural and healthy?

The Prototheory

At this point, where technical competence would be most relevant, science is lacking so these comments must be unduly general. First of all, do the behavioral patterns which Davies discusses reside in processes of neocortical memory or is it not equally likely or more likely that such behaviors as rage, anger, Erickson's distrust, Bowlby's separation, etc., are all articulated in different ways and may very fundamentally affect later lives of individuals —more so than the memories of past events stored in the neocortex? In other words, political behavior may result not only or so much from explicit experiences which are then stored and summated, but rather from other kinds of processes such as psychologists and psychiatrists traditionally, and more recently, ethologists have been busy describing. Nonetheless, Davies raises extremely important issues here insofar as a theory of memory is not widely agreed upon and the whole issue remains vexatious.

It may be unduly simple to assume that experience is governed cognitively by processes such as accumulation and discharging. One is impelled to ask, for example, what is the function of REM sleep and how does that deal with the kind of information which would be involved in the prototheory. What is the evolutionary or phylogenetic significance of dreams and particularly of nightmares? And, have these processes to do with the evolution of cortical tissue and its great growth? Indeed, dreaming may itself be a form of complex reorganization/accumulation/synthesis/review of information and I am not sure that this kind of process is easily assimilable into Davies' model. There is the suggestion here that the model is based on a relatively old-fashioned computer scheme of input, storage, and then use, rather than on a more complex pattern in which information is constantly reassessed—possibly, through such mechanisms as dreaming. As Jonathan Winsum of Rockefeller University has noted (personal communication), if the brain did function according to the old-fashioned computer model in which bits of information were stored and then used when necessary, given the complexity of its output, the human brain would be so large it would have to be carried around in a wheelbarrow. So I am asking if the condenser concept is complex enough to deal with the volume and nature of cortical activity.

Some Further Specific Points

Why does intensification necessarily lead to activation and violence? Empirically, violence is usually corporate and structural, rather than individual and it most commonly occurs as a result of the interaction of one group of males who are familiar with each other and another group also familiar with each other. Furthermore, there is usually some history of violence and cognitive symbolism which justifies or at least organizes it in space and time. Thus occurrences of violence may not depend on the process of accretion and condensation which Davies describes, but rather on a socially stimulated reverberating process. As to the specificity and coordination of responses to aggression—and Davies' reference to the experience of war—this also may not depend on individual experiences but may be, in part, the outcome of the extremely long selection in our human species for hunting activity, warfare and coordinated aggression of various kinds. (One of the most powerful memoirs about the effects of warfare on the unwilling individual is J. Glenn Gray's *The Warriors* (1959) which suggests the power of the contagious or mimetic processes of warfaring). Rather than individual life histories stimulating aggressive outcomes, perhaps there is among members of the species a shared set of neurological processes; in the way that the linguist Chomsky has argued that all human beings share a capacity to use the "universal grammar" necessary for human speech, a behavioral grammar, what Fox and myself call a bio-grammar (Tiger / Fox, 1971), may also reflect a capacity of the species to respond in relatively organized fashion to certain stimuli. And it is worthwhile to recall the psychiatrist David Hamburg's formulation (1963): that one consequence of evolutionary selection is that some things are easier for animals to learn than others and it is possible that the kind of coordinated response to aggressive behavior which Davies discusses is relatively easy to learn—perhaps easier to learn than an uncoordinated or anti-coordinated response. Not only is what is easy to learn important here, but also what is easy to teach. Accordingly, in addition to the individual's biographical neurological record there may be some "design features" of the cognitive system which keep the organism in a state of greater readiness to do one thing as opposed to another. And one aspect of the argument which could be developed more fully is discussion of the mechanisms which trigger sociability in the model, i.e., why does gregarious behavior occur? To what extent does it depend upon individual experience? And to what extent is it predictable and likely?

One must ask why does aggressive behavior seem to occur, in Davies' model, principally in response to frustration? Surely the frustration-aggression hypothesis as a broad theory has been shown to be too simple. For example, some aggression is clearly the result of the opposite of frustration-success and power. Racist activity is one example, as are the processes of scapegoating and creating "pseudospecies;" this is Eric Erickson's term and refers to the process by which human beings are turned through definition into less than humans and thus may be killed abused, pillaged, etc., without invoking any of the inhibitions violence to conspecifics may evoke.

The Question of the Right Kind of Social Stimulation

Davies suggests that it is the advanced parts of the brain which lead to demands for change. This I take to be a value judgment insofar as it may be implicitly assumed that change is desirable and may also represent a dynamic rather than a static bias. This is a particular conception of social systems; one could, in fact, argue the reverse—that it was the more "primitive" parts of the brain with their urgent concerns for status, reproduction, security, etc., which could the more urgently provide the stimulus for social change.

There is also the problem of integration of the three main areas of the brain as defined by Paul McCleon suggesting that Davies is, of course, correct that the involvement is extensive between the advanced parts of the brain and the earlier ones (1962). The question remains: how accurately does information flow from one area of the brain to another? Is there the likelihood of systematic distortion as cognitive events are affected by and affect the other systems of the brain. And, again as McCleon has asked, are there "paleopsychic processes" which take precedence over more obvious contemporary ones in affecting behavioral outcomes? Davies successfully poses these questions and, furthermore, by placing the discussion in the historical context of the controversy between Walter Cannon and William

James suggests that major forms of explanation are involved and formidable forces in scientific work.

Davies' extremely detailed and searching analyses of the phenomenon of stress and its cognitive and neurological implications is very valuable. Here we begin to see the outlines of the new territory which Davies is exploring and can also foresee means by which this extremely molecular approach can be integrated with existing global ones. For example, Lévi-Strauss has made the observation, at some length, that just as human beings find that some things are good-to-eat, similarly there are some things that are good-to-think. This is an extremely interesting observation because it suggests that certain thoughts in themselves offer pleasure and must affect the interaction of the brain and the other systems of the body in a benign and comforting fashion. While Davies focuses on the opposite of what is *good-to-think,* nevertheless, his emphasis on this kind of problem is a crisp contribution to clarifying the issue and defining its dimensions.

Along these lines, it may also be possible to differ from Davies' supposition that it is absurd to think there is "any such thing as a broadly social nationally communal or political area of the neocortex." Why shouldn't there be? Given the immense enthusiasm of human beings for gregarious contact and for transactions of various kinds, there may be indeed a localized, specialized center. Of course, I do not mean this literally but if we are to take seriously, for example, the conclusions of Lévi-Strauss about the functions of reciprocity and exchange in human groups and the commonly observed importance of exchange, even among relatively small children (the lively trade in baseball cards, hockey pictures, football picture cards is a case in point), then there may well be structures in the brain which accompany this extremely important element of the human behavioral repertoire. And it is important for social scientists to become involved with such matters because as Davies notes, neurophysiologists may not themselves appreciate the significance or implications of the issues that they deal with at the physiological level; when these become involved with behavioral processes the contribution of the social scientist is essential.

In general, this is a provocative and thoughtful contribution which is, from my point of view, weakened by an undue simplification of the mechanism proposed to explain political change. Again, this judgment may depend on my lack of information and competence rather than on any real characteristic of the prototheory. The analytical courage, the technically informed humanism and the sense of serious quest which mark the paper may impel others to augment Davies' work and, of course, also sustain him as he continues to grapple with issues, the very probing of which is not only daring in itself, but immediately useful to students of human political behavior.

Bibliography

Davies, J. C.
1963 *Human Nature in Politics: the Dynamics of Political Behaviour,* New York, Wiley.

Gray, J. G.
1959 *The Warriors,* New York, Harcourt Brace.

Hamburg, D.
1963 "Emotions in Perspective of Human Evolution," in: Knapp, P. (ed.), *Expression of the Emotions in Man,* New York, International Universities Press.

Heyer, P.
n. d. "Marx and Darwin: A Related Legacy on Man, Nature and Society," Rutgers University, Ph. D. Thesis, unpublished.

Marx, K.
1972 *The Economic and Philosophical Manuscripts of 1844,* Struik, D. V. (ed.), New York, International Publishers.

Tiger, L.
1973 "Biology, Rhetoric and Reform: The Allure of Low-Born High Ideals," *Social Science Information* 12 (5), pp. 51-60.

Tiger, L. / Fox, R.
1971 *The Imperial Animal,* New York, Holt, Rinehart and Winston.

JOHN H. CROOK

Ethology and Biopolitics: A Discussant's View

IN THIS BRIEF ARTICLE based on comments I made as a discussant to Roger Master's paper "The Impact of Ethology on Political Science" I would like to contribute a few ideas about the role of ethology in the social and behavioral sciences that have not received much of an airing at this meeting. The main themes concern *a)* the relation between physical and biological reductionism in approaching biopolitics; *b)* the limitations of Konrad Lorenz's views in relation to the wider reach of contemporary ethology; *c)* the problem of relating observational human ethology to psychodynamics; and *d)* the significance of some themes in socio-ecology and social ethology to the biopolitical perspective. I shall deal with these questions in only a very abbreviated form here. The following references will provide additional information for those wishing to explore further (Eisenberg / Dillon, 1971; Hinde, 1974; Washburn / Dolhinow, 1972; Crook, 1970; White, 1974; Jorgenson, 1972; Rowell, 1972; Tiger 1969).

I. The Appeal to Reductionism

Most behavioral scientists will probably agree on the following points: *a)* that the general phenomena of social behavior and some social organization of species populations are widespread in the animal kingdom; *b)* that human society is not conventional or contractual but at basis a special case, as Aristotle argued, of a natural system; *c)* that genetic roots for performance and capacity underlie behavioral expression in Man as well as other animals; *d)* that many functions of human rituals have their analogies in the phylogenetically inherited ritualized behaviors of animals and that in our species some of these functions may be assumed by institutionalized rituals mediated through learning, language and educational transmission; and *e)* that the study of non-human primate societies gives clues as to the evolution of the social processes of our species. It is important to realize that these views are the result of intense ethological and particularly primatological research since about 1960 and that they represent a new paradigm replacing, on the one hand, the behaviorist paradigm in psychology and, on the other, the sociologists denial of the relevance of biology both to his proper science and to social anthropology (Kroeber, 1952).

Roger Masters points out that the attempt to model the behavioral sciences on the approach of physics has been inappropriate to the material of these sciences. Yet psychologists and sociologists alike remain prone to seek some underpinning for their subject in "harder" sciences than their own. The latest candidate is Biology. Yet the appeal to biology is no less a reductionism than was the appeal to physics. In particular the search for something basic to human behavior in Ethology, especially in the ethology of Lorenz, represents a renewed search for a fundamentalist determinism, this time based in behavioral genetics and ethological nativism rather than in physics. The irony contained in the physical idealism of earlier behavioral metholodogy applies in this new form to much of the content of Master's paper. While many ethologists are finding the analysis of fixed action patterns useful in approaching problems of human communication, others are noting the flexibility of both syntax and semantics in the use of call notes by birds (W. John Smith, 1968) and the subtle intergraduation of primate vocal and visual signals that makes precise categorization and enumeration a purely descriptive device to be applied with caution to material that is variable along several dimensions. Likewise in the study of social organization, phenomena such as agonistic buffering (Deag / Crook, 1971), intergenerational replacement using social subterfuge (Crook, 1971), and the use of "rules" in governing interactions (Kummer, in press), reveals complexities that imply the operation of processes additional to simple genetic replication and selection in the historical transmission of behavior across generations and through time.

Biological reductionism, like its physical forebear, has an intellectual function in permitting early "closure" in the explanation of complex phenomena. In this it repeats in sophisticated form the appeal to instinct among nativist thinkers of earlier periods. Such "closure" amounts to an avoidance of the complexities of the issues raised by human behavior, especially when studied by human beings themselves. While the appeal to ethology is important and adds both new material and approaches to the exploration of human behavior, there is a danger of simplistic and scholastic dogmatism if it is applied too vigorously without recognition of the complexities it does not explain. In a parallel case, Richard Gregory (1975) has pointed out how the excesses of behaviorist naivety in psychology prevented an adequately vigorous examination of the perceptual and cognitive processes mediating stimulus and response. Only recently have the issues raised by the Gestalt school of psychology once more become a prime focus of attention. Likewise a biopolitics based too narrowly on an ethological reductionism may miss the implications of human complexity much in the way a Martian scientist would fail to understand computers were he to examine the hardware of the machinery alone.

II. Problems of Lorenzian Ethology

Masters points out that political scientists in the United States came to know ethology primarily through the popular, rather than the scientific works of Konrad Lorenz. The basic Lorenzian position emphasizes the innate component in animal behavior, the evolution of behavior through natural selection, and makes numerous generalizations from animal to human behavior based essentially on studies of fish and birds. The significance of Lorenz' contribution to ethology in the 1930s is well recognized. In many respects he is the "father" of modern ethology. However, many contemporary ethologists disagree with him fundamentally both with respect to the content of his motivation theory and with respect to the nature of his generalizations to Man. Critical evaluations of Lorenz' theories may be found in the following texts: M. F. Ashley Montagu (ed., 1973), D. Lehmann (1953), R. A. Hinde (1967), B. Chatwin (1974). Contemporary ethologists acknowledge Konrad Lorenz' historical position and particularly the importance of his work as an antidote to behaviorism but his ideas cannot be considered representative of the current creativity of the science. In particular the authoritarian stance adopted in, for example, *Civilized Man's Eight Deadly Sins* (1974) is highly personal to Lorenz and although an important work, it does not represent the views of ethologists generally. The contemporary reach of ethology is wide, extending from precise neurophysiological and endocrinological studies of behavioral determinants through abstract modeling of motivational processes to research on communication, behavioral genetics, behavioral evolution and the analysis of animal society. The field thus contacts sociology, psychology and anthropology at many points. Ethology is of interest to biopolitics primarily at the points where it contacts these human disciplines. The lines of thought are however varied and ethologists should not be considered as speaking with a single voice, let alone a Lorenzian one.

Probably it is Lorenz' emphasis on the innate determinants of behavior coupled with his ideas on aggression and Robert Ardrey's views on territorial behavior that have attracted most attention. While there is no doubt that in the evolution of behavior Man has shared with other species a common psychogenetic inheritance, the way in which the "biogrammar" interacts with behavioral programming based on acquisition of culturally transmitted traits has in no case been adequately worked out. Most of the views in this field are representative only of their author's personal choices. Thus the weight to be given to the innate component in universal social patterns other than elementary signal composition remains in doubt especially where the *use* of these components is affected by cultural traditions. However the issue is an important one and has been neglected by a behavioristically inclined psychology and sociology. For example, the role of the endocrine system in differentiating male and female behavioral and social behavior in our species has been ineffectually examined by those intent on proving the importance, which need not be doubted, of culturally programmed sex-roles and identity stereotypes (Crook, 1974). The lack of balance in our present understanding of sex roles in contemporary Western society arises from a lack of objective consideration of the way in which gender behavior is determined by biology and society in interaction.

III. Ethology and Psychodynamics

One of the main contributions of ethology to the human behavioral sciences lies in its methodology. The observational study of animal behavior has been largely based on the relative fixity of behavior patterns which permits classification and categorization of behavior into particulate responses forming a repertoire of expressive and communicative acts. Precise analysis of human expression in this light has contributed much to our understanding of non-verbal communication in our species. Furthermore the particularization of behavior is extremely useful for quantitative computer-based analysis of interaction processes. The study of the relationship between behavioral expressivity and human subjectivity has however barely begun. It is at this point that certain unique features of the human individual need consideration, for they seem to reveal the boundary at which ethological approaches need to be reconsidered in relation to underlying processes of human cognition and motivation not represented elsewhere in the animal kindgom.

It seems to me that the extended space and time binding properties of human language are the basis for the subject's knowledge of himself as an object. The differentiation of the "I" and the "me" lies at the root of an individual's self conception as an identity. The process of identity formation through the introjection of parental and social values means that human interaction is fundamentally an interplay of "persons" rather than organisms, and that the self-esteem system and personality of the subjectivities in relationship are crucial determinants of behavioral outcomes. This fact underlies the whole field of psychodynamics and psychopathology for, when an individual introjects a fundamentally maladaptive view of himself, the elements of interior conflict basic to neurotic behavior are established. Generalizations from animal to human studies may or may not have to take this fact into account. Thus while biological factors such as the endocrine determinants of behavior, the body chemistry underlying addiction to tobacco or the effects of lead intake from city water supplies on behaviors are approachable as direct biological effects on behaviorism (and indeed political) issues, other problems in which biological factors interact with those of personality formation through familial and societal programming cannot be treated in so direct a manner. Here complex systems analysis of interactions of factors of contrasting provenance is clearly needed—and for the most part these lie in the future.

It follows that enthusiasts for the biopolitical approach need to consider carefully the nature of the inference that is possible from different sorts of study and behavioral context. While the current vogue for biology is understandable as a reaction to the lack of productivity in behaviorist psychology, the need for careful consideration of the realities of human personal development and the existential nature of subjectivities in interaction is clear. A naive reductionism is only appropriate in very clear cut instances.

IV. Social Ethology and Biopolitics

It seems worthwhile concluding with a reference to certain themes in contemporary social ethology which have potential relevance for biopolitics. The comparative study of social organization in related species of birds or mammals has repeatedly revealed the way in which patterns of population dispersion, reproductive behaviour and vocal and visual communication form co-adapted systems in relation to particular constellations of ecological factors comprising the contrasting habitats of the species concerned. Thus explanations for contrasting conditions of solitary or gregarious dispersion, of monogamy or polygamy and communication patterns can be suggested in terms of systemic adaptation to ecology. The radiation of societal structure in birds (Crook, 1965; Lack, 1968), in primates (Crook, 1970), in carnivores (Kleiman / Eisenberg, 1973), in antelope (Jarman, 1974), and other groups (see Crook / Ellis / Goss-Custard, 1975), has been interpreted in this light and contemporary work focusses on a more exacting analysis of the actual causal factors involved (e.g., Clutton-Brock, 1974). In addition the contrasting types of mating and rearing systems have been elegantly interpreted in terms of differing types of parental investment in relation to environmental and social selection processes (Trivers, 1972; Alexander, 1974). Slowly a comprehensive understanding of the evolution of bird and mammal societies is emerging. Anthropologists have been quick to attempt the utilization of this thinking in interpreting the earliest stages of human social evolution as ecological adaptation proceeding the agricultural revolution (e.g., Fox, 1967; C. Jolly, 1970). In an analogous study Maitland-Bradfield (1973) interprets the transitions between hunter-gatherer to pueblo village societies in terms of shifts in relation to the ecology of differing environments. The comparison of the systemic relations between this and the non-human mammalian cases is very striking.

Societal structuring is mediated by the patterned interactions of individuals. Social dynamics (Crook / Goss-Custard, 1973) comprise complex patterns of relations in socially advanced mammals such as group-hunting carnivores and troop-living primates. Among Macaques intergenerational replacement involves the gradual ascent of young males into social positions of greater dominance and hence of improved access to females for mating. Among the complex devices used in this process young males may use babies to permit friendly approaches to older more dominant males. These interactions include a ritual presentation of the baby to the older animal and fondling. The behavior appears to be relatively local and, although of repeated occurrence to differing degrees in different species, largely a consequence of local traditions of interaction. During the life trajectory of an advanced primate an individual may occupy several different positions in the social structure, the sequence forming a pattern in relation to attempts at reproduction using social skills with varying degrees of success. These complexities clearly foreshadow the human processes of a similar type but in which the presence of language and time-space structuring in communicative symbolization adds a flexibility and cognitive complexity not attained in other animals.

V. Conclusions: Ethology and Biopolitics

Biological factors play roles of importance in contemporary political problems. This is manifest not only for example in the probable involvement of pollutants (such as lead) in affecting human brain processes and behavior in certain cities, but also in the widespread concern with ecology and the depreciation of the planet's surface resources and natural life (see for example Corning's article in this volume). Ecology and ethology are closely related sciences, as my remarks above on the socio-ecology of population dispersion and reproductive behaviors indicate. The socio-ecological principles currently emerging in contemporary research undoubtedly operate in our own species, but in that grossly more complex way that is an outcome of socio-cultural structuring through history and the operation of economic rather than purely ecological processes as prime determinants of the social system.

Clearly political scientists need to be aware of these issues in their reflections on the contemporary situation. I would however caution against too easy an acceptance of any poorly substantiated viewpoint in ethological theory. Behavioral sciences are peculiarly prone to a dialectical process in which the emphasis shifts from one pole to the other of the old nature-nurture controversy. Although we are now well into the computer age and sophisticated tools for the analysis of systems processes are available, the old habits of dualistic thinking do not die easily. In the last resort an enthusiasm for either an extreme behaviorism (nurture) or an extreme nativism in ethology (nature) will do political science little good except to shuttle potentially useful thinking into outdated *culs-de-sac*. The ground to occupy is that narrow ridge between these viewpoints where the interaction of system determinants is examined with dispassionate care and with due regard to any empirical findings that may be available. Much in human ethology and in the inferences to be drawn from animal to human studies (Crook, 1975) remains highly theoretical and, exciting though these ideas may be, caution in their application is needed. The uncritical acceptance of the ideas of the non-scientist Robert Ardrey by many who preferred skilled writing and glib thought to solid academic statement sounds a warning. To know the importance of a neighboring science to one's own requires a reading of the original materials.

An issue that illustrates the complexity of the interplay between biological, behavioral and political issues concerns the decreasing age of onset of puberty in the western world. Nutritional and educational factors interplay causally and produce major effects in school, in the development of industry catering for a monied adolescent class, in taste and sexual mores and a tendency to focus on adolescent modes of relating rather than on those of adulthood. It is probable that biological maturity occurring so much earlier than in primitive peoples is not necessarily associated with other psychological aspects of maturation. Likewise the turmoils of adolescence may receive undue attention in television and screen performances primarily due to a switch in the market combined with excessive concern by adults with those younger than themselves. The result is a kind of neurotic attitudinising that keeps adults playing adolescent roles long after this is appropriate. The difficulties that this process produces may be in part responsible for the "growth movement" as a grass roots phenomenon especially in the North American scene. The encounter group is well suited to the working out of emotional problems of this sort and allows a renewed effort after appropriate maturity. Clearly here cultural, biological, economic and political issues are closely interwoven. To my mind it is to cases such as this, in which to walk the ridge is the only way to an adequate understanding, that Biopolitics needs to direct a close attention.

Bibliography

Alexander, R. D.
1974 "The Evolution of Social Behaviour," *Ann. Rev. Ecol. Syst.* 5, pp. 325-383.

Chatwin, B
1974 "Man the Aggressor," *Sunday Times Magazine*, London, December 1st.

Clutton-Brock, T.
1974 "Primate Social Organisation and Ecology," *Nature* 250, pp. 539-542.

Crook, J. H.
1965 "The Adaptative Significance of Avian Social Organizations," *Symposia of the Zoological Society of London* 14, pp. 181-218.
1970 "The Socio-Ecology of Primates," in: Crook, J. (ed.), *Social Behaviour in Birds and Mammals*, London, Academic Press, pp. 103-166.
1971 "Sources of Cooperation in Animals and Man," in: Eisenberg / Dillon (eds.), *Man and Beast: Comparative Social Behaviour, op. cit.*, pp. 235-260.
1974 "Darwinism and the Sexual Politics of Primates," *Social Science Information* 12 (3), pp. 7-28.
1975 "Problems of Inference in the Comparison of Animal and Human Social Organizations," in: von Cranach, M. (ed.), *The Logic of Inference from Animals to Man*, in press.

Crook, J. H. / Ellis, J. / Goss-Custard, J. D.
1975 "Mammalian Social Systems: Structure and Function," *Animal Behaviour*, in press.

Crook, J. H. / Goss-Custard, J.
1973 "Social Ethology," *Ann. Rev. Psy.* 23, pp. 277-312.

Crook, J. H. (ed.)
1970 *Social Behaviour in Birds and Mammals*, London, Academic Press.

Deag, J. / Crook, J. H.
1971 "Social Behaviour and Agonistic Buffering in the Wild Barbary Macaque," *Folia Primat.* 15, pp. 183-200.

Eisenberg, J. F. / Dillon, W. (eds.)
1971 *Man and Beast: Comparative Social Behaviour*, Washington, DC, Smithsonian Institution Press.

Fox, R.
1967 "In the Beginning: Aspects of Hominid Behavioural Evolution," *Man* 2, pp. 415-433.

Gregory, R.
1975 "The Physical Attributes of the Oldest Profession," *Psychology Today* 2, May, pp. 62-65.

Hinde, R. A.
1967 "The Nature of Aggression," *New Society* 9 (231), March 2, pp. 302-304.
1974 *Biological Bases of Human Social Behaviour,* New York, McGraw Hill.

Jarman, P.
1974 "The Social Organization of Antelope in Relation to their Ecology," *Behaviour* 48, pp. 215-267.

Jolly, C.
1970 "The Seed-Eaters: A New Model of Hominid Differentiation Based on a Baboon Analogy," *Man* 5, pp. 5-26.

Jorgensen, J. G. (ed.)
1972 *Biology and Culture in Modern Perspective,* San Francisco, Calif., Freeman.

Kleiman, D. / Eisenberg, J.
1973 "Comparison of Canid and Felid Social Systems from an Evolutionary Perspective," *Animal Behaviour* 21, pp. 637-659.

Kroeber, A. L.
1952 "The Superorganic," in: Kroeber, A. L. (ed.), *The Nature of Culture,* Chicago, Ill., University of Chicago Press, pp. 22-51.

Kummer, H.
in press "Rules of Dyad and Group Formation among Captive Gelada Baboons," *Proc. V. Internat. Primat. Congress Nagoya,* 1974.

Lack, D. L.
1968 *Ecological Adaptations for Breeding in Birds,* London, Methuen.

Lehmann, D.
1953 "A Critique of Konrad Lorenz's Theory of Instinctive Behaviour," *Q. Rev. Biol.* 28, pp. 337-363.

Lorenz, K.
1974 *Civilized Man's Eight Deadly Sins,* London, Methuen.

Maitland-Bradfield, R.
1973 *A Natural History of Associations,* London, Duckwaith.

Montagu, Ashley, M. F. (ed.)
1973 *Man and Aggression,* 2nd ed., New York, Oxford University Press.

Rowell, T.
1972 *Social Behaviour of Monkeys,* London, Penguin.

Smith, J. W.
1968 "Message Meaning Analyses," in: Sebeok, T. A. (ed.), *Animal Communication: Techniques of Study and Results of Research,* Bloomington, Ind., Indiana University Press.

Tiger, L.
1969 *Men in Groups,* London, Nelson.

Trivers, R.
1972 "Parental Investment and Sexual Selection," in: Campbell, B. (ed.), *Sexual Selection and the Descent of Man,* Chicago, Ill., Aldine.

Washburn, S. L. / Dolhinow, P. (eds.)
1972 *Perspectives on Human Evolution,* vol. 2, New York, Holt, Rinehart and Winston.

White, N. F. (ed.)
1974 *Ethology and Psychiatry,* Toronto, University of Toronto Press (for the Ontario Mental Health Foundation).

Appendix

STEVEN A. PETERSON

Biopolitics: A Bibliographical Essay

THE FOLLOWING BIBLIOGRAPHY is divided into three sections: biopolitics, biological linkages in other social sciences, and a supplementary set of readings on various areas in biology. The first two sections provide an indication of the amount of effort in the social sciences toward developing a biological perspective. The final section includes a selected series of works which can give one a general acquaintance with the work of professional biologists and their kin. The first section of the bibliography is as close to comprehensive as possible, while the other two sections are designed to provide a taste of the literature in the subsumed substantive areas.

*

The *first section* of the bibliography is a compilation of works in biopolitics. It is hoped that there have been no omissions in this listing. Works are included here if they have been done by political scientists or if interested non-political scientists have delivered papers on politics at political science meetings (e.g., Tiger, 1972). This is perhaps a crude way of defining the area, but one which seems to do the least violence toward categorization.

Elsewhere, Somit has classified the biopolitics literature into four general categories (Somit, 1972):

1) The case for a more biologically-oriented political science.
2) Ethological aspects of political behavior.
3) Physiological and pharmacological aspects of political behavior.
4) Issues of public policy raised by recent advances in biology.

There are a host of works in the first area. The authors attempt to demonstrate in a general way that a more biologically-oriented political science is necessary and will prove profitable to the discipline. Some characteristic works here might include those by Somit, Corning, Thorson, Hummel, and, from the present set of papers included in this volume, Schubert (Somit, 1968; Corning, 1971a; Thorson, 1970a; Thorson, 1970b; Hummel, 1971).

The second area has also been a popular one for biopolitical works. Here, analysts use ethological concepts or substantive knowledge from ethology (broadly defined) to try to explain human political behavior. Among those who have written in this subarea are Pranger, Adrian, Willhoite, Tiger, Masters, and Schubert (Pranger, 1967; Adrian, 1970; Willhoite, 1971; Tiger, 1972; Masters, 1969; Masters, 1972; Schubert, 1973).

Other biopolitical pieces grapple with the relationship between drugs and physiology on the one hand and politics on the other. It is in this area that most of the empirical work in biopolitics has taken place. Some of the representative efforts include those by Schwartz, Jaros, Lodge and associates, Peterson, and Stauffer (Jaros, 1972; Lodge, Tursky, and Tanenhaus, 1973; Wahlke and Lodge, 1972; Peterson, 1973; Stauffer, 1969; Schwartz, 1970; Schwartz, 1973). Among the papers published in this volume, the works by Davies, Lodge and Tursky, and Schwartz fall into this category.

Finally, a number of political scientists have investigated the public policy implications raised by developments in the life sciences. Behavior control is perhaps the most familiar example. Some of those who have pursued this question include Somit, Caldwell, Lepawsky, and Corning (Somit, 1968; Somit, 1970; Caldwell, 1964; Lepawsky, 1967; Corning, 1971c; Corning, this volume).

The *second part* of the bibliography includes a number of works which suggest linkages between biology and other social sciences. This trend appears to be growing. One example of this is the preparation of a panel on "Biosociology" for the 1975 American Sociological Association convention.

Perhaps the most familiar works in this category are the popularizations of ethological material as applied to man. One immediately thinks of Morris, Ardrey, and Lorenz (Morris, 1967; Morris, 1969; Morris, 1972; Ardrey, 1961; Ardrey, 1966; Ardrey, 1970; Lorenz, 1966; Lorenz, 1974). Others who have achieved wide currency with their work but who have maintained a more rigorous approach include Tiger and Fox (Tiger, 1969; Tiger and Fox, 1971). One clear trend among those who try to draw linkages between biology and social science is the use of ethology. Examples abound, e.g., see Ambrose, Blurton-Jones, Bowlby, Count, Mazur, Tinbergen, Tiger and Fox (Ambrose, 1973; Blurton-Jones, 1967; Bowlby, 1957; Freedman, 1967; Mazur, 1973; Tinbergen, 1974; Tiger and Fox, 1966).

The biological analysis is reflected in many different areas of the social sciences. In sociology, there is the work of Mazur and Eckland (Mazur and Robertson, 1972; Eckland, 1967). In anthropology, one thinks of Count, Chapple, Callan, and Kortmulder (Count, 1958; Chapple, 1970; Callan, 1970; Kortmulder, 1968). Many in psychology have essayed this approach, such as Campbell, Cortes and Gatti, Freedman, Gray, and Heise (Campbell, 1972; Cortes and Gatti, 1972; Freedman, 1968; Gray, 1958; Heise, 1973). In linguistics, one well-known example is Lenneberg (Lenneberg, 1964; Lenneberg, 1967). These are only a very few of the titles.

The *third section* of this bibliography is a selection of more technical pieces to supplement the first two sections. These works are useful tools in evaluating the applications of biology to the social sciences. Four categories are provided: theoretical issues; animal behavior; evolution and genetics; and physiological bases of behavior.

The key theoretical issues dealt with are the question of analogy vs. homology and the nature-nurture issue which is still around to bedevil the unwary. In the first section of the bibliography, Masters' paper on analogy is listed (Masters, 1973b). Atz approaches the question from a biological perspective and indicates clearly the profound difficulties in establishing behavioral homologies (Atz, 1970). Others who touch on the issue of inter-species comparisons include Dimond, Beer, and Lorenz (Dimond, 1970; Beer, 1968; Lorenz, 1974). With respect to the battle between the instinctivists and the more developmentally oriented, there are a number of key writings. Lorenz, of course, provides the basic argument for instinct theory (Lorenz, 1965). The most devastating assaults on the Lorenzian position come from Schneirla, Lehrman, and Kuo. When their arguments are considered along with those of Gottlieb, Jensen, and Hebb, the classical Lorenzian paradigm is clearly demonstrated to be deficient (Schneirla, 1966; Lehrman, 1953; Lehrman, 1970; Kuo, 1967; Maier and Schneirla, 1963; Jensen, 1961; Hebb, 1953; Gottlieb, 1968; Gottlieb, 1970).

Among the basic books on animal behavior, some leading texts immediately come to mind. The best of these include those by Alcock, Marler and Hamilton, Manning, Hinde, and Scott (Alcock, 1975; Marler and Hamilton, 1966; Manning, 1973; Hinde, 1970; Scott, 1972). Eibl-Eibesfeldt and Tinbergen have written texts that reflect the classical ethological position (Eibl-Eibesfeldt, 1970; Tinbergen, 1953). Useful collections of readings have been gathered by McGill and Klopfer and Hailman (McGill, 1965; McGill, 1973; Klopfer and Hailman, 1972). Finally, there are a number of good volumes on primate behavior (DeVore, 1965; Jolly, 1972; Jay, 1968; Morris, 1967).

Many of those involved in biopolitics assert the importance of evolution and genetics for understanding aspects of the political (e.g., Corning, 1971a; Corning, 1971c). The *magnum opus* on molecular biology is Watson's classic volume *Molecular Biology of the Gene* (Watson, 1970). Waddington presents a scheme for understanding the development of the genetic material into a functioning organism. Bonner provides a somewhat more technical perspective on the same process (Bonner, 1965; Waddington, 1957). General works on evolution which provide reasonably good understanding of this central unifying concept of biology include those by Dobzhansky, Lerner, Mayr (Dobzhansky, 1970; Lerner, 1968; Mayr, 1970). Behavior genetics has become increasingly important over the last decade as an area of intellectual endeavor. Some of the more useful pieces in this field have been written by Hirsch, McClearn and deFries, and Roe and Simpson (Hirsch, 1967; McClearn and deFries, 1973; Roe and Simpson, 1958). On human evolution, see Buettner-Janusch, Campbell, Dobzhansky (Buettner-Janusch, 1973; Campbell, 1966; Dobzhansky, 1962). The above provide at least a general perspective on evolution and genetics.

The final subsection of the supplementary readings deals with the relationship between physiology and behavior. Basic volumes proliferate. Two useful ones are by Deutsch and Deutsch and Teitelbaum (Deutsch and

Deutsch, 1973; Teitelbaum, 1967). Some important titles on the nervous system might include Eccles, MacLean, MacCleary and Moore, Stellar, Papez (Eccles, 1957; MacLean, 1958; McCleary and Moore, 1965; Stellar, 1954; Papez, 1937). For further information on hormones and the endocrine system, there are a number of worthwhile pieces, three of which are by Lewin, Michael, and Young (Lewin, 1972; Michael, 1968; Young, 1961). The question of the manipulation of behavior through physiological means has become salient to many (e.g., Somit, 1968; Somit, 1970). However, Valenstein provides a caution about overemphasizing present potential in these directions (Valenstein, 1973).

Selected Bibliography

I. Biopolitics

Adrian, C.
1969 "Implications for Political Science and Public Policy of Recent Ethological Research," Paper presented for delivery at 2nd International Sinological Conference, Taipei, Taiwan, The China Academy.
1970 "Ethology and Bureaucracy," Munich, IPSA.

Beck, H.
1970 "Politics and the Life Sciences: Notes Toward a Theory of Biobehavioral Ecology," Munich, IPSA.
1971 "The Rationality of Redundancy: Neurocybernetic Foundations for a Theory of Biopolitics," *Comparative Political Studies* 3, pp. 469-476.

Beck, H. / Stampfl, J.
1973 "On the Use of Trend-Surface Models in the Spatial Analysis of Political Crises as Biobehavioral Phenomena," Montreal, IPSA.

Bell, A.
1970 "The Reception in Finland of the Organismic Doctrine of Rudolf Kjellen," Munich, IPSA.

Bernstein, P. / Schwartz, D.
1973 "A Note on the Impact of Health on Presidential Decision-Making," Montreal, IPSA.

Caldwell, L.
1964 "Biopolitics: Science, Ethics, and Public Policy," *The Yale Review* 54, pp. 1-16.

Corning, P.
1971a "The Biological Bases of Behavior and Some Implications for Political Science," *World Politics* 23, pp. 321-370.
1971b "An Evolutionary-Adaptive Theory of Aggression," Chicago, Ill., APSA.
1971c "Evolutionary Indicators: Applying the Theory of Evolution to Political Science," Revision of IPSA paper, Munich.
1973 "Politics and the Evolutionary Process," in: Dobzhansky, T. / *et al.* (eds.), *Evolutionary Biology,* Appleton-Century-Crofts, vol. 3.
1974a "The Dynamics of Collective Aggression in Human Societies," Prepared for American Psychological Association, New Orleans, La.
1974b "On the Problem of Defining Aggression," Prepared for Second Biennial Meeting of International Society for Research on Aggression, Toronto.

Davies, J. C.
1969 "The Psychobiology of Political Behaviour: Some Provocative Developments," Prepared for Western Political Science Association Meeting, Honolulu.
1970 "Violence and Aggression: Innate or Not ?," *Western Political Quarterly* 23, pp. 611-623.
1971 "Biology, Darwinism, and Political Science: Some Old and New Frontiers," Chicago, Ill., APSA.

Dearden, J.
1973 "Sex-Linked Differences of Political Behavior: An Investigation of Their Possibly Innate Origins," Montreal, IPSA.

Ferguson, LeRoy / Ferguson, Lucy / Bouterline-Young, H.
1970 "An Attempt to Correlate Rate of Physical Maturation with Attitudes Toward Politics," Munich, IPSA.

Frank, R. G.
1973 "Biological Referents in Political Rhetoric: A Study in Verbal Kinesics," Montreal, IPSA.

Haas, M.
1969 "Toward the Study of Biopolitics: A Cross-Sectional Analysis of Mortality Rates," *Behavioral Science* 14, pp. 257-280.

Halliday, R. J.
1971 "Social Darwinism," *Victorian Studies* 14, pp. 389-405.

Hummel, R.
1970 "A Case for a Bio-Social Model of Charisma," Munich, IPSA.
1971 "Teaching Political Theory: The Impact of Biopolitics," Chicago, Ill., APSA.

Jaros, D.
1972 "Biochemical Desocialization: Depressants and Political Behavior," *Midwest Journal of Political Science* 16, pp. 1-28.

Lepawsky, A.
1976 "Medical Science and Political Science," *The Journal of Medical Education* 42, pp. 896-908.

Locker, A.
1973 "A Systems-Theoretical View on the Origin and Decay of Political Movements: Some Analogies with Biological Evolutionary Processes," Montreal, IPSA.

Lodge, M. / Tursky, B. / Tanenhaus, J.
1973 "A Bio-Behavioral Framework for the Analysis of Political Behavior," Montreal, IPSA.
1974 "An Experimental Cross-Modal Framework for the Analysis of Political Attitude Scales: A Psychophysical Approach," Paper presented at NSF Conference, Delevin, Wisc.

Lodge, M. / Tursky, B. / Tanenhaus, J. / Cross, D.
1974 "The Development and Validation of Political Attitude Scales: A Psychophysical Approach," Paper presented at NSF Conference, Delevin, Wisc.

McAlpine, W. E.
1970 "Information Reduction Processes and Politics," Munich, IPSA.

Masters, R.
1967 "La redécouverte de la nature humaine," *Critique* (245), pp. 857-876.
1969 "Les racines biologiques d'une révolte," *Preuves,* February, pp. 74-81.
1970 "Genes, Language, and Evolution," *Semiotica* 2, pp. 295-320.
1972 "Political Behavior as a Biological Phenomenon," Washington, DC, APSA.
1973a "Functional Approaches to Analogical Comparison Between Species," *Social Science Information* 12 (4), pp. 7-28.
1973b "On Comparing Humans—and Human Politics—with Animal Behavior," Montreal, IPSA.

Mazrui, A.
1968 "From Social Darwinism to Current Theories of Modernization," *World Politics* 21, pp. 69-83.
1972 "Political Man and the Heritage of Hair: Some African Perspectives," *British Journal of Political Science* 2, pp. 1-20.
1973 "Phallic Symbols in Politics and War: An African Perspective," Montreal, IPSA.

Petchesky, R.
1971 "Biological Engineering as a Social Control Device," Chicago, Ill., APSA.

Peterson, S. A.
1973 "The Effects of Physiological Variables Upon Student Protest Behavior," Montreal, IPSA.

Phillips, C.
1973 "Biology, Cultural Evolution and the Political Process," Montreal, IPSA.

Pranger, R.
1967 "Ethology and Politics: The Work of Konrad Lorenz," Prepared for delivery at Southern Political Science Association Meeting, New Orleans, La.

Sanchez, A. L.
1970 "Biology, Politics and Society in Ortega y Gasset," Munich, IPSA.

Schubert, G.
1973 "Biopolitical Behavior: The Nature of the Political Animal," *Polity* 6, pp. 240-275.

Schwartz, D.
1970 "Perceptions of Personal Energy and the Adoption of Basic Behavioral Orientations to Politics," Munich, IPSA.
1973 "Health Processes and Body Images as Predictors of Political Attitudes and Behaviors," Montreal, IPSA.

Schwartz, D. / Zill, N.
1971 "Psychophysiological Arousal as a Predictor of Political Participation," Chicago, Ill., APSA.

Shubs, P.
1973 "Self Body-Image Correlates of Political Attitudes and Values," Montreal, IPSA.

Singer, D. / Luterbracher, U.
1970 "Crowding and Combat in Animal and Human Societies: The European State System, 1816-1965," Munich, IPSA.

Somit, A.
1968 "Toward a More Biologically Oriented Political Science: Ethology and Psychopharmacology," *Midwest Journal of Political Science* 12, pp. 550-567.
1970 "The Political, Philosophical, and Legal Problems Posed by Two Emerging Technologies," Prepared for delivery at International Futures Conference, Kyoto, Japan.
1972 "Biopolitics," *British Journal of Political Science* 2, pp. 209-238.

Stauffer, R.
1969 "The Biopolitics of Underdevelopment," *Comparative Political Studies* 2, pp. 361-387.
1970 "The Role of Drugs in Political Change," Munich, IPSA.

Stephens, J.
1970 "Some Questions about a More Biologically Oriented Political Science," *Midwest Journal of Political Science* 14, pp. 687-707.

Thorson, T.
1970a *Biopolitics*, New York, Holt, Rinehart and Winston.
1970b "The Biological Foundations of Political Science: Reflections on the Post-Behavioral Era," Munich, IPSA.

Tiger L.
1972 "Biology, Rhetoric, Reform: The Allure of Low-Born High Ideals," Washington, DC, APSA.

Tursky, B. / Lodge, M.
1971 "Compliance, Resistance, and Rebellion," Chicago, Ill., APSA.

Wahlke, J. / Lodge, M.
1972 "Psychophysiological Measures of Political Attitudes and Behavior," *Midwest Journal of Political Science* 16, pp. 505-537.

Welch, S. / Booth, A.
1973 "Crowding as a Factor in Political Aggression," Montreal, IPSA.

Wenner, M.
1970 "Symbiosis and Politics," Munich, IPSA.
1971 "Abstract: Biological and Environmental Factors as Political Variables," Chicago, Ill., APSA.

White, E.
1972 "Genetic Diversity and Political Life," *Journal of Politics* 34, pp. 1203-1242.

Wiegele, T.
1971 "Toward a Psychophysiological Variable in Conflict Theory," *Experimental Study of Politics* 1, pp. 51-81.
1973 "Decision Making in an International Crisis," *International Studies Quarterly* 17, pp. 295-335.

Wiegele, T. / Plowman, S. / Catey, R.
1973 "International Crisis, Cardio-Respiratory Health, and Political Attitudes," Montreal, IPSA.

Willhoite, F., Jr.
1971 "Ethology and the Tradition of Political Thought," *Journal of Politics* 33, pp. 615-641.
1975 "Equal Opportunity and Primate Particularism," *Journal of Politics* 37, pp. 270-276.

II. Biological Linkages in Other Social Sciences

Ambrose, J.
1973 "The Concept of Critical Period for the Development of Social Responsiveness," in: Foss, B. (ed.), *Determinants of Infant Behavior,* London, Methuen, vol. 2.

Ardrey, R.
1961 *African Genesis,* New York, Atheneum.
1966 *The Territorial Imperative,* New York, Atheneum.
1970 *The Social Contract,* New York, Atheneum.

Bajema, C. J. (ed.)
1971 *Natural Selection in Human Populations: The Measurement of Ongoing Genetic Evolution in Contemporary Societies,* New York, Wiley.

Ball, D.
1973 "The Biological Bases of Human Society," in: Douglas, J. D. (ed.), *Introduction to Sociology,* New York, The Free Press.

Barchas, P. / Fisek, M. H.
1969 "Rhesus and Freshmen; Studies in Status Order," Paper prepared for ASA meeting, San Francisco, Calif.

Bever, T. G. / Chiarello, R. J.
1974 "Cerebral Dominance in Musicians and Non-musicians," *Science* 185, p. 537.

Bischof, N.
1972 "The Biological Foundations of the Incest Taboo," *Social Science Information* 11 (6), pp. 7-36.

Blurton-Jones, N. G.
1967 "Some Aspects of the Social Behaviour of Children in Nursery School," in: Morris, D. (ed.), *Primate Ethology,* Garden City, NY, Doubleday.

Bowlby, J.
1957 "An Ethological Approach to Research in Child Development," *British Journal of Medical Psychiatry* 30, pp. 230-240.

Burnet, M.
1972 *Dominant Mammal: The Biology of Human Destiny,* New York, St. Martin's Press.

Callan, H.
1970 *Ethology and Society,* London, Oxford University Press.

Callan, H. M. W. / Chance, M. R. A. / Pitcairn, T. K.
1973 "Attention and Advertance in Human Groups," *Social Science Information* 12 (2), pp. 27-41.

Campbell, D.
1972 "On the Genetics of Altruism and the Counter-Hedonic Components in Human Culture," *Journal of Social Issues* 28, pp. 21-38.

Carey, G.
1972 "Density, Crowding, Stress and the Ghetto," *American Behavioral Scientist* 15, pp. 495-510.

Chapple, E. D.
1970 *Culture and Biological Man,* New York, Holt, Rinehart and Winston.

Cohen, Y.
1974 *Man in Adaptation: The Biosocial Background,* Chicago, Ill., Aldine.

Coleman, A. D.
1968 "Territoriality in Man: A Comparison of Behavior in Home and Hospital," *American Journal of Orthopsychiatry* 38, pp. 464-468.

Cortes, J. / Gatti, F.
1972 *Delinquency and Crime: A Biopsychosocial Approach,* New York, Seminar Press.

Count, E. W.
1958 "The Biological Basis of Human Sociality," *American Anthropologist* 60, pp. 1049-1085.
1973 *Being and Becoming: Essays on the Biogram,* New York, Van Nostrand.

Darlington, C. D.
1969 *The Evolution of Man and Society,* New York, Simon and Schuster.

Debré, R.
1969 "La biologie aide-t-elle à comprendre la jeunesse révoltée ?," *Revue de Paris* 76, pp. 25-32.

Eckland, B.
1967 "Genetics and Sociology: A Reconsideration," *American Sociological Review* 32, pp. 173-194.

Eysenck, H. J.
1967 *The Biological Bases of Personality,* Springfield, Ill., Charles C. Thomas.

Felipe, N. H. / Sommer, R.
1966 "Invasions of Personal Space," *Social Problems* 14, pp. 206-214.

Fox, R.
1967 "In the Beginning: Aspects of Hominid Behavioral Evolution," *Man* 2, pp. 415-433.

Freedman, D. G.
1967 "A Biological View of Man's Social Behavior," in: Etkin, W. (ed.), *Social Behavior from Fish to Man,* Chicago, Ill., University of Chicago Press.
1968 "Personality Development in Infancy," in: Washburn, S. L. / Jay, P. (eds.), *Perspectives on Human Evolution,* New York, Holt, Rinehart and Winston.

Galle, O. / Gove, W. / McPherson, J. M.
1972 "Population Density and Pathology: What are the Relations for Man ?" *Science* 176, pp. 23-30.

Glueck, S. / Glueck, E.
1956 *Physique and Delinquency,* New York, Harper.

Gottesman, I.
1968 "Biogenetics of Race and Class," in: Deutsch, M. / Katz, I. / Jensen, A. (eds.), *Social Class, Race, and Psychological Development,* New York, Holt, Rinehart and Winston.

Gray, P.
1958 "Theory and Evidence of Imprinting in Human Infants," *Journal of Psychology* 46, pp. 155-156.

Harlow, H. / Gluck, J. P. / Suomi, S. J.
1972 "Generalization of Behavioral Data Between Nonhuman and Human Animals," *American Psychologist* 27, pp. 709-716.

Hediger, H.
1969 "Biological Glimpses of Some Aspects of Human Sociology," *Social Research* 36, pp. 530-541.

Heise, D. R. (ed.)
1973 *Personality: Biosocial Bases,* Chicago, Ill., Rand McNally.

Hill, J.
1972 "On the Evolutionary Foundations of Language," *American Anthropologist* 74, pp. 308-317.

Hinde, R.
1962 "The Relevance of Animal Studies to Human Neurotic Disorders," in: Richter, D. / *et al.* (eds.), *Aspects of Psychiatric Research,* London, Oxford University Press.
1974 *Biological Bases of Human Social Behaviour,* New York, McGraw-Hill.

Imanishi, K.
1965 "The Origin of the Human Family: A Primatological Approach," in: Altmann, S. (ed.), *Japanese Monkeys,* Edmonton, S. Altmann.

Jay, A.
1971 *Corporation Man,* New York, Random House.

Jensen, A.
1969 "How Much Can We Boost IQ and Scholastic Achievement?," *Harvard Educational Review* 39, pp. 1-123.

Keiter, F.
1965 "Human Behavioral Biology (Ethology): A Modern Aspect of Cultural Anthropology," *Social Research* 32, pp. 357-374.

Kimble, D. P.
1973 *Psychology as a Biological Science,* Pacific Palisades, Calif., Goodyear.

Kortmulder, K.
1968 "An Ethological Theory of the Incest Taboo and Exogamy," *Current Anthropology* 9, pp. 437-449.

Lenneberg, E.
1964 "A Biological Perspective of Language," in: Lenneberg, E. H. (ed.), *New Directions in the Study of Language,* Cambridge, Mass., MIT Press.
1967 *Biological Foundations of Language,* New York, John Wiley.

Lorenz, K.
1966 *On Aggression,* New York, Bantam.
1974 *Civilized Man's Eight Deadly Sins,* New York, Harcourt, Brace, Jovanovich.

McGrew, W. C.
1972 *An Ethological Study of Children's Behavior,* New York, Academic Press.

Maclay, G. / Knipe, H.
1972 *The Dominant Man: The Pecking Order in Human Society,* New York, Delacorte.

Mazur, A.
1973 "Cross-Species Comparison of Status in Small Established Groups," *American Sociological Review* 38, pp. 513-530.

Mazur, A. / Robertson, L. S.
1972 *Biology and Social Behavior,* New York, Free Press.

Means, R. L.
1967 "Sociology, Biology, and the Analysis of Social Problems," *Social Problems* 15, pp. 200-212.

Montagu, A. (ed.)
1973 *Man and Aggression,* 2nd ed., New York, Oxford University Press.

Morgan, E.
1972 *The Descent of Woman,* New York, Stein and Day.

Morris, D.
1967 *The Naked Ape,* New York, Dell.
1969 *The Human Zoo,* New York, McGraw-Hill.
1972 *Intimate Behavior,* New York, Random House.

Papousek, H.
1967 "Genetics and Child Development," in: Spuhler, J. (ed.), *Genetic Diversity and Human Behavior,* Chicago, Ill., Aldine.

Ploog, D.
1970 "Neurological Aspects of Social Behavior," *Social Science Information* 9 (3), pp. 71-97.

Reynolds, V.
1966 "Open Groups in Hominid Evolution," *Man* 1, pp. 441-452.

Sheldon, W. H.
1940 *The Varieties of Human Physique,* New York, Harper and Row.

Tiger, L.
1969 *Men in Groups,* New York, Random House.
1970 "Dominance in Human Societies," *Annual Review of Ecology and Systematics* 1, pp. 298-301.

Tiger, L. / Fox, R.
1966 "The Zoological Perspective in Social Science," *Man* 1, pp. 75-81.
1971 *The Imperial Animal,* New York, Holt, Rinehart and Winston.

Tinbergen, E. A. / Tinbergen, N.
1972 "Early Childhood Autism: An Ethological Approach," *Zeitschrift für Tierpsychologie* 10, pp. 1-53.

Van den Berhge, P. L.
1973 *Age and Sex in Human Societies: A Biosocial Perspective,* Belmont, Calif., Wadsworth.
1974 "Bringing Beasts Back In: Toward a Biosocial Theory of Aggression," *American Sociological Review* 39, pp. 777-788.

Vine, I.
1973 "Social Spacing in Animals and Man," *Social Science Information* 12 (5), pp. 7-50.

Wickler, W.
1973 *The Sexual Code: The Social Behavior of Animals and Men,* Garden City, NY, Doubleday.

III. Supplementary Reading

1. *Theoretical Issues*

Atz, J. W.
1970 "The Application of the Idea of Homology to Behavior," in: Aronson, L. / *et al.* (eds.), *Development and Evolution of Behavior,* San Francisco, Calif., W. H. Freeman.

Beach, F. A.
1955 "The Descent of Instinct," *Psychological Review* 62, pp. 401-410.

Beer, C. G.
1968 "Ethology on the Couch," *Science and Psychoanalysis* 12, pp. 198-213.

Dimond, S.
1970 *The Social Behavior of Animals,* New York, Harper, chap. 9.

Gottlieb, G.
1968 "Prenatal Behavior of Birds," *Quarterly Review of Biology* 43, pp. 148-174.
1970 "Conception of Prenatal Behavior," in: Aronson, L. / *et al.* (eds.), *Development and Evolution of Behavior,* San Francisco, Calif., W. H. Freeman.

Hebb, D. O.
1953 "Heredity and Environment in Mammalian Behavior," *British Journal of Animal Behavior* 1, pp. 43-47.

Jensen, D. D.
1961 "Operationism and the Question: 'Is This Behavior Learned or Innate?'," *Behaviour* 17, pp. 1-8.

Klopfer, P. / Hailman, J. (eds.)
1972a *Function and Evolution of Behavior,* Reading, Mass., Addison-Wesley.
1972b *Control and Development of Behavior,* Reading, Mass., Addison-Wesley.

Kuo, Zing Yang
1967 *The Dynamics of Behavior Development,* New York, Random House.

Lehrman, D. S.
1953 "A Critique of Konrad Lorenz' Theory of Instinctive Behavior," *Quarterly Review of Biology* 28, pp. 337-363.
1970 "Semantic and Conceptual Issues in the Nature-Nurture Problem," in: Aronson, L. / *et al.* (eds.), *Development and Evolution of Behavior,* San Francisco, Calif., W. H. Freeman.

Lorenz, K.
1965 *Evolution and Modification of Behavior,* Chicago, Ill., University of Chicago Press.
1970, 1971 *Studies in Animal and Human Behavior,* vol. 1 and 2, Cambridge, Mass., Harvard University Press.
1974 "Analogy as a Source of Knowledge," *Science* 185, pp. 229-234.

Lorenz, K. / Leyhausen, P. (eds.)
1973 *Motivation of Animal and Human Behavior,* New York, Van Nostrand.

Maier, N. R. F. / Schneirla, T. C.
1963 *Principles of Animal Psychology,* New York, Dover Press.

Moltz, H.
1965 "Contemporary Instinct Theory and the Fixed Action Pattern," *Psychological Review* 72, pp. 27-47.

Schneirla, T. C.
1959 "An Evolutionary and Developmental Theory of Biphasic Processes Underlying Approach and Withdrawal," in: Jones, M. R. (ed.), *Nebraska Symposium on Motivation*, Lincoln, Neb., University of Nebraska Press.
1966 "Behavioral Development and Comparative Psychology," *Quarterly Review of Biology* 41, pp. 283-302.

Tinbergen, N.
1963 "On Aims and Methods in Ethology," *Zeitschrift für Tierpsychologie* 20, pp. 410-429.

2. *Animal Behavior*

Alcock, J.
1975 *Animal Behavior*, Sunderland, Mass., Sinauer Associates.

Altmann, S. A. (ed.)
1967 *Social Communication Among Primates*, Chicago, Ill., University of Chicago Press.

Crook, J. H. (ed.)
1970 *Social Behaviour in Birds and Mammals*, New York, Academic Press.

DeVore, I. (ed.)
1965 *Primate Behavior*, New York, Holt, Rinehart and Winston.

Eibl-Eibesfeldt, I.
1970 *Ethology*, New York, Holt, Rinehart and Winston.

Ewer, R. F.
1968 *Ethology of Mammals*, New York, Plenum Press.

Hinde, R.
1970 *Animal Behaviour*, New York, McGraw-Hill.

Jay, P. (ed.)
1968 *Primates*, New York, Holt, Rinehart and Winston.

Jolly, A.
1972 *Evolution of Primate Behaviour*, New York, MacMillan.

Klopfer, P. / Hailman, J.
1967 *An Introduction to Animal Behavior*, Englewood Cliffs, NJ, Prentice-Hall.

Lorenz, K.
1952 *King Solomon's Ring*, New York, Thomas Y. Crowell.

Manning, A.
1973 *Introduction to Animal Behavior*, Reading, Mass., Addison-Wesley.

Marler, P. / Hamilton, W. III
1966 *Mechanisms of Animal Behavior*, New York, Wiley.

McGill, T. (ed.)
1965 *Readings in Animal Behavior*, 1st ed., New York, Holt, Rinehart and Winston.
1973 *Readings in Animal Behavior*, 2nd ed., New York, Holt, Rinehart and Winston.

Morris, D. (ed.)
1967 *Primate Ethology*, Garden City, NY, Doubleday, Anchor Books.

Schiller, C. H. (ed.)
1957 *Instinctive Behaviour*, London, Methuen.

Scott, J. P.
1972 *Animal Behavior*, Chicago, Ill., University of Chicago Press.

Simonds, P.
1974 *The Social Primates*, New York, Harper and Row.

Southwick, C. (ed.)
1963 *Primate Social Behavior*, Princeton, NJ, Van Nostrand.

Tinbergen, N.
1951 *The Study of Instinct,* London, Oxford University Press.
1953 *Social Behavior in Animals,* London, Chapman and Hall.
1968 "On War and Peace in Animals and Man," *Science* 160, 1411-1418.

3. *Genetics and Evolution*

Bonner, J.
1965 *The Molecular Biology of Development,* New York, Oxford University Press.
Buettner-Janusch, J.
1973 *Physical Anthropology,* New York, Wiley.
Campbell, B.
1966 *Human Evolution,* Chicago, Ill., Aldine.
Campbell, B. (ed.)
1972 *Sexual Selection and the Descent of Man,* Chicago, Ill., Aldine.
Dobzhansky, T.
1962 *Mankind Evolving,* New Haven, Conn., Yale University Press.
1970 *Genetics of the Evolutionary Process,* New York, Columbia.
Hirsch, J. (ed.)
1967 *Behavior-Genetic Analysis,* New York, McGraw-Hill.
Lerner, I. M.
1968 *Heredity, Evolution, and Society,* San Francisco, Calif., W. H. Freeman.
Mayr, E.
1970 *Population, Species, and Evolution,* Cambridge, Mass., Harvard University Press.
McClearn, G. / DeFries, J. C.
1973 *Introduction to Behavioral Genetics,* San Francisco, Calif., Freeman.
Roe, A. / Simpson, G. G. (eds.)
1958 *Behaviour and Evolution,* New Haven, Conn., Yale University Press.
Simpson, G. G.
1967 *The Meaning of Evolution,* New Haven, Conn., Yale University Press.
Waddington, C. H.
1957 *The Strategy of the Genes,* London, Allen and Unwin.
Watson, J. D.
1970 *Molecular Biology of the Gene,* 2nd ed., New York, Benjamin.
Wilson, E. O.
1974 *Ecology, Evolution, and Population Biology,* San Francisco, Calif., Freeman.

4. *Physiological Bases of Behavior*

Delgado, J.
1969 *Physical Control of the Mind,* New York, Harper and Row.
Deutsch, J. A. / Deutsch, D.
1973 *Physiological Psychology,* Homewood, Ill., Dorsey Press.
Ecclès, J. C.
1957 *The Physiology of Nerve Cells,* Baltimore, Md., Johns Hopkins Press.
Lewin, R.
1972 *Hormones,* Garden City, NY, Doubleday, Anchor Books.
MacLean, P. D.
1958 "The Limbic System With Respect to Self-Preservation and Preservation of the Species" *Journal of Nervous and Mental Diseases* 127, pp. 1-11.
McCleary, R. / Moore, R.
1965 *Subcortical Mechanisms of Behavior,* New York, Basic Books.
Michael, R. (ed.)
1968 *Endocrinology and Human Behaviour,* New York, Oxford University Press.
Money, J. / Ehrhardt, A.
1972 *Man and Woman, Boy and Girl,* Baltimore, Md., Johns Hopkins Press.

Moyer, K.
1971 *Physiology of Hostility,* Chicago, Ill., Markham.
Papez, J. W.
1937 "A Proposed Mechanism of Emotion," *Archives of Neurology and Psychiatry* 38, pp. 725-743.
Quarton, G. C. / Melnechuk, T. / Adelman, G. (eds.)
1970 *The Neurosciences,* New York, Rockefeller University Press.
Stellar, E.
1954 "The Physiology of Motivation," *Psychological Review* 61, pp. 5-22.
Teitelbaum, P.
1967 *Physiological Psychology,* Englewood Cliffs, NJ, Prentice-Hall.
Thompson, R. (ed.)
1972 *Physiological Psychology,* San Francisco, Calif., Freeman.
Valenstein, E.
1973 *Brain Control,* New York, John Wiley.
Young, W. C. (ed.)
1961 *Sex and Internal Secretions,* Baltimore, Md., Williams and Wilkins.

ALBERT SOMIT

*Review article: Biopolitics**

THE IDEA THAT BIOLOGICAL concepts are helpful in explaining political phenomena, and that biological factors play a significant role in political behavior, has a long history in Western political thought. In one version, the state is portrayed as a living organism with the various parts of the state (or government) functioning much as do the parts of the body, i.e., head, heart, arms, etc.[1] Alternatively, when attention turns to the external rather than the internal life of states, the resulting relationships are treated as the inevitable outcome of the "struggle for survival" to which all living organisms are presumably condemned. In both cases, the language employed is rich in biologic metaphor—*lebensraum,* birth, death, growth, decay, youth, age, sickness and health.[2]

Another biologically-oriented approach treats "human nature" as a limiting, as well as an explanatory, factor in political life. Here we need only recall that distinguished school of political philosophy which holds that the behavior of political man springs from "human nature" and that selfishness, avarice and ingratitude are among the more outstanding, if unlovely, attributes of that nature. This line of thought can be traced from Greek speculation through eighteenth-century conservatism and, appropriately restated, contributed to, and drew sustenance from nineteenth-century Social Darwinism.[3]

Social Darwinism, and especially the tenet that social behavior was in significant measure the product of man's biological make-up, soon became intellectually disreputable. While there were many reasons for this, not all of them scientific, the fact remains that, by the early 1900s, the theory that human nature could more satisfactorily be understood in terms of learned responses was gaining ascendancy; by the thirties, it was the accepted wisdom. With few exceptions, social scientists trained after the First World War simply took it for granted that they could safely ignore man's genetic legacy.

Recent advances in biology have challenged this comfortable belief. Freeing themselves from the shackles of the earlier notion of "instinct"

* The editor and the publishers wish to express their gratitude to the *British Journal of Political Science* for permitting the publication of this article which appeared in vol. 2, April 1972, pp. 209-238.

(see below, p. 215), ethologists have argued that a good deal of our behavior has its roots in our biological make-up and that some social and political phenomena can be explained only by resorting to biological concepts. Concurrently, research in neurobiology, physiology and psychopharmacology has demonstrated beyond doubt that it is possible to influence—and even control—human behavior by altering the physiological functioning of the body.

Largely as a result of these several stimuli, there has lately appeared in political science a body of literature which concerns itself, from one vantage point or another, with the interrelationships between biology and political science. These several approaches are usually subsumed under the heading of "biopolitics" and, though the term is not altogether fortunate (see below, p. 212), I will use it for lack of a better. The purpose of this essay, then, is to review the burgeoning literature of biopolitics, a literature which now includes some forty items. I will try to identify the major areas of interest, to assess what has been done in each of these areas and, in the almost obligatory concluding section, suggest what might be the most promising future lines of exploration.

As is often the situation in a nascent field, the literature ranges widely and irregularly. After experimenting with other alternatives,[4] the following seemed to me to provide the most satisfactory rubric:

1) The case for a biologically-oriented political science.
2) Ethological aspects of political behavior.
3) Physiological and psychopharmaceutical aspects of political behavior.
4) Issues of public policy raised by recent advances in biology.

A couple of contributions, naturally, did not fall gracefully into any single category and a few arbitrary allocations were inevitable; further, since this is a subject-oriented rather than an author-oriented classification, some papers will be discussed under more than one heading.

Now, three final introductory comments: first, in selecting the items to be reviewed, I have limited myself to the writings of *political scientists*. My interest here is with biologically-oriented political scientists, not with politically-oriented biologists. The latter have been exceedingly generous, almost intolerably so, in their willingness to explain (or prescribe) political behavior, but to have dealt with them, except as absolutely necessary, would have taken this review in a totally different direction.

Next, the perceptive reader will note that, with few exceptions, the items are from English-language sources. Fellow political scientists in France, Germany, Italy, Spain, Scandinavia and Japan assure me that nothing else of this sort yet exists in their professional literature at the time of my "closing date," September 1971. Given my linguistic limitations, it was an assurance I was delighted to receive.

Lastly, as a matter of convenience for those who may wish to consult some of the literature mentioned below, I have provided two bibliographies. The first of these contains all of the items which I have subsumed under the heading of biopolitics; the second contains all other items mentioned in the course of the discussion. In the great majority of instances, I doubt that the reader will have any difficulty in deciding in which bibliography a given paper—or book—is to be found.

I. Toward A More Biologically-Oriented Political Science

Political scientists were relatively slow in grasping the relevance for their discipline of post-1945 developments in biology.[5] In the early 1960s, Harold Lasswell[6] very briefly touched on the possibility of using chemical and biological techniques to control political behavior; James C. Davies' *Human Nature in Politics* (1963) argued, *inter alia,* the relevance for political science of Abraham Maslow's theory of "organic" human needs; W. J. M. Mackenzie (1967) suggested that the work being done in ethology might be of more than passing interest to political scientists; and Lasswell (1968) urged greater attention to the "genetic predispositions of man and their possible impact on human politics and society" (p. 10).

The first general statement of the case for a closer linkage between political science and biology was, I believe, my own article in 1968. After briefly summarizing the ethological viewpoint and after treating, in more detail than any reader probably desired, research in psychopharmacology, I drew what seemed to be the obvious conclusion:

> In short, the ethologists insist that important aspects (but by no means all) of human behavior are rooted in man's biological (i.e., genetically transmitted) constitution; the psychopharmacologists have been able to induce profound behavioral changes by altering the physiological or biological functioning of the human body. Psychopharmacology thus confirms, I think it fair to say, the ethological contention that there is a direct link between biology and behavior in man as well as in other forms of life (p. 560).

The closing half-dozen pages examine some of the implications for public policy and for political science itself, which might follow from a more biological orientation.

The following year,[7] three more papers sounded the same theme. After reminding his colleagues that they had not yet taken full advantage of some of the psychological theories already available to them, in April, Davies (1969) briefly described recent research in psychobiology, neurophysiology and endocrinology. Advances in these and related areas give promise that in the near future "we will be able to specify far more precisely than hitherto just what human nature consists of" (p. 24) and that "we will then know what are the innate tendencies for which the environment must be restructured in order to realize man's fullest potential, in and out of politics" (p. 25). Several months later, Charles R. Adrian (1969) also proposed a closer tie between biology and politics but, as his paper focused exclusively on ethology, I have reserved it for discussion in Section 2 below.

The longest, and in some respects most detailed, statement of the case for biopolitics will be found in an article by Peter Corning (1971a), a greatly abridged version of a paper delivered some eighteen months earlier.[8] Corning's article has three main sections: a description of the "modern, 'synthetic' theory of evolution;" a review of the "current research on behavior from a biological and evolutionary point of view;" and a discussion of "some possible implications for political theory and political research." Of the three, the first, entitled "The Evolutionary Model of Society," is probably the most valuable for someone not familiar with the general literature, for it provides both a compact summary and an excellent bibliographic coverage. The next section, entitled "The Nature of Human Nature," is valuable for much the same reasons, although it retraces in considerable measure some of the ground covered by earlier papers. The final pages on "Some Implications" tend to be somewhat theoretical and are, I think, less useful, an opinion which will not necessarily be shared by others.

The only book-length contribution to this literature to date is Thomas L. Thorson's *Biopolitics* (1970a). I wish the book had been given some other title, as the actual subject matter has relatively little in common with the other biopolitical writings. There is, moreover, the danger that the term "biopolitics" will be identified with Thorson's philosophic position, one hardly shared by most of those working in the area. For neither of these conditions, of course, is Thorson to blame.

With some assurance, I can say that the book does not deal in any organized fashion with physiology, neurophysiology, psychobiology or psychopharmacology; the treatment of ethology is, at best, cursory (the term does not appear in the index); and there is no attempt, *à la* Corning, to present a systematic overview of modern evolutionary thought. The reader who looks for these topics will certainly be disappointed.

Thorson's mode of argument is such that, while I am reasonably sure of what is *not* in the volume, I am less confident that I grasp what he *is* trying to say.[9] Unless I grossly misread him, Thorson's basic concern is not with "biopolitics," as I have used the term in this essay, but with something quite different—philosophy of science. *Biopolitics* is basically an attack on the contemporary conception of scientific method and an argument for what Thorson believes to be a truly "evolutionary" approach to science. Whereas present-day scientific thought ("the Newtonian universal-generalization paradigm") regards scientific laws as "valid everywhere for all time," Thorson (espousing "the Darwinian evolutionary-development paradigm") treats such laws as temporarily and spatially limited. That is, a scientific generalization might well be valid at one period or place—and invalid at others. If this is true—and here Thorson does turn to political science—then the quest for general laws of political behavior is *a priori* doomed to failure.

Several comments would seem to be in order. Thorson's exposition of contemporary, evolutionary thought is at least questionable. He relies heavily on Teilhard de Chardin, whose ideas are regarded as less than orthodox by most professional biologists, and Thorson's references to the standard evolutionary literature, as mentioned above, are at best sketchy. Second, while it is "thinkable" that the laws of physics, for instance, are not today what they were eons ago (or will be eons hence), I doubt that many readers will find the notion any more plausible after reading Thorson than they did before. Finally, one could agree that we are not likely to discover "general laws of political behavior"—but this conclusion can be far more satisfactorily defended on quite different grounds.

In all fairness to Thorson, a subsequent paper directed specifically to political science and biology (1970b) seems to take a more moderate position. True, he again sharply criticizes a political science which apes the "model of 19th century physics" with its notions of "detached observer science" and its "is-ought separation." Political science, he insists, should be recast in terms of contemporary, evolutionary theory *and* information transmission theory. This would yield a "post-behavioral" political science informed by a concern for human values and by a revitalized and scientifically restated conception of "natural law." What this means, and which is the real Thorson, I can only leave for the reader to decide.

So far, all of the writers cited have advocated a closer tie between biology and politics even if the arguments have been convergent rather than identical. Stephens (1970) has considerately met the need for a dissident voice, although he is actually more cautionary than negative. Using the Somit (1968) paper as his point of departure, Stephens warns against an uncritical effort to utilize biological concepts in social science. His paper, which focuses largely on the general systems theorists of the post-1945 period, provides an excellent summary of these attempts and the difficulties encountered. Other than for the single item noted above, Stephens does not deal with the biopolitical literature itself. Most of the contributors to that literature, I think it safe to say, would endorse his caveat that:

> Before moving too rapidly toward a more biologically oriented political science, it will be well to consider the problems that others have confronted in attempting to apply theories developed in one area to another, quite diverse, area. The beneficial returns are often less than the confusion that is introduced by such borrowing. And in the group sciences where there is still a great deal that is not known about the relationship of individual behavior to group behavior, the addition of a biological level must be approached with the greatest methodological and empirical caution. The problem of reduction, which is already enormous, will be even more paramount in a more biologically oriented political science (p. 707).

Finally, mention should be made of a very recent paper in which Hummel (1971) discusses the significance of biopolitics for "conceptualization at all levels of political science where theory plays a role: in political philosophy, behavioral explanations of politics, methods of research and macro-models for political research" (p. 1). Taking each of these areas in turn, the author briefly considers the import for that area of one or two biopolitical concepts. Under political philosophy, for example, he considers the implications of ethology for the traditional view of the nature of man as found in Hobbes and Locke. Or, to take another example, what does ethology signify for such "operative political values" as the belief that "all men are created equal ?"

Regrettably, none of these areas is explored in any depth; a good part of what is said can be found elsewhere; the coverage of the biopolitical literature is extremely sketchy; the judgements are sometimes idiosyncratic;[10] and the tone is apocalyptic, rather than analytical.[11] Even more troublesome, I would say, is the reluctance to recognize that many of the biopolitical concepts treated are so ambiguous or controversial that it is still too early to do more than speculate about their long-run significance for political theory.

II. Ethological (Genetic) Aspects of Political Behavior

So effectively have the ethologists presented their case in the past decade that no more than a capsule summary of their general position is needed. It is a two-stage argument. To begin, they point out that natural selection[12] fashions, for each species, behavioral patterns which identify that species almost as readily as do its physical characteristics. Developed over literally millions of years, genetically transmitted modes or responses shape the manner in which the species conducts the functions essential to its survival—nesting, hunting, defense, care of the young, etc. These patterns vary in the degree and rigor to which they influence behavior, being most controlling among the lowest forms of life and allowing greater flexibility of response as we move up the "Great Chain of Being." From worms to primates, however, they play an important part in the behavior of all organisms. By defining "instinct" as a tendency, the ethologist thus not only concedes but stresses the fact that behavior, at almost all levels, can be influenced by learning, as well as by heredity.

To this point, the ethological argument is, in principle, acceptable to almost all parties, ethologists or not. Controversy arises, though, when we turn to human behavior. Does Homo sapiens, as well as all other animals, come into life with "genetically programmed" predispositions to respond to certain types of stimuli in certain types of fashion? Most social scientists would probably say either "no" or "not to any significant degree;" almost all ethologists and most biologists would say either "yes" or "to an important degree."[13] Since the matter cannot yet be empirically resolved, I would like to withhold further discussion of the issue and turn instead to the ethologically-oriented political science literature. I will note first the papers which attempt some sort of general overview[14] and then those which focus on specific ethological concepts.

1. *Overviews of Ethology*

The term "overview" is strictly relative. While the papers dealt with here are broader in scope than those in the next section, none of them offers more than a partial coverage of ethology. Pranger restricts his attention to Lorenz; Masters is concerned with Ardrey; and though Adrian casts a wider net, his treatment is at best summary. The fact remains—and as one who has contributed to this corpus, I can say this with reasonably good grace—that a really adequate critical analysis of the implications of ethology for *political science* remains to be written.

Pranger (1967) assesses Konrad Lorenz (1965, 1966) as biologist and as political theorist. On the former, Pranger wisely withholds judgement, observing that the biologists themselves are in disagreement. On the latter, he concludes that Lorenz makes, on balance, "a notable contribution to contemporary political thought," although he admits to some reservations about Lorenz' "solutions" for the problems of nationalism and war. Pranger tends to concern himself primarily, I should add, with Lorenz' treatment of aggression.

Masters (1967) reviews three books—Davies' *Human Nature in Politics*

(1963), Ardrey's *African Genesis* (1961) and *Territorial Imperative* (1966). The comparison is quite favorable to Ardrey, whose use of biology yields, in Masters' opinion, a more accurate conception of, and a more promising approach to, human nature than does Davies' "specialist" social science methodology. A "rediscovery of human nature" along the line projected by Ardrey, Masters quite correctly observes, could have profound impact on contemporary political thinking. It is interesting to note that Ardrey's most recent book (*The Social Contract,* 1970) attempts, with dubious success, to restate Rousseau in terms of Lorenz—an undertaking for which Masters served as intellectual godfather.

Adrian (1969) ranges more widely and his bibliography includes not only the familiar Ardrey-Lorenz-Morris-Tiger volumes but also a handful of items intended for a more professional readership. Organized in rather informal fashion, the paper moves through such topics as aggression, man's predatory origins, territorial behavior, the social structure of animal groups and crowding. The discussion is descriptive, rather than critical, and Adrian's general evaluation is pretty well reflected in his comment that

> The findings of ethologists would seem to indicate that Hobbes, as to man's basic nature, was more right than Locke; Madison more than Jefferson. Similarly, they suggest that the intellectual conservative, the political reactionary, and the contemporary radical, all of whom emphasize community over society, are more right than were nineteenth- and twentieth-century liberals, who reversed the emphasis (p. 22).

The concluding pages are devoted to questions of public policy, and I will briefly return to this aspect of the paper in Section 4 below.

In a very recent article, Wilhoite (1971) attempts "to explore tentatively some of the ways in which biological studies and interpretations of animal (including human) behavior may come to be useful to our understanding of man and political life, and to suggest some parallels and contrasts between ethological perspectives and a variety of theoretical conceptions within the tradition of Western political thought" (p. 641). The brief discussion of ethological perspectives is largely in terms of Lorenz; the Western tradition of political thought is dealt with almost *en passant ;* and the consideration of the relationship between the two does not move significantly beyond previous contributions. They may be related to the fact that, with two exceptions, Wilhoite seems unacquainted with the literature reviewed here.

2. *Specific Ethological Concepts*

There are also a few items which seek to apply—or at least make reference to—specific ethological concepts. Some of these concepts, of course, are also touched on in the papers just discussed.

Aggression

This is probably the most controversial of the notions advanced by Lorenz and popularized by Ardrey; not all ethologists, needless to say, share their views. Contemporary man, the argument goes, evolved from a predatory, carnivorous ancestor. Early man survived by hunting and killing—and these behavioral tendencies became and remain not only part of "human nature" but an important part. In one critical respect, however, man differs from other predators. By some ghastly evolutionary miscarriage, we failed to acquire that genetically transmitted inhibition against killing fellow speciates which normally characterizes other predators.[15] Homo sapiens is so constituted that he not only kills members of his own species but, even worse, has an "innate tendency" to do so.[16]

Considering the role of violence in contemporary life, the loss of life in past wars, declared and informal, and the danger that we may destroy ourselves in the next war, it is hardly surprising that the notion of man's "inherent aggressiveness" was soon picked up by political scientists.[17] The topic received brief mention in the earlier biopolitical commentaries (Somit, 1968; Corning, 1971a; Davies, 1969; Pranger, 1967) and two papers have subsequently dealt with it at greater length.

Davies (1970) does not deal directly with the arguments from ethology, though he does refer to Koestler's "para-ethological" *Ghost in the Machine* (1968), but confines himself primarily to the evidence from neurophysiology. Whether men are inherently aggressive or violent depends in part on the definition of terms and Davies defines violence and aggression in terms of "intent"—a definition which may be logically defensible but which poses near-insoluble problems for scientific inquiry. His conclusion is, I think it fair to say, inconclusive:

> Until we do know adequately, which may be centuries from now, we would do better to rest content with the tautology: man is man. He is good, he is evil. He is aggressive, he is cooperative. He is sensitive, he is callous. He is selfish, he is generous. He hates, he loves. And each of these characteristics when seen in the actions of men is without exception the result of a very complicated organism interacting with a very variegated environment. So man is man (p. 623).

In an ambitious undertaking, Corning (1971c) attempts to develop an "adequate theory of aggression." Such a theory, he holds, "must provide a satisfactory explanation of the origins, genetic bases, and survival consequences of aggressive behaviors." Furthermore, it must be consistent with the principles of evolution, must permit systematic cross species comparisons, must satisfactorily link individual and group behaviors and, perhaps above all, "must be able to account for and integrate the evidence generated in support of other theories of aggression" (p. 2). The three major "other" theories of aggression are, of course, the "biological instinctual," the "social learning theory," and the familiar "frustration-aggression hypothesis."

After reviewing the three competing theories, and the evidence adduced for each of them, Corning concludes that

> Aggressive behaviors are a product of evolution (of natural selection) and can be understood in relation to their survival consequences for particular species; because aggressive behaviors are evolved traits, they must *on balance* (at least in the past) have had adaptive value for any species in which such behaviors are common. . . (p. 13).

This hypothesis, as he himself admits, "is, if anything, a commonplace among evolutionary biologists and ethologists" (p. 13). Having formulated his hypothesis, Corning identifies the seven "expectations" about the adaptive value of aggression which should follow if the hypothesis is valid, and then examines the evidence for each of the seven. In an effort to test one of the expectations (the adaptive value of maternal aggression) he devised and carried out an experiment involving lactating female mice, pups and male mice.

Corning's coverage of the literature, as always, is meticulous; his review of existing theories, as usual, lucid and fair; and it is refreshing to find a political scientist at work in the laboratory, even with mice. Having said this, some reservations must also be voiced: his conception of the adaptive nature of aggression is hardly novel; the seven "expectations" are, at best, only partially fulfilled; and the relevance, for a theory of *human* aggression, of research on mice is at least debatable. Finally, while it may be that aggressive behavior has survival value for all other species, and may even also have had such value for man himself, it does not follow that it today serves the same adaptive function—or at least not in the forms in which it manifests itself in contemporary society.

Territoriality

According to some ethologists, man may share with many other species a "sense of territory" or, to put it somewhat differently, man may exhibit "territorial behavior." Such proclivity could manifest itself, in theory, by a deep attachment to a particular plot of land, to one's place of birth, to one's town or village, or in more recent times, by what we call "patriotism" or "nationalism."[18]

If it exists, human territorial behavior could have direct bearing upon any proposed solution to international strife; it could also raise grave questions about the viability of any political or economic system which envisages the abolition of private property. Curiously enough, the references to territoriality occur almost entirely in discussions of administrative behavior. Adrian (1970) suggests that the concept might be helpful in accounting for some types of organizational phenomena, but does not go beyond the suggestion. In the "non-biopolitical" literature, Anthony Downs (1967) earlier advanced the same idea and Sir Geoffrey Vickers (1965), drawing on a familiar ethological example, has commented that

> . . . the territorial behavior of stickle-backs is a distressingly exact description of a type of inter-departmental behaviour which is endemic in many institutions; and I have often felt, when taking part in wage negotiations, that I was participating in interaction which would be more easily described in terms of ethology than of economics (pp. 18-19).

Male Bonding

Faced by an implacably hostile environment, Lionel Tiger (1969, 1971) has argued, primitive man's chances of survival hinged in good part on his willingness and capacity to co-operate with fellow males in hunting food and in defending himself against predators—including, possibly, fellow speciates. Those who had this tendency survived and reproduced themselves; those who did not. . . Thus, over the ages, natural selection produced the "male bond," an innate tendency among men to join with other men for what we would now call political purposes, but which may also extend to sporting activities, secret societies and, to be sure, that classic male avocation. A corollary of this proposition, it should be mentioned, is that while the tendency toward "political" behavior is bred into the male, it was bred *out* of the female. Very brief references to this idea, but as yet no attempt to apply it, can be found in Adrian (1969), Masters (1970) and Corning (1971a).

Crowding and Personal Space

Many, if not all, species are so constituted biologically that they require a certain "personal space"—i.e., an area not occupied by a fellow speciate—if they are to function normally.[19] Calhoun's (1962) famous experiments with rats graphically demonstrated what happens when overcrowding occurs. In a way, the concept is akin to that of territory, though a species which does not display what we would technically call territorial behavior may nonetheless manifest a strong need for personal space.[20]

Singer and Luterbracher (1970) have sought to ascertain whether a relationship can be established between "crowding" (defined in terms of population, population change and population density) and "combat" (defined in terms of battle-connected deaths) for the European state system from 1816 to 1965. They found a gross positive correlation when they looked at the total state system and the total time period; the correlation decreased, and in some cases became negative, when the data were examined in terms of smaller subsets of nations and/or of time. As Singer and Luterbracher put it, "on the one hand, we cannot dismiss the general association that we found when population was the independent variable, but neither can we draw any confident inferences from these results" (p. 26). An obvious point is that the European nations with the greatest population tended to be the "great powers;" great powers, almost by definition, are more likely than minor powers to become involved in warfare; and there is likely to be a positive correlation between frequency of war and battle casualties.

To my knowledge, the Singer-Luterbracher paper is the only instance in which political scientists have attempted to utilize this particular biological concept. Those concerned with problems of urban areas, especially the recent upsurge of violence in our cities, may find the idea worth exploring.

3. *Obiter Dicta*

So far, I have dealt with the ethological concepts which, if only in modest fashion, have already been incorporated in the biopolitical literature. There remain at least two other concepts which will probably soon be "picked up" by political scientists and which merit passing mention.

Xenophobia

Drawing in part on evidence gathered from the observation of very young children, Keith (1948) argued that "to account for the babe's behavior, we have to assume that it has been born with a mental bias—an inclination as well as an aptitude to love the known but to turn away from the strange or unknown" (p. 65). Others, possibly less charitable, utilize evidence drawn from the observation of cattle, baboons, rats and ants to broach, if not actually support the proposition that men have a "natural" inclination toward xenophobia.

Imprinting

The tendency to distrust that which is foreign and strange might also be explained in terms of what the ethologists call "imprinting." Technically speaking, this refers to the manner in which the very young of some species learn to respond in a filial manner to an object. Under normal circumstances, the "object" to which they imprint is one of their parents, but many readers will recall Lorenz' delightful story of the baby duckling for whom he reluctantly became surrogate mother. Closely related to imprinting is the idea of a *critical period*—that relatively brief span of time during which, for most species, imprinting is possible and after which it is not likely to occur.

At least one eminent biologist, Renée Dubos (1968), believes that imprinting has relevance for social and political science. According to Dubos:

> Experiments in animals have shown that the young organism is particularly susceptible to the effects of conditioning during certain so-called critical periods of early development. It is unfortunate that scientific knowledge concerning these critical periods is extremely scanty, because the same biological law certainly applies to human beings. Most slum children, unfortunately, continue to conform to the ways of life of their destitute parents, despite intensive efforts by skilled social workers to change their habits and tastes. By the third or fourth year of life their behavioral patterns have already been environmentally and culturally determined. Furthermore, there is much reason to fear that they will in turn imprint similar patterns on their own children. It is not accurate to state that slum children are culturally deprived; the more painful truth is that slum life imprints on them a culture from which they are usually unable to escape (pp. 79-80).

4. *Social Darwinism*

The growing awareness of developments in biology generally and ethology in particular generated a modest renaissance of interest in Social Darwinism, old and new. Halliday (1970), seeking "to restore Darwinism as a critical discourse," sets out to offer a definition of Social Darwinism which, hopefully, "will facilitate a closer and more accurate identification of its practitioners" (p. 4). After a critical examination of previous usages, he concludes that Social Darwinism is best defined as "that discourse arguing for eugenic population control; an argument requiring a complete commitment to an exclusively genetic or hereditarian explanation of man's evolution" (p. 12). While Halliday has a third objective, that of taking up "whatever questions about man and society we can find which seem to raise other than ideological problems and which suggest more than trivial conclusions," the paper is most valuable for its coverage of British Social Darwinism during the 1880-1914 period.

According to Mazrui (1968), two of the underlying assumptions of nineteenth-century Social Darwinism have unwittingly been incorporated in contemporary theories of modernization. The conviction that the white man represented the most highly evolved "race" of Homo sapiens led, quite naturally, to the classification of cultures as "higher" or "lower" and to equally invidious distinctions among political regimes, with the "Anglo-American type" being regarded as the most advanced and "types that are found in Africa and Asia" as the least. But the Darwinians also held that states and societies, as well as animal life, underwent a "progressive evolution toward greater complexity and greater sophistication," and this second, more optimistic, tenet is also found in present-day thinking about comparative politics. Mazrui draws liberally on the modernization literature to support his thesis and, on balance, the idea seems quite plausible. Still, Hebrew, Greek, Roman and even later writings testify that Social Darwinism is not a necessary pre-condition either for the belief that some cultures are more advanced than others or for faith in a better political world to come.

A pair of essentially "historical" papers should also be noted. In a very brief essay, Luis Sanchez Agesta (1970) holds that Ortega's political philosophy derived, in large part, from an organic ("vitalistic"?) conception of man, society and State. Professor Sanchez Agesta is a specialist on Ortega, and it would be sheer sciolism for me either to agree or disagree with him. In a very, very long paper, Aaron Bell (1970) describes the influence of Rudolf Kjellén, author of *The State as a Form of Life* (1919), on Finnish political thought. Here again, I am in *terra incognita* and can say only that Mr. Bell's paper reflects a prodigious amount of scholarly labor.

By far the most ambitious effort in this genre is Peter Corning's "Evolutionary Indicators: Applying the Theory of Evolution to Political Science" (1971b). "This article," the author states in his abstract, "explores some of the conceptual and methodological problems involved in operationalizing an evolutionary-adaptive paradigm in political science; an approach is proposed based on a synthesis between the theory of evolution and the concept of social indicators. The core hypothesis upon which such a paradigm would rest is that the basic and continuing problems of all species, including man, is biological survival (defined ultimately as long-run reproductive efficacy)." "It should be possible," Corning declares, "to develop from an evolutionary paradigm an 'empirical theory' of politics capable of satisfying the requirements of a full-fledged explanatory theory. . . Equally important, we should be able to analyze and explain contemporary political artifacts functionally—that is, in terms of their survival consequences (as ethologists have been doing for many years with animal behavior)" (p. 3). The manner in which these objectives are to be achieved is explained in sections entitled "The Problem of Operationalizing Evolutionary Theory," "Toward a Framework for Evolutionary Macro-Analysis" and "Evolutionary Indicators: Measures of Adaptiveness."

Just as in Corning's other essays, the amount of work is prodigious, the ideas often stimulating and the paper an invaluable bibliographic guide to anyone interested in this subject. Nonetheless, I would have to disagree with him—not so much in terms of his objectives, although I find these a bit grandiose, but in terms of how he proposes to achieve them. Let me illustrate: Corning's paradigm rests, he himself emphasizes, on the notion that biological survival can be defined ultimately as "long-run reproductive efficacy." After repeating this statement almost verbatim in the section entitled "Toward a Framework for Evolutionary Macro-Analysis," he quite rightly asks: "But what does that mean? How does one go about operationalizing this criterion?" (pp. 14-15). Five pages later, the notion of "optimum number" is offered as a solution—but after another half dozen pages Corning concedes that the "concept of the optimum number cannot at present be operationalized in a very rigorous way" (p. 22).

What to do, then, if a paradigm's "core hypothesis" cannot "as yet" be operationalized? Corning's solution is to talk about "what we might be able to do" *if* this problem (and several others) were to be solved. Surely, he has every right to proceed in this fashion and, in our discipline, he keeps distinguished company. My own view, however, is that at such a juncture, efforts would be better invested in trying to make the concept empirically meaningful.

Davies' (1971) discussion of Social Darwinism moves in two directions, one rather familiar, the other not. A retrospective section provides a brief summary of previous efforts to build a social science on Darwinian notions, with particular attention to Spencer, Sumner and Henry Starr Jordon. When Social Darwinism became unfashionable, he observes, so did the search for linkages between social science and biology. In the latter half of his paper, Davies postulates the existence, among mankind, of a "species-wide" desire for equality, suggesting that the frustration of this desire may give rise (via as yet unknown biological mechanisms) to aggressive behavior and thus account for some of our present-day social disorders. This may well be true, but it is relevant to note that the ethological data derived from studies of other forms of animal life by no means unambiguously supports the thesis of an innate desire for "equality." If anything, the weight of the evidence is on the other side.

III. Physiological Aspects

If human behavior is to a significant degree influenced by man's biological make-up, and the ethologists point to our genetic "programming," some other very interesting possibilities arise. One is that changes in human physiological functioning are associated with, or can even be deliberately induced to bring about, concomitant changes in social and political behavior. There is the further possibility that, given the appropriate technology, indices of physiological states (blood pressure, body posture, eye-blink rate, etc.) can be utilized to facilitate political research. These are the two main areas which the physiologically-oriented practitioners of biopolitics have been exploring.

1. *Physiological Influences on Behavior*

Nutrition and Health, Disease, Exercise

Mention has already been made of two papers which identified, respectively, psychopharmacological (Somit, 1968a) and neurophysiological (Davies, 1969) influences. A different line of attack was attempted by Stauffer (1969) who hypothesized a relationship between physiological conditions in developing countries (measured by nutritional level, health rate, etc.) and political attitudes. Stauffer took his point of departure from earlier work by Keys and associates (1950), work called to the attention of political scientists by Davies (1963). (Keys found that systematic starvation, in an experimental situation, produced two clusters of attitudes—apathy and irritability.) On the basis of not entirely complete data, Stauffer concluded that "the rural population of the underdeveloped world faces a recurring annual period of undernutrition that borders on semi-starvation" (p. 369). He was not able, however, directly to link this condition of starvation to political attitudes.

The article explores a number of other parameters—malnutrition, death rates associated with malnutrition, the incidence of parasitic diseases, etc. The findings, at most, are tentative. Conceivably, deficiencies in both quantity and quality of food may be associated with "the willingness passively to accept fate, and other similar attitudes attributed to the people of underdeveloped societies. . ." (p. 377). More important is Stauffer's insistence that students of political development take into consideration biological variables which they have so far ignored in studying the politics of developing nations. This is the sort of article of which it is safe to say that it suggests more than it immediately achieves.

Probing for possible biological causes of internal and external social violence, Haas (1969) speculated *a)* that it is possible empirically to isolate those forms of death due to psychogenic causes, and *b)* that there is a correlation between death rates from such causes and levels of foreign and domestic violence. Data bearing upon 72 states were examined for 1960; death rates by cause were computed; and so were casualty rates arising from violence. Haas found that "there are *no* identifiable psychogenic causes of death cluster, *per se* ; there is *no* significant relationship between deaths due to either internal or external violence and any other cause of death. . . there is no relation between a psychogenic cluster, whatever its composition, and either prior foreign war or civil strife" (pp. 277-278). He suggests, though, that longitudinal analysis based on more refined data might yield more positive results. The essentially negative findings aside, the study represents an interesting attempt to apply highly sophisticated statistical techniques to data of somewhat uncertain quality.

Wiegele's essay (1970) resists both ready classification and succinct summarization. In essence, Wiegele postulates a correlation between physical fitness and the "incidence of tension and anxiety in individuals and groups." The "fact of general physical unfitness in modern, technologically advanced societies" could be an important factor in contemporary "societal tension, turmoil and instability" (p. 9.) Since his supporting evidence is largely conjectural ("rigorous empirical investigation may find that university students are relatively inactive physically," p. 11), Wiegele devotes almost ten pages to a "suggested experiment" which might yield the requisite corroborative data. The paper concludes with the observation that "if the findings from such an experiment would confirm even a small number of the stated hypotheses, then perhaps an important dimension would have been added to the study of political science." Final judgment must be reserved, obviously, until the experimental findings are in hand.

Neuronal Networks

Beck (1970), drawing on "that body of knowledge which has crystallized at the intersection of semiotics, the general theory of signs or messages, and etiology, the biological study of behavior," attempts to explain the present crisis in the American political system by applying a model based on Sokolov (1963) and Wallace (1956). The former, with its concept of a "neuronal model" is essentially physiological in orientation; the latter, with its notion of "mazeway," essentially psychological. Beck argues that an individual seeks a "concordance" between his cognitive map, on the one hand, and the incoming stimuli from his sensory mechanism, on the other. Where there is dissonance between the model of reality and the incoming information, the map will change to accord with the newly perceived reality.

Such a model, Beck feels, might explain both American student behavior in the 1960s and the recent political behavior of American blacks. White college students have moved from a concern with the future and with instrumental values to a greater concern with the present and with consummatory values; among blacks, the opposite has taken place. In both instances, he suggests, the strains on the cognitive map have led to a mazeway reformulation. As Beck himself points out, his data about student movements and about black political behavior are no more than anecdotal. The merits of his theory cannot be tested until some rather severe research problems are resolved.

In a subsequent article which employs much the same model, Beck (1971) suggests that, in designing political institutions, we seek to duplicate the "redundancy" which seems so useful in animal nervous systems. This redundancy (i.e., a system of overlap and duplication which makes possible the "construction of reliable systems from unreliable components") enables the organism "*to attend to other aspects of its environment (other messages), thereby increasing the range of incoming messages to which it can respond*" (p. 472). Or, as Beck later puts it, the identification of redundancies enables the system to respond to still more varied stimuli, to wander, as it were, in ever-increasing circles about the phase-space which defines its universe" (pp. 475-476). This line of reasoning leads him to such questions as: What sorts of institutions are necessary for human behavioral communities? What sorts of redundancies must be provided in these institutions? To what degree do existing institutions further the actualization of their members and the larger community? To answer these questions, Beck concludes, "political science as a vocation and political science as a way of life must come together" (p. 476).

McAlpine (1970) turns to neurophysiology for the concept of the organism as an information processing system. To cope with their problems, societies as well as individuals must develop an "information reduction process"—"a procedure whereby a community or an individual is enabled to render its behavior 'more simple' both with respect to the information which needs to be gathered from the environment in the course of performing an act, and with respect to the processes of decision through which 'responses' within the behavior are determined" (p. 1). So far, so good—but he then introduces the notion of "controls" and, after three readings, I am still unable to decide whether he has in mind the control of the environment (i.e., response to problems), the control of informational flow, some mixture of the two, or something else. In any event, McAlpine next suggests that judicial systems may perform the function of "information reduction." At this juncture, I concluded that my capacity to summarize his paper was not equal to the challenge and that I should let the author speak for himself:

> The main focus in this paper, however has not been on the role of a constraint on control as a parameter in decisions affecting the design of central institutions of a society but on broad "states of control" which may appear within a dynamic process of development, and particularly the possibility of "states" characterized by "demands" on information processing for control which are so severe that performance begins to deteriorate. The activities of government may conceivably both exacerbate and moderate such conditions. Where a constraint on the capability for control is manifest the relevant question might be "What general processes are available in society which tend to reduce the demand for information processing and control?" The approach taken within this paper has been to ask the question "What basic biological information reduction processes might serve as models and thus as aids in approaching more complicated processes of society?" (p. 2 of Summary).

Pubertal Age

Ferguson, Ferguson and Boutourline-Young (1971) report on a study of 300 young men whose grandparents were originally inhabitants of southern Italy. At the time of the study, one-third of the subjects lived in Palermo, another one-third in Rome and the rest in Boston. Questionnaire responses yielded a correlation between age at pubertal maturity and political attitudes. Late maturers were "less likely than their early and middle-maturing peers to report they had a strong interest in politics, that they frequently became excited or indignant as a result of political events, that they generally kept up with what's going on in politics." In addition, "late maturers were also less likely than their peers to be high in the sense of political efficacy" (p. 3). Since the findings are reported in very tentative fashion, no critical comment is really necessary. I should mention, however, that a distinguished French biologist has recently advanced a not too dissimilar hypothesis to explain recent manifestations of student unrest.[21]

Drugs and Politics

Criticizing the "near total lack of concern by political scientists with the possible political consequences of drug use" (p. 2), Stauffer (1970) seeks to "convince political scientists that drugs constitute a possible source of influence on politics sufficiently demonstrable to warrant serious study" (p. 3). The article provides both a commentary on the use of drugs (the term is used to include alcohol) and a survey of recent research describing the effects of drugs on individual behavior. The title of the concluding section "Drugs and Political Change" is not altogether accurate since the discussion, after dealing with "Drug-Centered Cults as Factors in Change," centers on the difficulties governments encounter in trying to arrive at a rational "drug" policy (national and international). While Stauffer offers no solution to this latter problem, he does insist that this is a matter to which political scientists should address themselves. For those who find his argument compelling, the paper provides a useful guide to the literature.

A quite different approach to the impact of drugs on political behavior was taken by Jaros (1970). Political socialization, he observes, has been commonly regarded as something accomplished by "learning and social conditioning." But—"can political behavior, particularly that relevant to support for regimes, be explained in terms of the pharmacological action of substances in the human body ?" (p. 4). On the assumption that the answer is in the affirmative, four hypotheses are advanced : *a)* that "depressants (by decreasing discriminative ability) should modify the substantive nature of political choices made by citizens" (p. 8); *b)* that "depressants should increase the quantity of political responses which individuals manifest" (p. 10); *c)* that "drugs induce dependency which results in increased probability of manipulations of political acts" (p. 11); and *d)* that "ingestion of depressants is associated with erosion of consensus on the dimensions of conflict and thus with an increase in the frequency of preference orderings which do not satisfy a common dimension in collective decision-making situations" (p. 13).

To test these hypotheses, Jaros devised an ingenious set of questionnaires and an elaborate five-stage experiment which involved responses once before and twice after the administration of a drug (pento-barbital). Subjects, including both a control group and an experimental group, were seventy-seven volunteer University of Kentucky undergraduates. The control group received placebos; the experimental group either a dosage of 100 mg or 60 mg in the usual "double-blind" setting. The resulting data ranged from "modest relationships" to "some directly contrary findings."

Jaro's hypotheses are by no means self-evident and the reasoning by which he arrived at some of them may not be altogether persuasive; he deals, furthermore, not with "real behavior" but with a contrived situation, and his results can only be described as inconclusive. Nonetheless, the article is worth careful study. The research design shows extraordinary inventiveness and imagination; Jaros has not been satisfied simply to talk about possible relationships between drugs and political behavior but has actually devised and carried out a means of putting his ideas to experimental test. I can only endorse his hope that these "beginning findings will be preliminary steps to more elaborate investigation."

Pain

Tursky and Lodge (1971) report an attempt to influence behavior (i.e., responses on attitudinal scales) by physiological means. Some fifty subjects initially responded to a mail questionnaire. About two weeks later the questionnaire, or selected elements of it, were re-administered in a "base trial," "experimental trial," a "post trial" and a "post questionnaire." During the base trial and the experimental trial, subjects received electric shocks of varying degrees of intensity, depending on the nature of their responses. The objective, sadism aside, was that of linking the subject's "verbal response to a direct consequence by systematically punishing disagreement to a series of survey-type political statements" (p. 2).[22]

Subjects reacted in three basic fashions: some changed their responses during shock trials ("compliance"); others maintained their original responses ("resistance"); a third group voiced even stronger commitments, in full knowledge that such a response invited painful shock ("rebellion"). "The whys and wherefores for these attitudinal differences," the author admits, "remain obscure" (p. 19). The most that can be said is that this is an intriguing exploratory effort to link verbal responses with "operational consequences."

Other Approaches

i. Charisma. Hummel (1970) looks for a "bio-social model" by which "crisis charisma" can be explained. He defines "crisis charisma" as:

> an unusual or abnormal social relationship in which "charismanics"—those who suffer from an anxiety-producing tension caused by the insufficiency of existing ways of acting and thinking in the world—seek a reordering of their relation with the world through the proffered offices of an extraordinary man, or group of men, viewed as endowed with extraordinary powers and typically claiming monopolistic access to a vision of a new order. Such leaders are here called "charismats" (p. 5).

After a very lengthy and somewhat unfocused discussion, several "propositions" are advanced. Among these are : *a)* "the pre-charismatic crisis is the result of an attack through social change on fundamental human biological tendencies" (p. 36); *b)* the "pre-charismatic crisis is caused by stimulus overload in the cortical centers of the brain" (p. 45); *c)* "aggression is likely to precede charisma formation" (p. 53); and *d)* "the crisis is composed of two types of internal crisis behavior which are distinctive: cortical failure and affective orientation" (p. 53). Taking each of these in turn, Hummel *either* identifies the available evidence which seems to support the proposition, *or* suggests ways in which the proposition might be tested. Since the argument leans more heavily on the latter than the former, little can presently be said about the merits of the propositions.

ii. Symbiosis. Wenner (1970) explores the possibility that the concept of symbiosis might be applied to the study of political behavior. After identifying eight types of symbiotic relationships, Wenner observes that "although there may be some disagreement among scholars as to the precise nature of certain close interspecies relations, there is more than sufficient evidence that symbiotic associations are essential to the existence and welfare of a substantial number of species" (p. 14). Is man one of these species? As might be expected, the answer is in the affirmative. Offered as examples are the relationships between the United Automobile Workers and the Automobile Manufacturers Association; the Federal regulatory agencies and the industries they regulate; the Arab nations and Israel; the Soviet Union and the United States, etc. The author correctly observes, however, that "before one can conclusively claim that the application of the concept of symbiosis to the field of human social, and particularly political behavior is more than a heuristic device, it would seem necessary to show that it provides a theoretically meaningful alternative conceptual scheme for the analysis of specific political problems" (p. 23). Even more correctly, he concludes, "the present paper does not claim to have fully accomplished that purpose. . . only further research. . . will prove whether or not our suggestions contribute to our understanding of Man" (pp. 23-24).

iii. "Hair." Easily the most "offbeat" piece in the biopolitical literature is Mazrui's "Political Man and the Heritage of Hair: Some African Perspectives" (1970). The opening pages quickly sketch Darwin's theory of natural selection ("a gem of an idea"), subsequent attempts to account for physical differences among the "races" (hair, skin color, skull shape, etc.) and the difficulties which arose in using these characteristics to decide whether whites or blacks were more "reminiscent of the ape." Subsequent sections deal with "Hair and Politics of Beauty," of relationships between the rise of black nationalism and the increased aesthetic value placed by blacks on "Afro hair style," with "Hair and Modes of Protest" (long hair and beards as marks of uniqueness, manliness—and social protest), and "Hair and Family Ritual" (the shaping, cutting or shaving of hair to mark the transition from one role or status to another). I think this can profitably be regarded as "recommended reading."

2. *Physiological Measures of Political Attitudes and Behavior*

Schwartz (1970) holds that a feeling of energy level is both "an important variable, intervening between attitudes (or personality) and political behavior". . . "an important intervening variable between physiological factors and behavior" (p. 10). Drawing in part on Lucian Pye (1962) and in part on the psychological concept of "energy-levels," he hypothesized "that people who perceived themselves as comparatively high in energy level would tend significantly to adopt active behavioral orientations to politics (e.g., nonconformity, reformism and revolutionism)—probably because they saw themselves as physically capable of playing a rather active role in politics, whereas the perception of self as relatively low in energy-level should be significantly associated with more passive forms of political orientation (e.g., conformity and ritualism)" (p. 11). Data were compiled via questionnaires mailed to a random sample of students and faculty at the University of Pennsylvania. The results, in Schwart's phrase, "appear to provide some support for these hypotheses." The notion of "energy" is, of course, an interesting one, but it will take harder evidence than Schwartz has presented to make it more than that.

In a more recent (and much longer) paper, Schwartz and Zill (1971) argue that political participation can be conceptualized as the result of the interplay between "*a)* the individual's overall arousal level (i.e., that level of general psychophysiological activism, excitation and energy mobilization which typically characterized the individual) and *b)* the situational variations in arousal level which are associated with specific stimuli" (p. 2).

From this premise, several hypotheses were derived as to the relationship between "general" and "situational" arousal levels, on the one hand, and subsequent political participation, on the other. Two sets of psychophysical indicators were used to measure arousal levels. One was a Postural Arousal Index Form (based on Mehrabian, 1969) on which was reported the subject's overall posture, neck tension, arm position, hand tension, leg position, overall tension rating and a rating of gross bodily movement. The second set of indicators dealt with tape recorded speech variables—total words, reaction time, articulation rate, "ah" ratio, speech disturbance patterns, etc. Forty-some professors at the University of Pennsylvania served as subjects and were administered a questionnaire designed to plumb a variety of political topics. The resulting correlations suggested that the postural index was more useful as a measure of general arousal level, and that the speech measures were more useful as indicators for situational arousal, although the authors stress the extremely tentative nature of their data. More important than the specific findings, I would say, is the attempt to apply the techniques developed by students of what is generally called "non-verbal behavior." The article provides, *inter alia,* a useful guide to what, for most political scientists, is probably an unfamiliar literature.

Noting, *sans* enthusiasm, that political scientists have increasingly relied on survey data, Lodge and Wahlke (1971) argue that survey research depends upon "verbal self-report" which is "both conceptually and empirically untenable." Badly needed, Lodge and Wahlke maintain, is a technique which provides an independent physiological measure of the *affect* attached by a subject to any particular response or set of responses. The article describes their efforts to explore the correspondence between verbal self-report of political attitudes and two physiological measures—heart rate and carotid pulse pressure.[23]

Some forty volunteer female subjects were shown a series of slides, each containing "political stimulus material," and their verbal responses and physiological reactions recorded. The subjects were then divided into an experimental and a control group. Each of the two groups was exposed to somewhat different sets of slides and the reactions again measured. For the group as a whole, and the responses as a whole, the correlations between the verbal and physiological measures for each stimulus were low "and seldom statistically significant." Statistically significant differences did emerge, however, when the women were analyzed in terms of general socioeconomic background.

The actual experimental findings ("all. . . tentative, some tenuous") are not impressive, although the authors do feel that "statistically significant rank-order correlations indicate that verbal and physiological indicators are structurally related" (p. 29). Lodge and Wahlke, nonetheless, have raised an important issue, and it would be unrealistic to expect an adequate solution immediately to be forthcoming. Posing the issue constitutes, in itself, a sizable step forward.

IV. Public Policy Issues

By implication, almost every one of the papers discussed so far raises questions of public policy since, to the degree that political behavior is influenced by biological factors, there exists the possibility that these factors might be manipulated to achieve social or political objectives. In more explicit fashion, a number of the items mentioned above touch, if sometimes very briefly, on the public policy implications of biopolitics (Adrian, 1969; Somit, 1968; Corning, 1971b; Davies, 1969, 1971; McAlpine, 1970) and, it may be recalled, Stauffer (1970) urged not only that political scientists study the impact of drugs on political behavior but also that they join in the effort to formulate a rational governmental policy toward drugs and drug users.

Beyond these, there is a small handful of items primarily, if not exclusively, focused on policy issues. Pride of place goes to Caldwell (1964), as well as credit for the first utilization of the term "biopolitics" in the sense in which it has since been utilized. "An explosion of biological knowledge and technology," Caldwell remarks, "is raising questions of public policy which until recently were hypothetic, and were therefore from a practical point of view unreal" (p. 2). These no longer unreal questions are of two sorts: those which are environmental (pollution, radio-activity, etc.) and those which relate to "individual human behavior in the use of cigarettes, tranquillizers, narcotics, and alcohol. . . the biochemical control of personality. . . social concern for the numbers and qualities of future population . . . [and] the issue of biological warfare" (p. 6). If they are to be solved, he argues, the natural and the social sciences will have to be brought into more productive communication; there must be a profound change in man's perception of his relationship to nature; there must be a better popular understanding of the true meaning of science; and there must be a vigorous leadership "toward a policy synthesis of scientific knowledge and ethical values." In short, there will have to emerge a *biopolitics* capable of achieving these several goals. If anything, Caldwell tends to understate the gravity and the urgency of the issues to be resolved, although I would heartily concur in his concluding sentence that the creation of a biopolitics equal to all of these tasks "requires an extraordinary fusion of understanding, audacity, and humility" (p. 16).

Although written from a different perspective, a paper by Lepawsky (1967) requires inclusion in this section. Lepawsky is concerned about the "estrangement between medicine and politics"—an estrangement which "has serious scientific as well as professional consequences" (p. 905). Noting the rise of interest in medicine on the part of our sister disciplines (sociology, economics, psychology and anthropology), Lepawsky urges a parallel development in political science. Medicine and political science have numerous "common interdisciplinary problems;" research and reform are closely interlined in both disciplines; and

> Above all, medicine and politics represent, within their parent fields of the biological and the social sciences, two interdisciplines or supra-disciplines which are centrally responsible for dealing with some of the most strategic "sore" spots of human society—medicine for integrally protecting the health of the individual, and politics as the essential sovereign arbiter among conflicting power groups in a pluralistic society (p. 914).

In essence, Lepawsky argues for the creation of what he calls "medical politics"—an "interdiscipline" which, he agrees, might require "a massive reordering of the whole structure of contemporary knowledge." Although the article is hardly in the mainstream of biopolitics, it may be of particular interest to political scientists concerned with national health policy and the administration of public health organizations.

I have no option here but to mention another paper of my own (Somit, 1970). Advances in biology, quickly to summarize, will soon make available two technologies. Of these, one "will eventually make it possible to manipulate man's genetic make-up;" the other, "already present in rudimentary form. . . will make it possible to influence, if not actually control, the social and political beliefs of individuals, of groups, and conceivably of large masses of persons" (p. 1). The paper briefly examines the kinds of social and political problems posed by recent advances in genetics and psychopharmacology; takes a rather bleak view of the optimistic hope that "the same science which spawns this troublesome knowledge may also provide us with some means of preventing its misuse" (p. 9); admits that the problems posed "are problems for which we today have no answers;" concedes that "in fact, the foregoing analysis notwithstanding, it may be too early to be entirely sure that we know the right questions to ask;" but nonetheless insists that "it may not be too early, though, to start thinking about what these questions might be" (p. 9). The paper has, if nothing else, the merit of brevity.

One more item should be mentioned, an informal "paper"[24] by Petchesky (1971) in which she reviews the several areas in which "biological engineering" might be used to control social behavior (i.e., eugenics, mind control, etc.), examines some rather pessimistic "scenarios" for the directions in which society might move, and raises the question of whether more attractive alternatives might be found. The bibliography provided is quite thin and is almost devoid of references to the political science, let alone biopolitical, literature.

V. A Look Ahead

If past experience is any guide, "new "approaches and constructs in political science have a life expectancy of perhaps a decade.[25] Whether or not biopolitics will "catch on" remains to be seen; a number of seemingly excellent ideas die aborning or in early infancy. If it does survive, it will have to yield something substantial in the way of results within the next seven or eight years. To this point, I have attempted to describe where biopolitics stands at what is still the beginning of the journey. In the remaining few pages, I will hazard (and the verb is appropriate) some comment as to where this road might take us and which directions, to continue the metaphor, might profitably be pursued.

1. *The Case for a Biologically-Oriented Political Science*

There have already been written at least a half-dozen papers discussing "the possible implications of biology for politics." Such papers are necessary to launch a new idea, but their utility is soon exhausted. After the first couple of years, the real task is to demonstrate that the approach is truly a productive one; continued speculation about "possible implications" becomes less and less persuasive. I would hope that the biopoliticians will declare a moratorium on this kind of writing until they (we) have something substantive to report. The pressure to publish being what it is, the hope will likely be disappointed.

2. *Ethology and Politics*

What could be more exciting and tempting to a political scientist than the prospect of being able to answer the ultimate question—what is the nature of political man? Ethology holds out this promise but, I fear, the promise will be a long time in fulfillment. The immediate benefits to political science from ethology, I would predict, are likely to take quite another form. Let me expand briefly upon these judgments.

There are several reasons why ethology is not likely to add greatly to our understanding of political man *in the short run*. First, ethology itself is in a state of considerable intellectual disarray, with its practitioners often in disagreement both about the meaning of their concepts and the validity of the data being advanced to support conclusions on a number of important issues. They do not agree upon the meaning, for example, of so critical a term as "innate."[26] Until these and other large issues are resolved, political ethology simply will not have a sufficiently solid scientific base on which to build. Second, political scientists do not, as yet, have the technical competence to work in an area where the experts themselves disagree. This competence cannot be acquired by reading Lorenz, Ardrey, Tiger or Morris—it requires hard, intensive training. None of the "first generation" of biopoliticians originally had this competence. Some are now painfully acquiring it. Until this shortcoming has been corrected, there remains the risk of "describing the behavior of one group of organisms in language appropriate to another." Third, for all that has been accomplished in the study of animal behavior over the last twenty years, ethology is still compelled to argue from *analogy* when dealing with humans. Ethology has yet to demonstrate that these analogies are scientifically legitimate and/or to devise some experimental technique which can avoid this difficulty.

Let me hasten to add that even in the short run, political science may nonetheless benefit from ethology in two important respects. Ethology has made great advances by looking at actual animal behavior. This may stimulate political scientists, or at least some of them, to return to the study of actual political *behavior*. One of the great ironies of contemporary political science is that so many of those marching under the banner of "behavioralism" have turned their backs upon behavior itself. Political science may derive some profit by applying ethological concepts or even some ethological findings. It will profit to a much greater degree, I believe, if the ethological inspiration reorients us to the study of what people do, and how they behave, in day-to-day political life.

Hardly less important is the fact that the ethologists are forcing us to consider (or reconsider) the possibility that *some* political phenomena are due in *some* measure to our genetic programming. To the extent that this is accomplished, students of bureaucracy, of international affairs, of political socialization—in fact, of all aspects of political behavior—will begin to view these phenomena in a different light. The great achievement of Karl Marx, in the final analysis, was that he compelled non-Marxist social scientists to consider the possibility that economic and material factors influence political behavior. Ethology will have performed a similar service if it forces upon us the same "open-mindedness" with regard to biological factors.

3. *Physiological Factors in Political Behavior*

Two areas have been explored so far. One looks at the influence of physiological and biological factors on political behavior; the other seeks to use measurements of physical state as clues to either political attitudes, the emotional intensity (affect) actually attached to the expression of political attitudes, or as indications (clues) to likely political behavior.

The possibility that malnutrition, disease, crowding, noise, drugs, fatigue or any of the half-dozen other physiological influences have an impact upon political behavior opens a broad expanse for future investigation. Initially, most of this research will be attempted within a laboratory setting; eventually, some means will have to be found of moving out into the real world. In either case, the results attained will hinge on the willingness of political scientists working in these areas to acquire a technical competence which they presently lack. In theory, the need could be met by having political scientists join forces with physiologists and biologists; in actuality, if this kind of collaborative research is going to be truly fruitful, the political scientists will need a solid foundation in biology. (Two or three American political science departments, I should mention, are already moving to provide this kind of graduate training.) Still, no matter how this research is done, there remain some massive methodological problems to be solved if biopolitics is to go beyond the kind of "correlational" studies which have already been attempted.

Perhaps the line of attack which holds out the brightest immediate promise is the quest for physiological indices which will afford an independent measure of "verbal self-report." Such a tool is urgently needed if survey research data, beyond simple expressions of preference, are to be taken seriously. But to perfect such techniques for laboratory application is one thing; to refine them for field use is quite another. Perhaps those working in "non-verbal behavior" will be able to demonstrate that eye-blink rate, body posture, gesture mannerisms, etc., provide reliable clues to the validity of survey responses, the sincerity of political orators, or even the likely voting behavior of a congressman on an immediately forthcoming roll call.

4. *Public Policy Implications*

Laymen undoubtedly assume that political scientists are intensely interested in questions of public policy. Those within the discipline are aware, however, that many of their fellow practitioners have, in effect, abjured policy questions. If biopolitics establishes itself as a significant aspect of political science, such a development would almost automatically entail a great concern with policy questions.[27] Almost every aspect of biopolitics, as we have seen, has policy implications. The great issues already upon us are largely biological in nature—pollution, atomic and biological warfare, population control, drugs and, if some of the ethologists are correct, so are race relations and social violence. Looming on the horizon are two more extraordinarily troublesome issues: one, public policy in the field of eugenics; the other, the need somehow to regulate the use of the "mind-controlling "drugs and techniques which an advancing science will soon make available.

Few of these issues, it will be observed, are purely scientific; few, purely political. Biopoliticians, assuming the breed survives, should be able to make a twofold contribution: possessing at least a basic grasp of the biological issues involved, they may be able to play a helpful role in the formulation of policy; in other areas, they may be most useful in devising ways whereby the desired scientific objectives can be made politically acceptable. Eugenics offers an instance of the former; pollution policy of the latter.

Finally, and this should be particularly important in countries with democratic governments, biopolitics can contribute significantly to the formulation of public policy by improving and refining the ways whereby public opinion is ascertained. Present-day survey techniques are useful, but they also have serious limitations. Should biopolitics yield nothing more in the long run than a sensitive and reliable technique for sounding public opinion on complex as well as simple issues, it will have served the discipline well.

To be sure, this is a far cry from, and a much more modest aspiration than, many of the pronouncements found in some of the biopolitical writings. A firmer grasp on the real nature of human nature could ultimately bring us to the point where we can deal more successfully than heretofore with the perennial problem of international war; conceivably, it could suggest more effective means of reducing to a more tolerable level the degree of internal violence which now seems endemic within twentieth-century society; even more remotely, in time it could lead to that restructuring of the state and reordering of society which has been the dream of utopian theorists from Plato on. But all of these possibilities still lie far in the future. It will be far better for biopolitics if it eschews, at least for the present, these larger objectives for more modest and, hopefully, more attainable goals. Browning's plea that "a man's reach should exceed his grasp" may be sound sermonizing, but it is dubious research strategy.

Notes

1. John of Salisbury probably offers the most familiar instance of this usage, but other examples come almost as readily to mind. For a brief summary, see Sorokin (1928), pp. 197-207.
2. Two of the most distinguished American political scientists may be cited as cases in point. Woodrow Wilson (1908) declared that government ". . . is not a machine, but a living thing. It falls, not under the theory of the universe, but under the theory of organic life. It is accountable to Darwin, not to Newton. It is modified by its environment . . . shaped to its functions by the sheer pressure of life." And the American Political Science Association Presidential address of A. Lawrence Lowell (1909) was entitled "The Physiology of Politics." Similar examples, no doubt, abound in the British political science literature.
3. Donald G. MacRae has a useful essay on "Darwinism and the Social Sciences" in Barnett (1958). For another interpretation, see Banton (1961).
4. One was to group the articles by the traditional "fields" of political science; another to organize them according to biological "concept." Neither scheme, whatever its abstract merits, was very fruitful.
5. As might be expected, psychologists were closely attuned (McClearn, 1962, 1969 and Hebb, 1966); for anthropology, see Count (1958) and Callan (1970); for sociology, see Means (1967), Wrong and Gracey (1967) and Alland (1969). For a cross-disciplinary assessment (i.e., sociology and history), see Glass (1968). Not all forays were equally felicitous. Thus, Hediger (1969) felt free to write that "the most biological, that is the most natural, way of dwelling is without doubt to live in a self-contained house surrounded by a garden" (p. 532).
6. In Farber and Wilson (1961).
7. Mention should be made here of a paper delivered by John Wahlke (1968). Although the paper was concerned with the need for a more theoretically-based approach to the study of political behavior, Wahlke urged that, for the elements of such a theory, political science "must look beyond its own boundaries, particularly into such new fields of inquiry as biology and ethology" (p. 40).
8. The earlier version is, in some respects, the more interesting of the two but, since it is not likely to be available to the reader, I will limit my comments to the article published.
9. The shortcoming may be mine rather than Thorson's, although I find this explanation unattractive on intuitive, as well as empirical, grounds.
10. Hummel declares, for instance, that "Corning's major insight of profound philosophical consequence is that arbitrary philosophies and policies have led Man right up against the edge of possible existence: ideas do have limits and one of these limits is the destruction of brains that bear ideas" (p. 4). Equal restraint is manifested in describing the inconclusive Lodge / Wahlke experiment (see pp. 312-313 below) as having "possible monumental impact on the future use of attitudinal surveys" (p. 18).
11. Thus, his concluding passage : "If the social consequences of the apparent revolution in the life sciences are not predictable, the only basis on which we can now press for the education of generalists broadly at home in the life sciences and the social and political sciences is that of our capacity to predict the consequences of not developing a unified science of Man."
12. As the ethologist sees it, those species survive best which are best fit for their particular environment. Fitness is defined, in turn, as the ability to leave behind a "satisfactory" number of offspring under these particular environmental conditions. Slight variations in structural and/or behavioral characteristics are always present in any biological population and some of these are genetically based, i.e., they can be passed along to the organism's offspring. When such variations contribute to the species' fitness, they will appear increasingly in successive generations until they are literally bred into the entire species. The causes of these initial variations, it is relevant to add, are just now beginning to be adequately understood.
13. For a convenient summary of the essentially "anti-ethological" viewpoint, see Montagu (1968).

14. These differ from the papers mentioned in Section I in that the latter surveyed the broad field of *biology*, whereas those to which we now turn deal *only* with ethology.
15. Arthur Koestler (1968) arrives at the same conclusion by a somewhat different route.
16. In Ardrey's graphic phrase, "man is a predator whose natural instinct is to kill with a weapon" (1966, p. 316).
17. But by no means all of them. Gurr (1968) treats violence as essentially sociological, rather than biological, in origin; Marty (1971) discusses violence and non-violence without touching on the possibility that it may spring from man's "nature;" and Brodie (1971) manages almost the same thing in a discussion of "Theories on the Cause of War."
18. According to Keith (1948) the "territorial sense—a conscious ownership of the homeland, one charged with a deep emotion—is highly important factor in human evolution" (p. 5).
19. E. T. Hall (1959) has called attention to the apparent human need, observed in normal inter-personal relations, to maintain a given distance from other persons, and the manifest discomfort evidenced when others move either too close or too far. The space "needed" seems to vary from culture to culture and Hall treats the phenomenon as essentially socially derived rather than innate. It may be, of course, that the need for distance is innate but the manner in which the need manifests itself is a function of nurture, rather than nature.
20. See V. C. Wynne-Edwards (1961).
21. See Debré (1969).
22. A second objective was that of ascertaining "the correspondence between verbal and physiological measures of attitude during different coping [i.e., coping with electric shock] behaviors" (p. 3). For this purpose, a number of physiological measures (pulse, blood volume, heart rate, skin potential response, etc.) were taken in conjunction with verbal responses during the base and experimental trials. Regrettably, these latter data were not available at the time the paper was written.
23. There was also a second objective, that of determining "the direction and intensity of attitude change in response to the experimental manipulation of perceived threats to the polity" (p. 1). Since the latter objective is not directly relevant to this article—and the results were inconclusive—I will not deal with it further in this summary.
24. This was a summary of a doctoral dissertation, now nearing conclusion, delivered at the American Political Science Association convention in September, 1971. No formal paper was available; instead the author distributed a four-page outline plus a list of recommended readings.
25. The "group" approach to politics, "decision-making" theory, and "systems" theory (to name the most recent examples) took roughly this length of time to be proposed, popularized, tried and found wanting. In each case, much the same pattern can be observed: an idea or approach is suggested by one or two persons who thereby earn for themselves immediate professional fame, if not actual immortality. Next, a sizable part of the profession leaps onto the band-wagon and the literature burgeons almost overnight. Then come the critics, a few at first, but in increasing number and in louder voice thereafter. The penultimate phase sees the original progenitors arguing that they were really misunderstood and the erstwhile disciples insisting that they had their doubts all along. And, finally, the idea is quietly interred—a ceremony which tends to coincide with the birth of another bright "new" theory.
26. For some graphic illustrations see, for instance, Aronson (1970).
27. As this manuscript was about to be mailed, I received the abstract of a paper (Wenner, 1971) just delivered at the American Political Science Conference in Chicago, Illinois. The relevant sections of the abstract read as follows: "Abstract: Biological and Environmental Factors as Political Variables. After emphasizing that the essay does not seek to propose a concrete list of biological and environmental variables which may be substituted for, or added to, those lists of variables which political scientists already utilize for their studies of human social and political behavior, the paper suggests that: (1) there are biological and environmental factors which may be synergistic with the widely recognized variables which political scientists consider useful; and (2) many such biological and environmental factors may be highly important to the policy process and the kinds of decisions which have been, are being, or will be made. Material offered under (1) above is primarily a defense, unfortunately still necessary, of the thesis that a greater knowledge of the findings of the biological sciences may materially contribute to our knowledge of the sources and patterns of human social and political behavior. The primary emphasis of the paper, however, is a defense of the second contention (above)."

Bibliography

a) Biopolitical Literature

Adrian, C.
1969 "Implications for Political Science and Public Policy of Recent Ethological Research," Paper read at the Second International Sinological Conference, The China Academy, Taipei, Taiwan.
1970 "Ethology and Bureaucracy," Paper read at the 8th World Congress of the International Political Science Association, Munich (hereinafter cited as "IPSA").

Beck, H.
1970 "Politics and the Life Sciences : Notes Toward a Theory of Biobehavioral Ecology," IPSA.
1971 "The Rationality of Reduncancy : Neurocybernetic Foudations for a Theory of Biopolitics," *Comparative Political Studies* 3, pp. 469-476.

Bell, A.
1970 "The Reception in Finland of the Organismic Doctrine of Rudolf Kjellén," IPSA.

Caldwell, L.
1964 "Biopolitics: Science, Ethics and Public Policy," *The Yale Review* 54, pp. 1-16.

Corning, P.
1971a "The Biological Bases of Behavior and Some Implications for Political Science," *World Politics* 23, pp. 321-370.
1971b "Evolutionary Indicators: Applying the Theory of Evolution to Political Science," Revised version of a paper presented at IPSA and entitled "The Problem of Applying Darwinian Evolution to Political Science."
1971c "An Evolutionary-Adaptive Theory of Aggression," Paper read at the annual meeting of the American Political Science Association, Chicago, Ill. (hereinafter cited as "APSA").

Davies, J.
1969 "The Psychobiology of Political Behavior: Some Provocative Developments," Paper read at the annual meeting of the Western Political Science Association, Honolulu, Hawaii.
1970 "Violence and Aggression: Innate or Not ?," *The Western Political Quarterly* 23, pp. 611-623. The original of this paper was delivered as the presidential address at the Western Political Association convention, Sacramento, Calif.
1971 "Biology, Darwinism, and Political Science: Some New and Old Frontiers," APSA.

Ferguson, L. / Ferguson, L. / Boutourline-Young, H.
1970 "An Attempt to Correlate Rate of Physical Maturation with Attitudes Toward Politics," IPSA.

Haas, M.
1969 "Toward the Study of Biopolitics: A Cross-Sectional Analysis of Mortality Rates," *Behavioral Science* 14, pp. 257-280.

Halliday, R. J.
1970 "Social Darwinism," *Victorian Studies,* June 1971. The original of this paper was read at IPSA.

Hummel, R.
1970a " A Case for Bio-Social Model of Charisma," IPSA.
1970b "Teaching Political Theory: The Impact of Biopolitics," APSA.

Jaros, D.
1970 "Biochemical Desocialization: Depressants and Political Behavior," IPSA. *(Midwest Journal of Political Science,* February 1972.)

Lepawsky, A.
1967 "Medical Science and Political Science," *The Journal of Medical Education* 42, pp. 896-908.

Masters, R.
1967 "La Redécouverte de la Nature Humaine," *Critique* 245, pp. 857-876.

Mazrui, A.
1968 "From Social Darwinism to Current Theories of Modernization," *World Politics* 21, pp. 69-83.
1970 "Political Man and the Heritage of Hair: Some African Perspectives," IPSA. *(British Journal of Political Science* 2, 1972, pp. 1-20.)

McAlpine, W. E.
1970 "Information Reduction Processes and Politics," IPSA.

Petchesky, R.
1971 "Biological Engineering as a Social Control Device: Implications for Political and Social Theory," APSA.

Pranger, R.
1967 "Ethology and Politics: The Work of Konrad Lorenz," Paper read at the Southern Political Science Association meeting, New Orleans, La.

Sanchez Agesta, L.
1970 "Biology, Politics and Society in Ortega y Gasset," IPSA.

Schwartz, D.
1970 "Perception of Personal Energy and the Adoption of Basic Behavioral Orientations to Politics," IPSA.

Schwartz, D. / Zill, N.
1971 "Psychophysiological Arousal as a Predictor of Political Participation," APSA.

Singer, D. / Luterbracher, U.
1970 "Crowding and Combat in Animal and Human Societies: The European State System, 1816-1965," IPSA.

Somit, A.
1968 "Toward a More Biologically-Oriented Political Science: Ethology and Psychopharmacology," *Midwest Journal of Political Science* 12, pp. 550-567.
1970 "The Political, Philosophical and Legal Problems Posed by Two Emerging Technologies," Paper read at the International Future Research Conference, Kyoto, Japan.

Stauffer, R.
1969 "The Biopolitics of Underdevelopment," *Comparative Political Studies* 2, pp. 361-387.
1970 "The Role of Drugs in Political Change," IPSA.

Stephens, J.
1970 "Some Questions About a More Biologically Oriented Political Science, "*Midwest Journal of Political Science* 14, pp. 687-707.

Thorson, T.
1970a *Biopolitics,* New York, Holt, Rinehart and Winston.
1970b "The Biological Foundations of Political Science: Reflections on the Post-Behavioral Era," IPSA.

Tursky, B. / Lodge, M.
1971 "Compliance, Resistance, and Rebellion: The Strength of Political Convictions Under Stress," APSA.

Wahlke, J. / Lodge, M.
1971 "Psychophysiological Measures of Change in Political Attitudes," Revised version of a paper read at IPSA and entitled "Verbal and Psychophysiological Measures of Political Attitudes."

Wenner, M.
1970 "Symbiosis and Politics," IPSA.
1971 "Abstract: Biological and Environmental Factors as Political Variables," APSA.

Wiegele, T.
1971 "Toward a Psychophysiological Variable in Conflict Theory," *Experimental Study of Politics* 1, pp. 51-81. The original of this paper was delivered at IPSA.

Wilhoite, F., Jr.
1971 "Ethology and the Tradition of Political Thought," *The Journal of Politics* 33, pp. 615-641.

b) Related Items

Alland, A.
1969 "Darwinian Sociology Without Social Darwinism?," *Social Research* 36, pp. 549-561.

Ardrey, R.
1961 *African Genesis,* New York, Atheneum.
1966 *The Territorial Imperative,* New York, Atheneum.
1970 *Social Contract,* New York, Atheneum.

Aronson, L. / Tobach, E. / Lehrman, D. / Rosenblatt, J. (eds.)
1970 *Development and Evolution of Behavior,* San Francisco, Calif., Freeman.

Banton, M. (ed.)
1961 *Darwinism and the Study of Society,* London, Tavistock.

Barnett, S. A.
1958 *A Century of Darwin,* Cambridge, Mass., Harvard University Press.

Brodie, B.
1971 "Theories on the Causes of War," in Maurice N. Walsh (ed.), *War and the Human Race,* Amsterdam, Elsevier.

Calhoun, J. B.
1962 "Population Density and Social Pathology," *Scientific American* 206, pp. 139-148.

Callan, H.
1970 *Ethology and Society: Towards an Anthropolitical View,* Oxford Clarendon Press.

Count, E.
1958 "The Biological Basis of Human Sociality," *American Anthropologist* 60, pp. 1049-1085.

Davies. J.
1963 *Human Nature in Politics,* New York, Wiley.

Debré, R.
1969 "La Biologie Aide-t-elle à Comprendre la Jeunesse Révoltée?," *Revue de Paris,* December.

Downs, A.
1967 *Inside Bureaucracy,* Boston, Mass., Little, Brown.

Dubos, R.
1968 *So Human an Animal,* New York, Charles Scribner's Sons.

Farber, S. M. / Wilson, R. H. L. (eds.)
1961 *Control of the Mind,* New York, McGraw-Hill.

Glass, D. (ed.)
1968 *Genetics,* New York, Russel Sage.

Gur, T.
1968 "A Causal Model of Civil Strife: A Comparative Analysis Using New Indices," *American Political Science Review* 62, pp. 1104-1124.

Hall, E. T.
1959 *The Silent Language,* Connecticut, Fawcett.

Hebb, D.
1966 *A Textbook of Psychology,* Philadelphia, Pa, W. B. Saunders.

Hediger, H.
1969 "Biological Glimpses of Some Aspects of Human Sociology," *Social Research* 36, pp. 530-541.

Keith, Sir A.
1948 *A New Theory of Human Evolution,* London, Watts.

Keys, A. / *et al.*
1950 *The Biology of Human Starvation,* Minneapolis, Minn., University of Minnesota Press.

Koestler, A.
1968 *The Ghost in the Machine,* New York, Macmillan.

Lasswell, H.
1968 "The Future of the Comparative Method," *Comparative Politics* 1, pp. 3-18.

Lorenz, K.
1965 *Evolution and Modification of Behavior,* Chicago, Ill., University of Chicago Press.
1966 *On Aggression,* New York, Harcourt, Brace and World.
Lowell, A. L.
1909 "The Physiology of Politics," *American Political Science Review* 4, pp. 1-15.
Mackenzie, W. J. M.
1967 *Politics and Social Science,* Baltimore, Md., Penguin.
Marty, W.
1971 "Nonviolence, Violence and Reason," *Journal of Politics* 33, pp. 3-24.
McClearn, G.
1962 "The Inheritance of Behavior," in L. Postman (ed.), *Psychology in the Making,* New York, Knopf.
1969 "Biology and the Social and Behavioral Sciences," *Social Science Research Council Items* 23, pp. 33-37.
Means, R.
1967 "Sociology, Biology and the Analysis of Social Problems," *Social Problems* 15, p. 2.
Mehrabian, A.
1969 "Significance of Posture and Position in the Communication of Attitude and States Relationships," *Psychological Bulletin* 71, pp. 359-372.
Montagu, A. (ed.)
1968 *Man and Aggression,* New York, Oxford Press.
Morris, D.
1969 *The Human Zoo,* New York, McGraw-Hill.
Pye, L.
1962 *Politics, Personality and Nation Building,* New Haven, Conn., Yale University Press.
Sokolov, E. N.
1963 *Perception and the Conditioned Reflex,* New York, Pergamon Press.
Sorokin,
1927 *Sociological Theories,* New York, Harper.
Tiger, L.
1969 *Men in Groups,* New York, Random House.
1971 *Imperial Animal,* New York, Holt, Rinehart and Winston.
Vickers, G.
1965 *Art of Judgment,* New York, Basic Books.
Wahlke, J.
1968 "Political Behavior Theory and the Study of American Political Behavior," Paper delivered in the Florida State University series on "The State of Scientific Theories in Political Science."
Wallace, A. F. C.
1956 "Revitalization Movements," *American Anthropologist* 16, pp. 23-27.
Wilson, W.
1908 *Constitutional Government in the United States,* New York, Columbia University Press.
Wrong, D. / Gracey, H.
1967 *Readings in Introductory Sociology,* New York, Macmillan.
Wynne-Edwards, V. C.
1961 *Animal Dispersion in Relation to Social Behavior,* London, Oliver and Boyd.

Biographical Notes

Peter A. Corning
Lecturer in human biology and political science, Stanford University. Author of two books, *The Evolution of Medicare: From Idea to Law* (1970) and *Evolution and Political Science* (in press), and of several articles in social biology, political theory and public policy. Currently conducting research in social indicators and forecasting and is preparing a monograph on *Survival Indicators: An Approach to Measuring the Quality of Life.*

Dr. Corning was born on June 5, 1935 in Pasadena, California. He received his B.A. degree from Brown University and his Ph. D. from New York University. He has also held post-doctoral fellowships from the Institute for Behavioral Genetics and the Hoover Institution on War, Revolution and Peace.

His professional affiliations include the Society for the Study of Social Biology, the Behavior Genetics Association and the American Association for the Advancement of Science. He is also newsletter secretary for the International Society for Research on Aggression.

John H. Crook
Educated at Sherborne School and Southampton University John Crook took his doctorate at the Zoology Department of the University of Cambridge in 1958. His thesis concerned the evolution of weaver bird society based on extensive travels in West Africa. Later work in Asia, Seychelles and East Africa produced a monograph on the subject (Behavior Monographs 10, Leiden, Brill) in 1964. The thesis that bird society could be interpreted in terms of adaptation to ecology became especially fruitful when, with Steve Gartlan, Crook applied the idea to Primates in 1966. Comparative surveys on other mammals have followed and inferences to Man made both by Crook and social anthropologists. Crook is especially interested in the implications of animal social ethology to human behavioural sciences and maintains a critical attitude to the often overpopularised works of ethologists and others in this area. Reader in Ethology at Bristol University, England, Crook directs a lively school of research

workers studying birds and mammals both in the laboratory and in the field in many parts of the world.

David V. Cross
Associate professor of psychology at the State University of New York at Stony Brook. All his degrees are from the University of Michigan: he received a bachelor's degree in 1958, masters degree in 1959 and a Ph. D. in mathematical psychology in 1965. He joined the psychology faculty at Harvard in 1965 and for seven years there was a member of the Laboratory of Psychophysics. He began his present appointment at Stony Brook in 1972. His main research interests are the identification and measurement of factors which influence judgment in perception and the expression of opinion. He has contributed to measurement theory and the development of mathematical models in psychophysical scaling.

James Chowning Davies
Professor of political science, specializing in theory of political behavior. Developed (1962) a psychological theory of political revolution based on the initial satisfaction and subsequent frustration of physical and mental needs (the J-curve). Posited (1963), on a general biological and psychological base, systematic relationships between the individual and the polity, to explain political behavior as a function of the interaction between the individual and the environment. Developed (1975) a systematic theory relating the priority of human needs to successive stages of political development, and vice versa. Author of *Human Nature in Politics* (1963) and of numerous articles. Editor of *When Men Revolt and Why* (1971).

Research Training Fellow of the Social Science Research Council (1950-1951), Carnegie Fellow in Political Science at the University of Michigan Survey Research Center (1951-1953), Faculty Research Fellow of the Social Science Research Council (1961-1962), Fellow of the Rockefeller Foundation (1962-1963), Senior Fellow at the University of Pennsylvania Foreign Policy Research Institute (1969). Taught at the California Institute of Technology 1953-1963, except for a year at the University of California at Berkeley (1959-1960), and since 1963 at the University of Oregon.

David Easton
Took his Ph. D. from Harvard University (1947) and holds honorary doctorates from McMaster University and Kalamazoo College. He presently holds joint appointments as Andrew MacLeish Distinguished Service Professor at the University of Chicago, and as Sir Edward Peacock Professor of Political Science at Queen's University. Professor Easton has been a Fellow at the Center for Advanced Study in Behavioral Sciences; a member of the American Academy of Arts and Sciences; and is past President of the American Political Science Association. He has served in addition, on numerous social science and political science committees at both the national and international level.

His major interests have been in political philosophy and in political socialization. Among his books are *The Political System* (1953, 2nd ed.

1971); a *Systems Analysis of Political Life* (1965); and (with J. Dennis) *Children in the Political System: Origins of Political Legitimacy* (1969).

J. A. LAPONCE
Was educated at the Institute d'Études Politiques in Paris (diploma 1947) and at UCLA (Ph. D. 1955). He taught at the University of Santa Clara in 1955 and has been in the Department of Political Science at the University of British Columbia since 1956. His works include the *Protection of Minorities* (1961), the *Government of France under the 5th Republic* (1962), *People vs. Politics* (1973); he is presently working on a study of Left and Right and more generally on the relating of the perception of space to political ideology.

Milton LODGE
Is currently Co-Director of the Laboratory for Behavioral Research at the State University of New York at Stony Brook and holds a joint appointment as Associate Professor in the Department of Political Science of SUNY at Stony Brook and in the Department of Psychiatry at the University Medical School. After receiving his Ph. D. from the University of Michigan, he taught courses in and published in the political behavior field at the University of Iowa. In 1970 he received a Social Science Research Council research training fellowship for study at the Psychophysiological Laboratories at Harvard Medical School. As Principal Investigator of a NSF grant studying the verbal, physical, and physiological measurement of political attitudes and behavior, his current research focuses on the psychophysical scaling of opinion and experimental studies of concept formation and political attitude change.

William MACKENZIE
Born in 1909, was educated in both classics and law in Edinburgh and at Oxford. Now retired, Professor Mackenzie has taught classics in Oxford; politics in Oxford, Manchester and Glasgow.

The author of many books and articles, and a member of various public bodies, he has always sought wide rather than special grounds for political explanation. Professor Mackenzie is particularly interested at present in developing long-standing interests in the relations between language and politics and between biology and politics.

Professor Mackenzie currently has two books in press: *Power, Violence Decision*, and *Political Identity*. He will shortly deliver in the Douglas Robb Lectures on Political Adaptivity in Auckland University, New Zealand.

Roger D. MASTERS
Comes to the study of biology and politics from the field of political philosophy. He took his B.A. at Harvard (1955, Summa cum laude) and did his M.A. (1958) and Ph. D. (1961) at the University of Chicago, where he wrote his doctoral dissertation on Rousseau under Leo Strauss. His publications include: *Rousseau's First and Second Discourses* (editor and co-translator; St Martin's, 1964); *The Nation is Burdened* (Knopf, 1967); and *The Political Philosophy of Rousseau* (Princeton, 1968). Based on work begun during a Guggenheim Fellowship in 1968-1969, he has more recently

published a number of articles on the relationship of the biological sciences to human behavior and politics, including: "Genes, Language, and Evolution" (1970); "Functional Approaches to Analogical Comparisons between Species" (1973); "Vers Une Science?" (1975); "Politics as a Biological Phenomenon" (1975); and "Of Marmots and Men: Human Altruism and Animal Behavior" (in press). He is currently a Fellow of the Institute of Society, Ethics, and the Life Sciences; and serves as Chairman of the Editorial Board of the "Biology and Social Life" section of *Social Science Information* (published under the auspices of the International Social Science Council and the Maison des Sciences de l'Homme in Paris). He is Professor of Government at Dartmouth College.

Steven A. Peterson
Is presently an Assistant Professor of political science at Alfred University. He received his Ph. D. from the State University of New York at Buffalo. His research interests include biopolitics, future studies, and judicial process. Among the papers he has presented at professional meetings are: "Science Fiction and Anarchist Theory" and "The Effect of Physiological Variables Upon Student Protest Behavior."

Glendon Schubert
University Professor of political science at the University of Hawaii-Manoa, where he has taught since 1971. His interest in biopolitical behavior developed out of his work in political (and particularly judicial) behavior, throughout the past quarter of a century. The principal subject of his studies of judicial process and behavior has been the social psychology of political ideology, attitudes, and decision-making. Much of his research since 1960 has been cross-cultural as well as multidisciplinary in orientation; and he has taught (at Syracuse University, the University of California at Los Angeles, Howard University, Rutgers University, Franklin and Marshall College, the University of Minnesota, Michigan State University, the University of Oslo, Norway, the University of North Carolina, Chapel Hill, and York University, Toronto, Canada), done research (his research fellowships included residences at the Center for Advanced Study in the Behavioral Sciences, Palo Alto, the East-West Center, Honolulu, and the Center for the Study of Democratic Institutions, Santa Barbara), traveled, and published (for a discussion of his major writing, see Fred Kort, "The Works of Glendon Schubert," *The Political Science Review* no. 4, 1974, pp. 193-227) widely in foreign as well as American Forums. His other writings on biopolitics include "Biopolitical Behavior: The Nature of the Political Animal," *Polity* no. 6, 1973, pp. 240-275, "Future Stress, Constitutional Strain, and the American Judicial System," ch. 12 in his *Human Jurisprudence* (Honolulu, University Press of Hawaii, 1975), "The Rhetoric of Constitutional Change," ch. 11 in his *Human Jurisprudence (op. cit.),* and "Biopolitical Behavioral Theory," *The Political Science Reviewer* no. 5, 1975.

David C. Schwartz
Professor and Chairman, Department of Political Science, Livingston College and Graduate Faculty, Rutgers, The State University of New

Jersey. He is Associate Editor of *Society* magazine, an associate of the Danforth Foundation and a consultant in research to several public foundations and agencies. Dr. Schwartz is the author of *Political Alienation and Political Behavior,* co-author of *New Directions in Political Socialization* and has published numerous scholarly articles on political psychology, biology and politics, alienation, revolution and related topics. Dr. Schwartz is presently studying the influence of people's health, personalities and life experience on their political attitudes and behaviors, under grants and contracts from the H. F. Guggenheim Foundation, Ford Foundation and Earhart Foundation.

Albert SOMIT
Professor of political science and Executive Vice President, the State University of New York at Buffalo, holds a Ph. D. from the University of Chicago (1947) and has previously taught at New York University and the Naval War College. His major interests, other than biopolitics, are political philosophy, intellectual history, organizational theory, and political futurism. He is co-author of *The Development of American Political Science: From Burgess to Behavioralism,* Allyn-Bacon (1967) and *American Political Science: A Profile of a Discipline,* Atherton Press (1964). His most recent book was *Political Science and the Study of the Future,* Dryden Press (1974).

Lionel TIGER
Born in Montreal in 1937 and studied at McGill University and the London School of Economics and Political Science. His Ph. D. dissertation in political sociology focused on changes in the Ghanaian civil service on independence. He is presently Professor of anthropology in the graduate school of Rutgers University and Research Director of the Harry Frank Guggenheim Foundation. He is the author of scholarly articles, of *Men in Groups* (1969), *The Imperial Animal* (with Robin Fox, 1971) and *Women in the Kibbutz* (with Joseph Shepher, 1975).

Bernard TURSKY
Co-Director of the Laboratory for Behavioral Research at the State University of New York at Stony Brook where he holds a joint appointment as Professor in the Departments of Political Science, Psychology, and Psychiatry. For many years he was associated with Harvard University and Harvard Medical School where he was the Principal Investigator in NIH supported psychophysiological research. He recently served as President of the Society for Psychophysiological research and is currently an associate editor of *Psychophysiology.* Currently he is the Principal Investigator of an NIH research grant to investigate factors relating to pain perception and Co-Investigator of an NSF grant to investigate a cross-modal approach to the study of political attitudes and behavior.

John C. WAHLKE
Born and educated in Cincinnati, Ohio, and received the degrees A.B., A.M, and Ph. D. from Harvard. He is Professor of political science at the University of Iowa, and has previously taught at the State University of

New York at Stony Brook, the State University of New York at Buffalo, Vanderbilt University, the University of California at Berkeley, and Amherst College. He is co-author with Milton Lodge of "Psychophysiological Measures of Political Attitudes and Behavior," *Midwest Journal of Political Science* 56, November 1972, pp. 505-537, and with Heinz Eulau / William Buchanan / LeRoy C. Ferguson of *The Legislative System* (New York, John Wiley, 1962). Besides articles and monographs on representation and legislative behavior, his most recent works include *The American Political System* with Bernard E. Brown (Homewood, Ill., Dorsey Press, 1967 and 1971), *Government and Politics: An Introduction to Political Science,* with Alex Dragnich / *et. al.* (New York, Random House, 1966 and 1971), and *Comparative Legislative Behavior: Frontiers of Research* with Samuel C. Patterson (New York, John Wiley, 1972).

IMPRIMERIE DARANTIERE — DIJON-QUETIGNY

Bei Fragen zur Produktsicherheit wenden Sie sich bitte an:
If you have any questions regarding product safety,
please contact:

Walter de Gruyter GmbH
Genthiner Straße 13
10785 Berlin
productsafety@degruyterbrill.com